T0260114

BIOINSPIRED DEVICES

BIOINSPIRED DEVICES

Emulating Nature's Assembly and Repair Process

Eugene C. Goldfield

Harvard University Press
Cambridge, Massachusetts
London, England
2018

Printed in the United States of America

First printing

Library of Congress Cataloging-in-Publication Data
Names: Goldfield, Eugene Curtis, author.
Title: Bioinspired devices : emulating nature's assembly and repair process / Eugene C. Goldfield.
Description: Cambridge, Massachusetts : Harvard University Press, 2018. | Includes bibliographical references and index.
Identifiers: LCCN 2017009706 | ISBN 9780674967946 (alk. paper)
Subjects: LCSH: Biomedical engineering. | Biomimicry. | Medical innovations.
Classification: LCC R856 .G66 2018 | DDC 610.28—dc23
LC record available at https://lccn.loc.gov/2017009706

For Beverly

Contents

Preface

Ameliorating the burden of the brain diseases and injuries that disrupt individual and interpersonal behavior is among the most difficult challenges of our time. The evolutionary, developmental, and ecological processes of building a human brain have, paradoxically, made us vulnerable at particular points in the lifespan to catastrophic injuries, such as cerebral palsy, stroke, and schizophrenia. In this book, I propose that through biologically inspired engineering—that is, emulating how nature builds and repairs biological and behavioral ecosystems—we may become better able to heal ourselves and our relationships with our planet and with each other. In laboratories around the world, disparate disciplines in biology, medicine, and behavioral science are adopting biologically inspired, or *bioinspired*, engineering to address some of the major questions of our time: How do brains work? How do complex systems repair themselves? How do hosts and microorganisms enter into symbiosis-like social relationships to build large-scale communities?

In the chapters of this text, I address these and other questions by considering how and what nature builds, how nature repairs aging or damaged systems, and how scientists and engineers may emulate nature's processes of assembly and repair. Not all approaches to biologically inspired technologies are the same. Here, I adopt as an organizing framework the respective approaches developed at the Harvard Wyss Institute for Biologically Inspired Engineering, where I have worked with engineers on biologically inspired devices for nearly a decade, and at the Boston Children's Hospital, where I have conducted pediatric research and have been a faculty member for twice as long. At the heart of the Wyss Institute approach is the process of emulating how living tissues self-organize and naturally regulate themselves, while the specialization in pediatrics at Boston Children's Hospital motivates an appreciation for the parallels between how systems are initially built and how they repair themselves.

The content of the text is seen through the lens of my personal experiences, but it goes well beyond them to consider the work conducted in many laboratories, clinical settings, and engineering facilities. I consider a wide range of topics in materials science, robotics, microfluidics, and bioprinting. These are all placed in the context of building devices for repairing injured nervous systems, as well as for creating neurorehabilitation environments that go beyond the individual patient and go outside the medical setting. I present an "insider's view" of ongoing work on devices conducted by Wyss Institute colleagues, while at the same time presenting cutting-edge work in other labs worldwide. I also provide a personal perspective by considering devices relative to my own experiences and interests in music, art, and natural history, including some of my favorite films and classic TV. Early visions of future technologies may be quite informative (and often humorous), not only for what they allow us to imagine, but also for what they omit.

Throughout the book, I update a dynamical systems perspective presented in my earlier book, *Emergent Forms* (Goldfield 1995; see also Thelen and Smith 1994). The study of dynamics provides a methodology for examining how systems evolve over multiple time scales—across transformations of species; within the life spans of members of a particular species; and in individual human performance over a period of hours, minutes, or seconds. My goal in the ensuing years has been to apply principles from dynamical systems to the development of medical devices for children with neuromotor disabilities, including devices for assisting infant feeding (Goldfield 2007) and for assisting the walking of developmentally delayed children (Goldfield et al. 2012).

This systems perspective goes beyond the individual to include supportive environments for maintaining and restoring function. It treats the individual and environment as part of a larger whole, with both partners in a sometimes-symbiotic relationship. Indeed, an appreciation of symbiosis, from host-bacteria consortia to adult-child interactions, is proposed as a model for building the next generations of robotic devices that will become partners to humans. Symbioses between neurons and glia, for example, may provide a perspective on building self-repairing bio-hybrid devices, from neuroprosthetics to microscale and nanoscale machines that live inside us. But there are many aspects of biological partnerships that we do not yet fully understand. As one illustration, biological systems are able to maintain functioning, even though their components are continuously turning over and being replaced. The replacement process itself is a great mystery because blood cells or skin cells are manufactured throughout our lives. And yet, it is only during certain developmental periods that neuronal synaptic connections may be assembled, disassembled, and reassembled in new ways.

In high school, I wanted to be a biologist, but I loved machines. As an undergraduate, I thought I would become an engineer, but I was intrigued by the brain, how the mind emerges, and how development takes so many different pathways. So, my graduate school training was in Developmental Psychology. Then, as a postdoc, I began a decades-long fascination with the relationship between the brain and behavior. In my academic career, I have tried to emulate nature and social relationships in building devices for developmentally delayed children. At the level of behavior, human skills and relationships appear to develop in a symbiosis-like social environment. Adults are not simply models to be imitated. Nor are they mere sources of positive and negative reinforcement. Instead, adults establish physical and informational coupling through repeated and varied sequences of touch, vision, and vocal communications. Children learn to walk, talk, and use tools through a process by which adults build, modify, and dissolve structured supportive environments for learning. It is in these relationships, perhaps, that we may be more successful at building robots that will become our social partners, and succeed in building devices that we can trust with our lives.

Why is there a need for this book? As we approach the third decade of the twenty-first century, devices that are smart, safe, and seamlessly integrated with the human body, and with each other, have become a high priority in our national health and science agendas. Part of the challenge of bioinspired engineering is that solutions require new alliances in the fields of biology, physics, materials science, neuroscience, cognitive science, mathematics and computer science, and biomechanics. The bridge between science, engineering, and clinic is reflected in the chapter organization of the three book sections. Part I presents scientific foundations for a new generation of bioinspired devices. Chapter 1 considers the many sources of biological inspiration in nature, from insect wings to smart receptor systems. Chapter 2 considers physical constraints on what nature builds, including scaling laws, mechanical influences on evolving form, and how nature harnesses physical law to perform work. Chapter 3 examines the emergence of the functional properties of materials from the formation of complementary structural patterns within a hierarchy of scales. It also highlights what devices nature builds: smart perceptual instruments for exploring structured energy arrays and special-purpose devices for performing distinctive modes of behavior, including locomoting and communicating. Chapter 4 introduces a bioinspired approach to fabricating materials, sensors, actuators, and soft robots. The emphasis is on technologies that emulate the processes of biological self-assembly, including origami and microfabrication techniques such as "pop-ups."

Part II of the book includes three chapters on the adaptive, but vulnerable, human nervous system, considered in a comparative context. Chapter 5 highlights the rich set of technologies being developed for recording neurons in progressively larger and more complex nervous systems. The chapter also considers the embodied brain: how network interactions function to harness the biophysical properties of the body acting in the environment. Chapter 6 continues the comparative theme and examines nervous system development and vulnerabilities. The chapter highlights how the advantages of human brain power come at a high metabolic cost, leaving the essential connecting hubs of the network organization vulnerable to injury. Chapter 7 is organized around comparative cases of remodeling and their implications for neuroprosthetic devices and neurorehabilitation: regeneration of neural circuitry in salamanders, metamorphosis from tadpole to frog, and synaptic plasticity in the acquisition of birdsong.

Finally, Part III considers the challenges of building neuroprosthetic and neurorehabilitation devices and how emerging technologies are changing our conception of the relation between human and machine. Chapter 8 introduces a perspective on neuroprosthetics that integrates brain, body, and device. I consider the requirements for a device to become truly embodied, turning again to the biology of symbiotic relationships. Where is the functional boundary between biological and synthetic organs? As part of their normal functioning, do nervous systems generate a "virtual body" that may persist even after loss of a part, or does it become distorted in diseases, such as schizophrenia? Chapter 9 introduces an approach to neurorehabilitation using devices capable of dynamic adaptive assistance, a process whereby control systems autonomously grow or shrink the coupling connections between available resources, or "primitives" (for example, biological muscles or synthetic actuators). The final chapter looks ahead to devices that may become part of us, at multiple scales, such as nanoscale robots that may provide surveillance and repair of our nervous system and robots that become a part of our family, including health care robots that act as caregivers and fulfill other prosocial roles.

The intended audience for the book includes students who become excited by new ideas in science and technology as they are immersed in the demanding content of biology, medicine, mathematics, physics, and chemistry. The book is meant to encourage these readers to persevere in the difficult challenge of mastering their studies and to themselves discover new approaches to clinical care. The book is also written for scientists and clinicians who think that they may have a disruptive, innovative idea that may revolutionize practice by adding value to development of advanced devices for clinical care. And, importantly, the book is meant for potential end-users of these devices as a glimpse of advanced technologies that will become part of our lives in the decades ahead. There is an increase in the number of individuals who are surviving debilitating brain injuries during the earliest and latest stages of life (for example, perinatal brain injury in preterm infants and stroke in the aging) as technology becomes better able to support and sustain lost or partial functioning. Together, our challenge will be to progress beyond survival and promote optimal health following injury.

I have been fortunate to have the opportunity to collaborate with wonderful colleagues and friends from graduate school onward. For more than twenty years, I have been inspired by the clinical work and groundbreaking research of my colleagues at the Boston Children's Hospital and at the affiliated hospitals of Harvard Medical School. For nearly a decade, I have also worked with scientists and engineers from a wide range of disciplines at the Wyss Institute for Biologically Inspired Engineering at Harvard. This book is one tangible result of what I have learned as part of a cross-fertilization of ideas from the clinic, scientific forum, and engineering labs.

My primary mentor at the Boston Children's Hospital has been Peter Wolff, MD, now retired at age 90, as of this writing. As a psychiatrist intrigued by the ideas of Jean Piaget, and with a genius for combining observations of infants with insights from neuroscience and music, Peter helped to guide me to questions about the relation between brain and behavior. My chief of service, David DeMaso, MD, a child psychiatrist, has been extremely generous over the years, recognizing as he does the vagaries of federal grant funding. Brian Snyder, MD, PhD, an orthopedic surgeon and director of the cerebral palsy clinic, has allowed me to learn first-hand about the challenges faced by patients with cerebral palsy seen in the clinic. Jane Newberger, MD, MPH, allowed me to enter the cardiology clinic to learn about the challenges of feeding in infants having open heart surgery, and Carlo Buonomo, MD, and the Boston Children's Hospital Feeding team collaborated in a number of research projects on infant swallowing. I thank them for allowing me to observe and record videofluoroscopic examinations and for helping me interpret recorded events that required many hundreds of hours experience. Ann Hansen, MD, provided access to the Boston Children's Hospital NICU and left me in awe at treatment of very sick premature infants. At Beth Israel Deaconess Medical Center (BIDMC), DeWayne Pursley, MD, Chief of Neonatology, has made it possible for me to conduct several studies of the prematurely born infant by providing access to well-baby nurseries and a wonderful NICU. Vincent Smith, MD, MPH, has been a supportive colleague and friend and has helped me conduct several research projects in the NICU. I am most grateful to all of these clinician-scientists for their support and for sharing their experiences during the writing of this book.

At the Wyss Institute, Founding Director Don Ingber, MD, PhD, invited me to join the Wyss faculty at its founding and to continue to the present as associate faculty. Over nearly a decade, Don has made it possible for me to pursue ideas that have been "outside the box" and has provided access to the wealth of technical resources at the Wyss Institute. My Platform leader, Jim Collins, PhD, has encouraged my work on wearable devices, and lead senior engineer Jim Niemi has made certain that I had access to engineering staff and materials needed to build these devices. My colleagues in robotics and computer science at the Wyss, especially Radhika Nagpal and Rob Wood, have dazzled me with the depth of their knowledge in their respective fields.

Several individuals, both professors and students whom I first met during graduate school, have continued to have an important influence on my thinking, and their ideas are evident throughout the book. Among these are the trio of Turvey, Shaw, and Mace, and our friend Elliot Saltzman. Michael Turvey and Bob Shaw introduced me to the role of physical law for understanding perceiving and acting, and Bill Mace has always shared his enthusiasm for the work of the Gibsons and is a model of collegiality. Elliot Saltzman has been a guidepost throughout my career, sharing "interesting stuff," providing tutorials on task dynamics, and making certain that I knew about grant opportunities.

I have been fortunate in receiving generous support over the years from the National Institutes of Health and the National Science Foundation. Over the past decade, David Corman, Helen Gill, Sylvia Spengler, and Wendy Nilsen of the NSF Cyberphysical Systems program have especially encouraged my attempts to bridge the gap between clinical science and advanced technologies, and I sincerely extend my thanks to them. Family and friends have been patient and supportive as I wrote this book. My wife, Beverly, and daughter, Anna, have always shared ideas, made certain we always ate well, and provided love and encouragement. Our friends have always opened their homes to us to celebrate holidays and to share vacations. This supportive environment has been a critical ingredient in my work.

I thank Michael Fisher, now retired from Harvard University Press. Michael first saw the possibility of my ideas for a book. Janice Audet, my current editor, has been instrumental in ensuring that the book got written. Janice encouraged my early efforts, and her enthusiasm for music and culture, as well as for the natural sciences, helped me to find a voice for the book. I am most grateful to both of them.

BIOINSPIRED DEVICES

PART ONE

Biological Inspiration for Nature's Building and Repair Process

Nature builds by self-assembly: millions of independently behaving termites construct meter-scale air-cooled homes. Nature's structures harness available energy resources: the thorax of flying insects is a resonant structure that stores energy subsequently recovered on each stroke to power wing flapping. Nature builds interconnected networks of receptors that function as smart instruments, including odometers and compasses, as well as solar and celestial navigation systems. And nature builds soft animals, such as caterpillars, capable of locomoting by means of sequencing the oscillations of multiple body segments that attach and release the animal from a substrate. Inspired by nature, scientists strive to understand how plants and animals embody physical law to perform work, as when fish exploit vortices that arise from fluid flow for locomotion or when the collective schooling of fish creates an informationally connected emergent "creature" capable of scaring away potential predators. The information communicated between ants induces both construction and repair of hive chambers, suggesting that remodeling in response to damage may engage the same fundamental processes by which nature initially builds structures. Engineers and computer scientists inspired by nature's methods of assembly and repair use teams of independent micro-robots, each with limited capabilities, to add bricks to a growing and changing structure. Insect scale flapping wing robots are manufactured from components that "pop up" and snap into place. Soft robots selectively inflate and deflate internal body compartments to achieve an undulating gait. The chapters of this first section (Chapters 1, 2, 3, and 4) consider these topics of how nature builds, what nature builds, and how to emulate nature's building and repair processes.

✦ 1 ✦

Bioinspired Devices as Parts of Complex Systems

Fictional writing and films are often a harbinger of new directions in technology. So, on a long flight from Boston to Japan, I was thrilled to watch the beautiful, award-winning 2014 animated film *Big Hero Six*. It provides a subtle and clever introduction to possibilities for building robots that help us, spanning a range of architectures from swarms to social robots. At one end of a continuum of robot architectures, the film depicts a swarm of microbots, capable of linking themselves together to form towering mobile structures, in a manner reminiscent of the way that ants in the rainforest connect themselves together to build bridges and rafts (Foster et al. 2014). But, for me and many other fans of this movie, the most compelling robot is Baymax, a healthcare companion. Baymax uses sensors to immediately detect a human who is injured and identifies symptoms predicting whether the individual is at risk of physical or psychological harm. The robot then implements some healing response (including medical advice and actual interventions). Baymax, thus, presents a prosocial message of possibilities for robotic companions.

The design of Baymax points to coming capabilities in robotics. The robot body is a hybrid, a trade-off of hard and soft components (Bartlett et al. 2015), capable of transformations in form and function, subject to vulnerabilities, and capable of different modes of behavior. The soft pillow-like outer body is a supernormal rendition of comfort, while the hard inner structures contain a support architecture and battery power. Because the film is meant to amuse and educate, this body design provides opportunities for wonderful slapstick when the soft outer body springs leaks and the internal batteries run down. Baymax becomes a wobbly, less functional, indeed comical, character. Baymax is capable of exhibiting different modes of behavior. When new programs are loaded—and the robot body is augmented with a hard shell, wings, fighting arms, and legs with rocket propulsion—Baymax is transformed to function in superhero "rescue" mode. But these modes of behavior are not mutually exclusive. Indeed, the prosocial and rescue modes begin to constitute an emergent and recognizable "personality." Baymax becomes an indispensable member of a social unit, part of a group of five young adult humans who, like this robot, are able to transform themselves to collectively function as superheroes for prosocial and rescuing acts, *Big Hero Six.*

Fiction gives writers license to violate physical law. But biology is real. Antman can be transformed to the size of an insect, and at the scale of human-sized bodies; Spiderman can carry loads many times the weight of his own body, as ants do at insect scale. But the forces that dominate at the scale of insects are not the same as at the scale of the human body. Other superheroes carry shields that deflect bullets and wear suits that stretch to many times the extent of their limbs. Materials science works within the constraints of physical law to push the boundaries of what is possible within those constraints. For example, in 1941, when Swiss engineer George de Mestral returned from a hunting trip, he noticed that there were burs clinging to his clothing and to his dog's coat. When he examined the seedpods under a microscope, de Mestral identified the means by which the burs so effectively attached to clothing and animal fur: hooked projectiles. He was eventually able to mold nylon into a fabric with hooks or loops that functioned in the same way as the natural burs. This marvelous invention, Velcro, was introduced into the United States in 1958. Now, in the twenty-first century, biomaterials scientists are fabricating adaptive adhesive materials by means of processes that emulate not only how nature *builds,* but also how it *repairs.*

Biological Inspiration

For centuries, philosophers, natural scientists, tinkerers, and engineers, have been intrigued and inspired by the wing veins and material properties that give insects the ability to outmaneuver predators (Tanaka et al. 2011) or the adhesive van der Waals force of bristles that allow the gecko's foot to provide support for gravity-defying acts on sheer vertical surfaces (Autumn et al. 2006; Hawkes et al. 2015; Heepe and Gorb 2014). Biologically inspired approaches to the design and fabrication of materials and devices share a recognition that natural systems have arrived at "good enough solutions" to survive in uncertain environments. By adopting these same principles for solutions to difficult engineering problems—such as building devices for repairing injured nervous systems—biologically inspired approaches strive to discover how to translate nature's principles into engineering practice. Biologically inspired devices are based on composite materials that typically include sensors, actuators, and control systems (Dunlop and Fratzl 2010; Fratzl and Barth 2009). Nature's sensors are "smart instruments," actively moved through structured energy fields to reveal complex variables, such as body-scaled information (Turvey and Carello 2011). Nature's actuators—muscles—serve multiple functions and leverage the body's physical properties, such as its compliance, to generate force (Dickinson 2000; Nishikawa et al. 2007). Finally, biological control systems are characterized by adaptive networks, consisting of a range of decentralized to centralized architectures, with feedback and feedforward loops (Revzen, Koditschek, and Full 2009).

Biological development is a process that spans multiple spatial scales, from molecular to macroscopic, to build hierarchically organized adaptive structures that are robust, multifunctional, and organized into decentralized networks. From a comparative perspective, remodeling of nervous systems is, itself, a developmental process that expresses itself in different ways (see Chapters 6 and 7): Salamanders regrow a severed tail or limbs with complete restoration of neural connections; the spinal cord of tadpoles reorganizes itself from a solely axial pattern generating system as the body resorbs the tail and sprouts legs; and human nervous systems initially produce an overabundant supply of synaptic connections and, through an activity-based competitive process, selectively eliminate weak connections. The multiple spatial scales of self-assembly, as well as the time scales of evolution, development, and plasticity (learning), have all inspired biological approaches to engineering and computer science: evolution is the foundation for morphological computing (see, for example, Bongard

Figure 1.1 Adaptive gait by a hexapod robot that has one limb experimentally removed. A robot automatically learns to keep walking after damage in the form of a broken front leg. Photo © Antoine Cully / UPMC-Université Pierre et Marie Curie (CC BY 4.0).

2011; Bongard, Zykov, and Lipson 2006); human ontogeny provides inspiration for humanoid robots, such as iCub (Natale et al. 2016; Shaw, Law, and Lee 2013); the process of neural circuit formation of animals interacting in structured environments has inspired animal-like robotic control systems, such as robotic lamprey and salamanders (Ijspeert 2014); and nature's self-repair capabilities have inspired robots that adapt to loss of a body part (see Figure 1.1).

How nature builds. A principled starting point from biological inspiration for *how* nature builds at all scales is the fundamental inseparability between individuals (cells, animals, communities) and the surrounding protective trophic matrix that transduces energy to seamlessly join the inner and outer worlds (for example, cells and their extracellular matrix, animals and their skin, and communities of people and their shared food, energy, and communication systems, respectively) (see Table 1.1). For example, at the cellular level, cell rearrangements generating lines of force within the extracellular matrix form tissues with bends and folds, tubes, branched architectures, and

Table 1.1 Animals and the networks they use to connect to environmental resources

Animal / network	Connection to resources
Slime mold *(D. discoideum)*/ microtubular network	Whole-body network changes shape to link food sources.
Spider / things that vibrate	Web used for predation; plants, leaf litter, rock, and sand are used for courtship.
Ants / chemicals and antibiotic bacteria	Chemicals used for navigation in foraging tunnels; antibiotics used for fungus farming.
Copepods, shrimp, and molluscs / coral network	Food web used for grazing, predation.

organs that loop. At the scale of insects, within a world of mechanical vibrations, the sensory system of spiders is attuned to the vibratory properties of its web. It is in this sense that at the scale of an individual insect, the web is a trophic matrix. Within the web, the actuators of the spider's body are neuromechanical systems attuned to use patterns of vibratory information to locate captured prey. In vertebrates, internal peripheral and central nervous systems move the body through richly structured informational fields. Humans are not like spiders that affix their webs within a single niche. Instead, we take our information-detecting and protective webs with us as we move through multiple niches. Our nervous system networks generate parallel loops of electrical activity within densely structured mechanical and biochemical fields. These parallel loops of activity are the means for signal transmission at multiple time scales, from ultrafast (zero delay) to time delays that may extend over an entire lifetime. Mechanotransduction occurs within organs for integrated sensorimotor control, including the muscles and its spindle and golgi tendon organs. The flow of blood through a dense network of blood vessels, lines of force mechanically transmitted at the speed of sound through tissue, and an exchange of mechanical potential energy and chemical energy at neuromuscular junctions all make each muscle an information-driven thermodynamic engine. Our social systems include standards that allow us to share the Earth's finite energy resources.

What nature builds. The developmental processes at all scales that sculpt living tissue and nervous system network structures to build special purpose devices and organ systems constitute a foundation for *what* nature builds (Chapter 3). The integrative approach of this book examines nature's devices, with a focused emphasis on perceiving and acting, communication networks, and adaptive behavior through development and learning. The functional and integrative approach to perceiving draws attention to smart perceptual instruments (Turvey and Carello 2011; Runeson 1977), sensory organs (including the eyes, ears, nose and mouth, and skin) that are capable of detecting rich patterns of available structured information—for example, in gradients and body-scaled affordances within particular econiches (Gibson 1986; Goldfield 1995). What makes these instruments smart is not only their ability to detect complex relational patterns, but also that they reveal available information through guided bodily motion. Action systems that orient and move the smart perceptual instruments through the environment (Goldfield 1995) have evolved through tinkering (Jacob 1977). Self-assembling muscle groups organize rapidly and temporarily to transform sensory organs into active perceptual systems (that is, eyes scan, the head orients the ears, the hand moves fingertips over surfaces). Indeed, these characteristics of flexible assembly, functioning via active movements through the environment, and multifunctionality present major challenges for building synthetic devices for seeing, hearing, touching, and sniffing. For example, we may be able to build a bionic eye, but how do we integrate it within a control system that nests eye, head, and body movements for active orienting within a structured environment?

Another way that nature makes perceptual instruments smart is by making them capable of physically and informationally connecting with other animals that perform the same functions (albeit with a range of embodiments). At the smallest scales of animal life, functions emerge from connectivity when bacteria form swarms, when slime mold amoebae form a "foot" for locomotion, when ants and bees forage, and when butterflies globally navigate thousands of miles across oceans and continents. At an intermediate scale of small vertebrates, birdsong regulates competition and cooperation for limited resources, and the collective swimming of fish schools creates an informationally connected larger "creature" whose size scares away potential predators of the individual small fish comprising the group. Nonhuman primates gesture with their hands and face, vocally hoot and howl, grunt their displeasure, and screech with excitement. This style of communication is well adapted for maintaining cohesion and cooperation within relatively small social groups and for providing boundaries with other social groups. For better or worse, human communications, with the aid of electronic devices, has allowed us to globally convey our ideas, as well as our ape-like ranges of emotion.

We now have the capability to build smart devices, such as robotic hands, robots that can augment the function of our injured or aging bodies, and social robots that communicate with us to find out how they may help (or how we may help them). A challenge in the coming decades will be to use biological principles to solve the extraordinary difficulties raised by integrating the human body with complex synthetic systems, including

- Can we repair, regrow, or replace partially, or even completely severed spinal cords?
- Can we restore sensorimotor function following stroke?
- Can we restore vision to blind individuals?
- Can we protect developing nervous systems from injury?

Part of the challenge of using principles from biology for engineered solutions to spinal cord injury, stroke, blindness, and developmental vulnerabilities is the recognition that the developing nervous system is part of a complex body with intrinsic mechanical properties and energy exchange with the physical world. Can we build devices that are part of this complex web of life?

Bioinspired Devices as Parts of Complex Webs

Bioinspired devices that are able to create a connection between the central or peripheral nervous system and the human body must be complex, heterogeneous, adaptive systems for replacing and / or restoring lost function. Attempts to build devices for repairing damaged nervous system networks may benefit from an understanding of some of the adaptive processes that nature uses to repair its networks. Consider the various types of network structures constructed by insects, and nature's solutions to repairing damage, as instructive for elucidating how bioinspired devices may be used to remodel injured nervous systems.

The spider's web is a kind of communications network, whose orb is constructed of silk material (Buehler 2010). The impact of a tasty flying insect on the web generates vibrations that are detected by sensitive mechanotransducers of the lyriform slit on its body surface (Barth 2004; Fratzl and Barth 2009). The non-linearity of the silk and orb architecture makes the web robust but still susceptible to local damage (Cranford et al. 2012). For the spider to be a successful predator, it must repair any local damage and does so by measuring the spacing of silk threads comprising the orb architecture with its mechanotransducing smart instrument (Turvey and Carello 2011). The means for repairing the spider's communications network is a spinning device for generating silk fibers. Fluids mixed via several glands in the abdomen are ejected from a spinning duct. Fluids from the different glands create silk with the specific properties required for different structural components of the web. As I discuss in Chapter 4, materials scientists are now able to emulate the spinning process to manufacture complex microfibers (Kang et al. 2011).

Tunnels within ant nests also exhibit network architecture, and ants use the tunnels for transportation and foraging within their widely distributed fungus farms. The farms must be maintained, for example, to prevent growth of harmful bacteria, and ants produce a natural antibiotic, mirmicacin, spread throughout the fungus farms (Currie et al. 1999). Nature also endows ants with smart perceptual instruments for measuring distances, by *odometry*, as they navigate their way through the foraging tunnels and with organs for producing and detecting chemical pheromones to communicate with nest mates (Wolf 2011). And finally, the hives of bees are hubs of large food-gathering and distribution networks. Bees use propolis (resin or bee glue) with antimicrobial properties, as part of an immune system-like network, to reduce disease within the hive (Simone-Finstrom and Spivak 2010). Bees also, famously, use vibration within the hive and waggle dances while in flight to communicate to their nestmates the location and distance of a food source (Menzel 2012; Frisch 1967). These lessons from the insect's use of smart perceptual instruments as well as adaptive repair of networks may be applied to neuroengineering of neuroprosthetic devices for repairing and rehabilitating injured nervous systems, the topic of Chapters 8 and 9.

The great neuroscientist Ramon y Cajal remarked that human nervous systems are like "impenetrable jungles," with 85 billion neurons and 100 trillion synapses (Azevedo et al. 2009). But it is not the sheer number of cells or connections alone that make a nervous system complex. The brains of parrots and songbirds contain twice as many neurons as primate brains of the same mass, giving them a higher neuron packing density (Olkowicz et al. 2016). Nervous system complexity is also the result of extreme heterogeneity of the parts and their organization into intricate and highly structured networks, with hierarchies and multiple scales (Carlson and Doyle 2002).

Our bodies as complex systems are comprised of communities of individuals (see Chapter 10). The formation of the placenta during pregnancy provides bidirectional communication between mother and fetus: the placenta delivers antibodies from the mother's immune system, and the fetal systems send chemical messages back to the mother (Guttmacher et al. 2014). Moreover, we share our body space with a community of microorganisms—bacteria, fungi, viruses, and archaea—that provide us with genetic variation and gene functions that human cells have not had to evolve on their own (Grice and Segre 2012; Sommer and Backhed 2013). We are holobionts, part of a microbial community (Charbonneau et al. 2016). Almost immediately after a human infant is born, a microbial ecosystem, a microbiome, is apparent in the gastrointestinal tract (Palmer et al. 2007; Koenig et al. 2011). Energy-regulating ecosystems within our bodies are apparent in interactions between the genome and microbiome (Grice and Segre 2012), in the relation between bone formation and energy metabolism (Karsenty and Oury 2012), and in the role of synaptic signaling complexes (molecular machines) in the exuberance and pruning of synaptic structure (Grant 2012).

An *ecological* meaning of biological complexity comes into clearer focus by shifting our perspective to a planetary scale. Life forms (cells, plants, animals) survive and reproduce by exchanging resources (energy, metabolites, information) with the environment and with other life forms as part of ecosystems whose resources may be abundant, stressed, or depleted. The fundamental macroscopic unit for life on earth is the ecosystem, such as rainforests, oceans, and coral reefs. To illustrate, nearly one-quarter of the earth's biomass, comprising 1.5 million species, consists of fungi that communicate via chemicals, using sophisticated extracellular signals and cellular responses (Leeder et al. 2011). In the rainforests that remain on our planet, fungi are diverse in appearance and lifestyle. Their ability to decompose plant fiber is harnessed by insects, such as leafcutter ants in South and Central America, who use fragments of leaves to grow a nutritious fungus that exists nowhere else (Marent 2006; Hölldobler and Wilson 2009).

Coral reefs are the "rainforests of the sea," with complex food webs, including fish, filter feeding invertebrates (sponges, ascidians), and a nutritional symbiosis with single-celled algae (Knowlton 2008). From the vantage point of microorganisms, the ocean is a vast, heterogeneous sea of gradients. Different species exploit this heterogeneity by foraging in different ways—for example, some marine bacteria propel themselves with corkscrew flagella, while others are nonmotile and harvest sinking detritus (Stocker 2012). Different means to detect gradient concentrations, and to follow a gradient toward its source, are apparent at multiple spatial scales, and in other types of ecosystems, including receptors for molecular cyclic AMP during formation of a locomotor pseudopod in the slime mold *Dictyostelium discoidum* (Camazine et al. 2001) (see Figure 1.2), insect receptor organs for detecting and sharing information about the location of trophic substances (Hölldobler and Wilson 2009), and cell membrane sites of climbing fibers during formation of the unique architecture of the human cerebellum (Hashimoto et al. 2009).

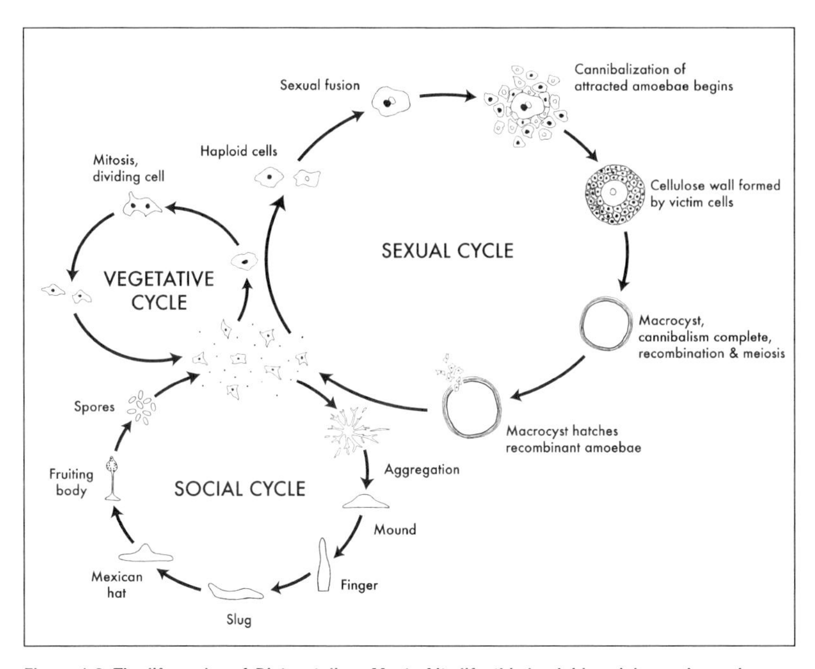

Figure 1.2 The life cycles of *Dictyostelium.* Most of its life, this haploid social amoeba undergoes the vegetative cycle, preying upon bacteria in the soil and periodically dividing mitotically. When food is scarce, either the sexual cycle or the social cycle begins. Under the social cycle, amoebae aggregate to cyclic AMP (cAMP) by the thousands and form a motile slug, which moves toward light and ultimately forms a fruiting body. Under the sexual cycle, amoebae aggregate to cAMP and sex pheromones, and two cells of opposite mating fuse and then begin consuming the other attracted cells. Before they are consumed, some of the prey cells form a cellulose wall around the entire group. When cannibalism is complete, the giant diploid cell is a hardy macrocyst, which eventually undergoes recombination and meiosis and hatches hundreds of recombinants. Not drawn to scale. Diagram by David Brown and Joan E. Strassmann (CC BY 3.0).

Self-Organization in Complex Systems

Both physical and biological processes are characterized by *self-organization.* According to Camazine et al. (2001), self-organization emerges locally from interactions among lower-level system components. Self-organizing systems are characterized by stochastic fluctuations, feedback loops, and symmetry breaking (Camazine et al. 2001). Symmetry breaking in biological systems is the process by which uniformity is broken (Li and Bowerman 2010). In self-organizing biological systems, such as the embryo, symmetry breaking may be initiated by stochastic fluctuations, which are then amplified through one or more feedback loops (Karsenti 2008; Wennekamp et al. 2013). The feedback loops for amplifying stochastic fluctuations are provided by interactions between biochemical and mechanical processes, which modify each other's behavior for generating and maintaining asymmetries (Li and Bowerman 2010). A well-studied example of biological symmetry breaking is the establishment of the anterior–posterior (AP) axis in one-celled C *elegans* embryos (Goehring and Grill 2013). Here, symmetry breaking depends on the interactions between a biochemical partitioning-defective (PAR) protein network and mechanical (actin-myosin) networks. The anterior and posterior PAR networks form mutually antagonistic negative feedback, resulting in a locally self-amplifying feedback loop. This amplification of fluctuations creates a cytoplasmic fluid flow that follows a mechanical gradient created by a cytoskeletal layer along the AP axis and eventual polarization (Wennekamp et al. 2013).

More generally, self-organizing patterns may persist for brief moments (insect swarms; Mirollo and Strogatz 1990), for decades (living tissue), or for longer periods (for example, structures built cooperatively, such as termite mounds, cities, and societies; Camazine et al. 2001; Turner 2010). Physical structures may simply erode or collapse. The connectivity of biological systems such as the brain may regress, partially collapse and remodel in response to species-specific genetic regulatory signals (for example, in apoptosis), or remodel as an adaptive response to mechanical stresses and injury (Volpe 2009). Another way to appreciate the self-organization of patterns in nature is to consider the irregularities of things that nature builds—its fractal geometry (Kelty-Stephen et al. 2013). A defining feature of fractal objects is self-similarity: the pieces resemble the whole. So, for example, a microscopic image of lung tissue reveals a resemblance between the branching structure of alveoli and the larger branches of the airway (Goldberger and West 1987). The fractality of the things that nature builds is also apparent in biological processes that unfold in time, such as the flow of currents through ion channels and proteins in the fatty membranes of animal cells that let sodium and potassium ions enter or exit the cell (Liebovitch, Jirsa, and Shehadeh

2006). Variability in the organization of human performance on tasks such as walking or reading reflects *fractal noise*, self-similarity in the temporal organization of component processes across multiple scales. As we walk, talk, read, and interact with others, fractal noise may be an indication of the coordination of control loops operative across multiple scales of temporal organization (Riley, Shockley, and van Orden 2012). Fractal noise at all temporal scales of control is an indication that the energy transactions during loops of nervous system and body in the environment are inexorably bound together over time.

Systems That Are Robust, Embodied, Anticipatory, Smart, Self-Repairing, Emergent, and Developmental

The preceding sections were oriented toward an appreciation of the complexity of biological systems. Biological organs are interconnected complex networks whose interactions make the component parts inseparable, lest they lose their emergent properties. How are we to ever develop devices that achieve a comparable level of complexity, making possible a seamless interface between biological and synthetic systems? Will we be able to emulate the biological processes that have evolved in living systems? Will synthetic wearable devices designed to assist behavioral functions—for example, locomotion—be able to achieve the same levels of adaptability, scalability, and fault tolerance as small vertebrates, such as the snake, lamprey, or salamander? In order to answer these questions, I next consider how nature has achieved seven fundamental characteristics of living systems upon which we may model synthetic devices: being robust, embodied, anticipatory (prospective), smart, self-repairing, emergent, and developmental.

Robust. Complex systems are *robust;* that is, they are able to remain stable despite fluctuations in the component parts environment (Carlson and Doyle 2002). Microbes are an ancient and dominant form of life on Earth due, in part, to their robustness. As noted by Wyss Institute faculty member Pamela Silver, a key to the robustness of microbes over eons is the formation of microbial communities, or consortia. Consortia allow microbes to touch each other, exchange genes, compete for or provide resources, and influence the growth of their neighbors (Hays et al. 2015).

Another illustration of robustness is the ability of neural circuits to maintain target circuit performance despite ongoing neuron channel and receptor turnover (Gjorgjieva, Drion, and Marder 2016; Marder and Taylor 2011). Nature builds robust systems characterized by performance trade-offs. Examples include how flying insects exhibit a trade-off between speed and stability in their take-off behavior (Card 2012); how reptiles use solar heating rather than an internal convective system; and how primate nervous systems exhibit a trade-off between wiring cost and topological value, as well as degeneracy (Edelman and Gally 2001). Natural networks are comprised of some functions that are highly conserved, while others may be tinkered with. Nature also achieves robustness by building control systems via exploration of many solutions to find one that works (Loeb 2012). This is based upon the view that solutions in nature need to be useful, not necessarily optimal. A diversity of potential solutions leads to system robustness. For example, nature has discovered a robust solution for hovering flight in both insects and hummingbirds.

Nature does not need to be robust to everything, just to the actual perturbations encountered (Lander 2007). Nature's solutions are a trade-off, a balance between the frequency of occurrence of a perturbation and the disadvantages associated with failing to compensate for it (Lander 2007). So, for example, in some flying animals such as flies (Card and Dickinson 2008a,b), nature may compensate for the predictable occurrences of certain predators in an econiche with sensor-activated escape mechanisms that are rapid and do not require the high energy cost of a highly centralized nervous system. Other animals (for example, the octopus) have evolved "hybrid" nervous systems that use widely distributed nets of neurons for responding rapidly to more predictable perturbations and reserve costly centrally organized networks of neurons for decisions for which the stakes are high (Zullo et al. 2009). Animals that occupy multiple econiches, such as the first land-invading tetrapods (Daeschler, Shubin, and Jenkins 2006)—as well as birds, animals that fly, and live on land and in the water—are perhaps more likely to encounter unpredictable perturbations. In these animals, energy-costly, highly centralized nervous systems promote active exploration for novelty and leave adaptations to more predictable perturbations of little consequence to local nets.

Embodied. Cellular processes have evolved over millions of years, resulting in the emergence of prokaryotic, eukaryotic, and multicellular morphologies, and a body plan (Davidson and Erwin 2006; Knoll and Carroll 1999; Raff 1996). The vertebrate body plan that may have emerged approximately 600 million years ago—and has remained unchanged, or conserved, to the present—is characterized by an elongated, bilateral, structure with openings at both ends connected by an internal passageway (Kirschner and Gerhart 2005). However, often missing from accounts in evolutionary biology is that the body plan is incomplete unless it is embodied within the substances, surfaces, and fields of its ecological niche (Gibson 1966; Goldfield 1995). In this book, therefore, *embodiment refers to a reciprocal and dynamical (energy-sharing) coupling among the body, nervous system, and environment of an agent (that is, animal or robot); it is a relationship in which control is distributed throughout the system components.* Living systems, and bioinspired synthetic ones, such as some robots (Pfeifer, Lungarella, and Iida 2007) incorporate the environment as an inseparable part of their functioning because they require an exchange of energy with the environment in order to adapt to its opportunities and challenges.

One way that nature expands the boundaries of individuality is by attaching environmental resources to the body. These resources may afford nourishment, protection, or functional tool-like capabilities that are not part of an animal's biological endowment. Consider the hermit crab, a decapod crustacean. The abdomen of most hermit crabs is not protected by a calcified skeleton and is vulnerable to dessication and attack by predators. For protection, hermit crabs occupy hollow objects, usually empty gastropod shells, which are then carried around as protective enclosures (Briffa and Mowles 2008). As they grow in size, hermit crabs often change between types of shells that vary in size, weight, and shape and may also attach sea anemones to the shell (Hazlett 1981).

Most hermit crabs have a curved abdomen and a clear asymmetry in the size of their chelae (claws), allowing them to grip and fit into the spiral of an empty shell. An occupied shell, thus, becomes an extension of the hermit crab body. Hermit crabs extensively explore each new potential shell by using their antennae to detect chemical information revealing shell condition by the amount of calcium detected. They also manipulate the shell with their chelae and walking legs to ascertain its size, weight, and species and may add sea anemone to help balance the shell (Briffa and Mowles 2008). Studies of *Coenobita rugosus,* a terrestrial hermit crab, have examined the process by which the animal learns to stabilize the motion of each new shell as it is walking within a semi-partitioned corridor (Sonoda et al. 2012). When a plate-like plastic extension is experimentally added to the shell, *Coenobita* rapidly learn to include the plate as an extension of their body in order to maneuver around the partitions, without touching the partitions with the plate. The legs of the hermit crab are adapted for different functions. The fourth and fifth pairs of pereopods are adapted for shell support, while walking is performed with the second and third periopods (Chapple 2012). During walking without their shell, crabs often topple over, indicating that the weight of the shell actually stabilizes crab locomotion.

A vision for bioinspired devices is that they participate seamlessly in the functioning of body organs and systems—that is, they become *embodied.* In the robotics literature, embodiment refers to a reciprocal and dynamical coupling among the brain, body, and environment, distributing control and processing throughout (Pfeiffer et al. 2007). Thus, *during the exchange of energy and information with the environment, an embodied device is one that shares control with a biological system in the performance of functional tasks.* But there are few existing robots or other devices that currently fit this definition. A challenge is to understand what may be required for an embodied synthetic device to truly share control with a biological one. The argument in this book is that *to share control with injured nervous systems, bioinspired devices must be designed to function more like mutualistic, multifunctional systems than as passive machines.* Future generations of bioinspired devices will live inside us, be worn like clothing, extend our body function in a seamless fashion, or watch over us to assist as needed (see Chapter 10). What can we learn about embodiment from biological systems?

A starting point for discussing embodiment is the work of Andy Clark, a philosopher whose ideas on embodiment have had a broad impact in cognitive science (see, for example, Clark 1997, 2003, 2008). Clark (2008) makes a key distinction among three degrees of embodiment: mere, modest, and profound embodiment. The distinction hinges on the link between control and command decisions and a body, in performing a goal-directed task. A "merely embodied" agent (biological or synthetic) is equipped with a body and sensors engaged in closed-loop interactions with the world but in which the body simply carries out commands. A "modestly embodied" agent is one in which a hard-wired architecture directs sensors to orient to some specific information and then directs it to approach the source of information. In neither of these cases is the body designed to anticipate unexpected events or change its configuration to adapt its current capabilities to new demands. An agent that is "profoundly embodied," by contrast, is guided by evolutionary and developmental routines to make open-ended internal and / or bodily adaptations to the unexpected. Minds based upon profound embodiment are "promiscuously body-and-world exploiting . . . systems continuously re-negotiating their own limits, components, data-stores and interfaces" (Clark 2008, 277).

Here, I consider Clark's taxonomy of embodiment along a continuum that distinguishes devices not only by their ability to make control decisions, but also according to their material composition, use of sensory information to provide feedback from the device to the body and nervous system, and the degree to which they share energy resources with the body. Examples of merely embodied devices include wheelchairs, prosthetic hands and feet with no sensory feedback to the body / brain, and exoskeletons with no sensory feedback to the body / brain. All of these simply execute preprogrammed control commands, use sensory feedback that is not directly provided to the user, and have battery power sources from which the user is isolated for safety. Wheelchairs are, fundamentally, motorized chairs not capable of supporting legged locomotion. Some prosthetics and exoskeletons are being designed with components that allow them to participate in the transfer of kinetic and potential energy between robot and body (Goldfarb, Lawson, and Shultz 2013; Rouse, Mooney, and Herr 2014) and use sensors to restore the experience of touch (Tabot et al. 2013), but they are still limited in using signals from the nervous system for control decisions (Thakor 2013).

Devices that cross the boundary to moderate embodiment include computer-controlled sequential muscle activation (for example, functional electrical stimulation, or FES; Corbett et al. 2013) and implanted synthetic organs such as a subretinal visual implant (Hafed et al. 2016). The current generation of merely embodied devices do not achieve profound embodiment because they lack the output from the device to multiple levels of the nervous system that provide the wearer with a psychological experience of "ownership." As I discuss later, this fundamental hurdle to achieving profound embodiment will require the kind of multisensory feedback provided by the active information pickup described in Chapter 3. In addition, profoundly embodied devices will be fabricated from materials that incorporate living cells into synthetic matrices, such as (mostly water) hydrogels (see, for example, Nawroth et al. 2012). For these hybrid devices to be viable, they will need to share cellular sources of energy, have their components repaired or replaced (perhaps by teams of nanoscale "gliobots"), and function robustly despite the constant turnover of their component parts. In other words, the boundaries between biological and synthetic components of these profoundly embodied devices will be computationally, materially, and experientially fuzzy and amorphous. Profoundly embodied devices will enter into a true symbiotic relationship with us. But we are not there yet (see Chapter 10).

There is now considerable evidence that sensory information from a body part is crucial for the experience of psychological ownership of that body part. However, it remains a difficult technical challenge to incorporate what we know about the process of information pickup by active perceptual organs—such as the skin surface, visual system, and auditory system—into profoundly embodied synthetic devices such as prosthetic limbs. The insights of James and Eleanor Gibson into the way that the body's perceptual systems detect the same amodal information through multiple means via a process of active exploration (see, for example, Gibson 1986; Goldfield 1995; Turvey and Carello 2011) provide one source of biological inspiration. Each perceptual system transduces patterned information revealed through active movements of the perceptual organs with respect to environmental structure. It is the correlation between information obtained from the multiple sensory modalities that may strengthen the synaptic connectivity of neuronal networks and provide a basis for the emergence of the psychological state of body ownership (see, for example, the perspective of Deisseroth, Etkin, and Malenka 2015, on the relation between neural activation states and the emergence of psychological states).

Anticipatory. Depictions of devices for predicting the future prototypically involve a figure peering into a crystal ball. Medical science has no crystal ball, but predicting the future is a vital need because effective treatment depends upon identification of disease before it causes irreversible damage or system failure. According to principles of homeostasis based upon traditional control theory, healthy systems are self-regulated to reduce variability and maintain physiological constancy (Cannon 1929). However, Ary Goldberger, a cardiologist and dynamicist at Beth Israel Deaconess Medical Center in Boston, has established a treasure trove of physiological data, called *physionet*, demonstrating that physiological time series contain "hidden information" characterized by nonlinearities (Goldberger et al. 2002). These data hint at a different homeostatic role for the body's regulatory systems—namely, to *maintain systems within a range of function where complexity lives: at the edge of chaos* (see Chapter 2). Living systems at the edge of chaos are healthy systems. Conversely, an individual on the verge of a catastrophic failure of one or more physiological systems (for example, a heart attack) exhibits a loss of complexity (Goldberger et al. 2002). Perhaps, then, it is possible to use this hidden information diagnostically to anticipate system failure, as I discuss in Chapter 9.

The brain itself is an organ that functions to anticipate actions. The future-oriented signals that report the ongoing activity of the body's motor systems to its sensory organs are called *corollary discharge* (Crapse and Sommer 2008). One illustration of corollary discharge comes from a study by Schneider, Nelson, and Mooney (2014) that measured the relative timing between movement initiation and movement-related signals to the auditory cortex of freely moving mice. Schneider et al. (2014) found that a wide range of natural movements strongly suppressed the synaptic activity of auditory cortical excitatory cells before and during movement. This means that intrinsic, action-specific signals prepare and guide information pickup by the auditory system, perhaps segregating body egomotion from environmental—that is, exteroceptive—sources of information. A second illustration with special relevance for the design of rehabilitation devices is the generation of cortical activity that anticipates action. For example, recent research on reaching behavior in monkeys reveals a preparatory period in which anticipation moves cortical population dynamical states to an initial value from which accurate movement-related activity follows (Shenoy, Sahani, and Churchland 2013). The motor system may use this starting point to initiate action, and then use sensory feedback to refine the activity.

As I show in Chapter 6, the brain itself develops according to a process by which immature circuits (for example, in the cerebral cortex) are initially prepared for functioning by earlier developing (for example, thalamic) circuits. A second period then follows, during which sensory input becomes available and the early circuits are refined according to the individual experiences of the animal (see, for example, Blankenship and Feller 2009). Thus, creating new circuits involves a preparatory period of intrinsic activity, with a subsequent mapping of extrinsic activity onto the existing patterns, in order to refine them. As I discuss in Chapter 7, post-injury neuroplasticity also seems to involve processes that return circuits to an earlier preparatory state so that they may re-differentiate available circuitry as an individual actively generates new post-injury experiences. Nervous system function, development, and plasticity, therefore, all inform the design of control circuits for robotic devices and other interventions for remodeling injured nervous systems and suggest a developmental framework for neurorehabilitation, presented in Chapter 9.

Smart. We are at the dawn of a new era of humanoid robots, wearable soft robots, and bio-hybrid robots—devices smart enough to assist us in activities of daily life, protect us from harm, and help to heal us when we are injured. But what makes a device smart? A bioinspired approach to robots and other smart assistive devices invites us to turn to nature for an answer. Where do we start? Because it lacks a nervous system—though it exhibits surprisingly sophisticated behaviors—a favorite starting point of experimental biologists and neuroscientists for looking at smart behavior is the slime mold *Dictyostelium discoideum,* affectionately called "dicty." *Dicty* is a social amoeba that lives in soils and feeds on bacteria and other microbes (Fets, Kay, and Velazquez 2010). Several decades of study have revealed that dicty is capable of (1) using receptors that detect chemotactic gradients and a propulsive mechanism to move toward or away from particular nutrients; (2) shifting between exploring and exploiting available nutritional resources; (3) making robust responses to perturbations, including performance trade-offs; (4) transforming its constituent cells into macroscopic organs, such as a fruiting body and a pseudopodium; and (5) building portable cases as shelter (a capability of the amoeba *Difflugia corona*) (Bonner 2010; Hansell 2005; Reid et al. 2012).

What do the impressive capabilities of *dicty* imply about the meaning of "smart"? Being smart may have emerged during the earliest evolution of animal life, with some fundamental characteristics highly conserved across all species. Subsequent evolutionary changes may have made animals better able to rapidly learn from members of their own, and other species, within more varied niches, over increasingly extended periods of time, over ever-larger and complex networks, and at further and further distances from "home." Animals are smart because their biological functions adapt at multiple time scales (that is, evolution, ontogeny, and behavior). For example, at the time scale of evolution, so-called evolutionary tinkering (Jacob 1977) is accomplished by a functional grouping of genes, a gene regulatory network

(Davidson and Erwin 2006). Gene regulatory networks consist of regulatory and signaling genes that establish which genes will be active or inactive. During the embryogenesis of animal bodies, regulatory genes regulate one another as well as other genes, and every regulatory gene responds to multiple inputs while regulating multiple other genes (Erwin and Davidson 2009). A relatively young field, called evolutionary developmental biology, or *evo devo* (Carroll 2008; Gilbert 2001; Hall 2006; Raff 1996), has emerged as a consequence of the discovery that evolutionary innovations occur through regulatory mechanisms that modify the relative timing of events during development (ontogeny) (Carroll 2008). Through cis-regulatory genes, nature transforms organs used for one function into new functions. An example is the diversification of arthropod segmental body parts (Carroll et al. 2005). Another example is the tree hopper (see, for example, Prud'homme et al. 2011), with its peculiar "helmets."

Animal behaviors are organized around the processes of actively obtaining fluctuating supplies of resources and of responding to unpredictable events in their econiches. Adaptive networks of molecules, cells, and neurons have evolved to achieve these goals in colonial organizations of insects as well as in animals. Here, the focus is on three characteristics of adaptive biological networks as inspiration for synthetic ones: degeneracy, context-aware intelligence, and distributed consensus for decision making. In a degenerate system, structurally different elements are able to perform the same function or yield the same output (Edelman and Gally 2001), a ubiquitous property of biological systems. Context-aware intelligence is characteristic of systems that obtain or actively explore the environment to extract meaningful information from available signals—that is, information that is useful for ongoing behavior. An example is the sensory receptor systems of animals, including a wide range of active somatosensory systems: whiskers that pick up vibratory information during whisking (Kleinfeld and Deschenes 2010), epidermal organs in the nose of burrowing animals (Catania 2012), and lateral line organs that detect hydrodynamic flow in fish (Coombs 2014). For biological decision-making systems in a world of choices, deciding upon ways to behave may emerge through a competition of influences at multiple levels in parallel throughout the system so that a final decision is achieved by what is called "distributed consensus" (Cisek 2012). This style of decision making is apparent not only in vertebrate nervous systems (see, for example, Cisek and Kalaska 2010), but also among large groups of social insects, such as honeybee colonies (Leadbeater and Chittka 2007).

An example of the way that animals are smart is that they are able to find life-sustaining resources, including food and water. The marine mollusc *Aplysia californica* illustrates how the earliest nervous systems adapted conserved anatomical structures for feeding (Nishikawa et al. 2007). This adaptation appears to involve changes in the sequential activation of the functionally complementary states of a feeding structure called the buccal mass. The buccal mass contains a muscular structure, the odontophore, or grasper, covered by a flexible, toothed sheet of cartilage, called the radula. The grasper has two complementary states: open-close and protract (move toward the jaws) or move toward the esophagus (retract). By regulating the timing of these states (open, close, protract, retract), the buccal mass can generate different functional responses (recall the preceding definition) for feeding, biting, swallowing, and rejection. Nervous systems also evolved to weakly couple the jaw with other structures for the pumping function of ingestion and swallowing, as in the first jawed fish. The earliest vertebrates were similar to contemporary lampreys, which are jawless fish. Embryological studies illustrate how the first jawed fish evolved through a regulatory process that changed the pathway of neural crest cell migration. The difference in allowing the formation of the jaw appears to be a difference in the timing of cellular migration events (Kuratani et al. 2001). In jawless fish, there is already a structure, called the naso-hypophyseal plate, that forms a barrier prior to neural crest cell migration. By contrast, in jawed vertebrates, there is an early separation of the naso-hypophyseal plate that allows cranial neural crest cells to enter the pharyngeal arches, and those neural crest cells migrating into the first pharyngeal arch form the mandible (lower jaw) and the maxillary process (roof of the mouth).

As aquatic animals, such as fish, became increasingly mobile and predatory, muscles became larger and more powerful. The bilaterally symmetric body plan may have promoted sequential activation of agonist-antagonist muscle groups to harness these powerful muscles for the sinusoidal motion of swimming. Increased metabolic demands for oxygen promoted the evolution of gills from existing tissue. When vertebrates that lived exclusively in water began to fill the ecological niches available in shallow water and marginal habitats, there was a dramatic change in the effects of physical forces on the body (Downs et al. 2008). Water is a neutrally buoyant environment, but a gaseous medium is not. In fish, extracting oxygen from water requires branchial and opercular pumping of water, a relatively viscous medium, across the gills. *Tiktaalik rosae*—a vertebrate that was transitional between fish and land-dwelling tetrapods—was adapted for these changes with appendages and ribs capable of supporting the body against the full extent of a gravitational load, a respiratory system less reliant on water breathing, and a head capable of independent motion provided by the neck (Daeschler et al. 2006). Fish are able to orient the entire body in order to position the mouth toward prey. The appendages of *Tiktaalik* were planted on the ground for support against gravity, but a neck provided an improved capability to orient the mouth toward food. The head of *Tiktaalik* was also well adapted for feeding because it was a more solid structure that formed a structural link among the branchial skeleton, operculum, palate, and lower jaw.

The transition from water to land, by animals such as *Tiktaalik*, may have resulted in selective pressure for internal lungs, and a system of internal shunts and valves to segregate airflow from flow of liquid and foodstuffs into the mouth. The changing oral anatomy, dentition, and diet of mammals may have promoted increased size and differentiation of their brain structures and the emergence of a chewing and swallowing mechanism for breaking down foodstuffs before swallowing. The requirement for using the same anatomical pathways for moving air to the lungs and food to the gut may have promoted a greater use of sensory information for regulating rapid transitions between one configuration of mechanical and muscular components and the emergence of mechanisms to protect the airway from penetration by foodstuffs. In the transition from four-legged to bipedal gait, larger-brained dexterous primates and early hominids, such as *Ardipithecus ramedus*, or *Ardi* (Potts 2012; White et al. 2009), became capable of transferring food from hand to mouth so that specific foodstuffs could be selected and ingested. The increased planning skills of these highly mobile bipedal primates allowed them to enter new niches and explore and select new food sources. Our primate cousins could eat in a more leisurely fashion within the safety of cohesive social groups, savoring the variety of flavors and textures of the different foodstuffs they encountered.

Self-repairing. The human body is comprised of an awe-inspiring set of internal networks for self-repair (see Table 1.2). Complex systems may use the same principles for repairing or replacing existing forms as for growing the initial ones. The possibilities for repair may be constrained by the same processes (for example, sensitive and critical periods) that limit new growth. Consider the processes involved in bone growth and bone remodeling. Bone is considered not only a skeletal organ, but also an endocrine organ that is critically involved in energy metabolism. Bone is also a mechanoreceptive organ that responds in specific ways to environmental stimulation (Karsenty 2003). For example, bone is the only tissue in the body that contains a cell type whose function is to resorb, or destroy, the host tissue (the osteoclast) under certain conditions (Harada and Rodan 2003). Destruction of bone is required during bone remodeling, a process that consumes a large amount of energy. Bone mechanoreceptors are exquisitely sensitive to physical strain, stress, and pressure and may influence bone remodeling through a set of complex feedback loops with

Table 1.2 Adaptive self-repair/remodeling of internal networks

Network	Adaptive repair process
Blood circulatory system	Vasculogenesis and angiogenesis
Central nervous system	Homeostatic regulation of neuroplasticity
Skin	Wound healing
Bone	Endocrine regulation of osteoblasts and osteoclasts
Muscle and tendon	Mechanical regulation of satellite cells and endomysium

many organ systems, including the brain (Ehrlich and Lanyon 2002; Thompson, Rubin, and Rubin 2012).

Bone serves many functions, a reflection of vertebrate evolution as well as the complex interactions between human bone, brain, and other body organs during growth and aging. Two distinct types of bone evolved in vertebrates: (1) cortical bone, consisting of concentric layers of mineralized hardened collagen, and mainly present in long bone shafts, and (2) cancellous bone, comprising a vast, interconnected, spongy trabecular network (Karsenty 2003). Our mineralized skeleton is a reservoir for calcium, with parathyroid hormone (PTH) as a regulatory stimulus for the tight control of plasma calcium (Ca^{2+}) (Zaidi 2007). Bone is, thus, an endocrine organ because the secreted protein (hormone) believed to control energy metabolism, osteocalcin, is made only by osteoblasts (Lee et al. 2007). Osteoblasts, derived from bone marrow mesenchymal cells, first construct and then model (shape) the skeleton for maximal resilience. A second type of cell, the osteoclast, derived from the hematopoietic stem cell, maintains mineral homeostasis by resorbing the extensive surface of mainly cancellous bone (Zaidi 2007). The careful balance between bone deposition and resorption is crucial for the proper development of bone size, shape, and integrity (Kronenberg 2003).

Emergent. According to Goldstein (1999), emergence occurs when novel and coherent structures, patterns, and properties arise during the process of self-organization in complex systems. Emergent behaviors have the following characteristics: (1) They have features that are not previously observed in the complex system under observation; (2) they are neither predictable nor deducible from lower or micro-level components; (3) they appear as macroscopic integrated wholes that tend to maintain some sense of identity over time, and this coherence spans lower-level components; and (4) they arise as a complex system evolves and develops over time (Goldstein 1999). Emergence in the relationship between individuals and supportive environment is apparent at the microscopic scale of bacterial "biofilms." Biofilms are microorganism aggregates in which the cells are often enmeshed in a self-produced extracellular matrix of polymer-based substances, and so adhere to each other and / or a surface (Flemming et al. 2016). The relation between bacteria and extracellular matrix exhibits emergent functional supports, including architecture and stability; a means for capturing resources, enzyme retention that provides digestive capabilities; and social interaction in formations of microconsortia (Flemming et al. 2016). The sponge-like biofilm and extracellular matrix, for example, provides sorption properties for resource capture, influencing the exchange of nutrients, gases, and other molecules between biofilm and environment. Moreover, the matrix is a kind of fortress: extracellular enzymes secreted by biofilm cells form stable complexes that are extremely resistant to dehydration (Flemming et al. 2016).

Human behavior embedded in a social matrix is also emergent. Consider speech. Its earliest forms include the repetitive trains of consonants and vowels, called babbling (Iverson 2010). My daughter, Anna, has two developmental scientists as parents, and so Beverly and I have a closet filled with audio and video recordings of Anna's early achievements. One of our favorites is the segment in which she produces her first babbling sounds, to our obvious delight. The sounds were briefly effortful and, then, as if a switch were turned on, the babbling began for minutes on end. Was Anna's babbling an emergent phenomenon? The babbling sounds certainly met all four of the preceding characteristics. And, moreover, because her babbling occurred in a social context, we responded to it excitedly and repeated and varied what she said. This was a foundational step for Anna's further speech and language development and her social advancement associated with increasingly sophisticated communication skills.

Emergence is also a fundamental concept for understanding the etiology of brain disease and psychopathologies, such as autism spectrum disorder (ASD) and schizophrenia. Unlike Anna, children with autism spectrum disorder exhibit an early deficit in social communication and speech (Frith and Happé 2005). Emergence and development go hand in hand, and the etiologies of psychopathologies reflect failures of emergence at particular critical periods in development. To come full circle from macro to micro, evidence reviewed in Chapter 10 suggests the possibility that interactions between microorganisms and the developing nervous system *in utero* may be a nexus for the initiation of cascades of events leading to psychopathologies.

Developmental. A fundamental species-specific characteristic of the primate nervous system is that cortical development is not simply additive, but rather begins with an overabundance of neurons and synaptic connections. Then, as a consequence of active exploratory behavior and social interactions, individual experience fine tunes connectivity so that some connections are strengthened, while others wither—a process that is regulated by the opening and closing of critical periods. This developmental remodeling of healthy nervous systems during the opening and closing of critical periods may hold significant clues about developmental psychopathologies, such as schizophrenia.

Complex nervous systems are robust, but errors in the developmental processes by which they are assembled are not always corrected by nature. Later behavior may be related to these very early developmental events in neurogenesis. Consider the consequences of disruptions in neurogenesis for a particular type of GABAergic interneuron, called the parvalbumin-positive interneuron (PVI). These are inhibitory, fast-spiking, and fast signaling (firing early after stimulation) interneurons, and their fundamental role in normal brain function is to regulate the activity of principal neurons (which are excitatory) (Hu, Gan, and Jonas 2014). However, while PVI are critically involved in balancing excitatory activity to generate oscillations, they are not simply network stabilizers. PVI also contribute to advanced computations, such as path integration in the hippocampus, as well as to regulation of plasticity and learning (Hu et al. 2014). Genetic mutations in a protein selectively expressed in PVI in several brain regions are frequently found in schizophrenic patients (Lewis, Hashimoto, and Volk 2005). These mutations may disrupt the process of providing molecular guidance for neurons migrating throughout the cerebral cortex (Ayoub and Rakic 2015). Moreover, PVI are metabolically "hungry," and so are vulnerable to oxidative stress. Mistimed plasticity causes damage to the interneuron itself, as well as to its surrounding perineuronal nets and myelin-forming oligodendrocytes, thus altering excitation-inhibition balance and destabilizing circuit function (Do, Cuenod, and Hensch 2015). PVI are critical for integration and segregation of neuron populations in network communities (Sporns 2013a,b), and regulate 30–80 Hz cortical gamma band oscillations involved in higher cognitive functions, such as attention and social cognition (Siegle, Pritchett, and Moore 2014). Destabilization of these circuits, over time, may eventually lead to the adult onset of the psychotic behaviors that are characteristic of schizophrenia (Green, Horan, and Lee 2015).

Biologically Inspired Devices: Bridging the Gap

A cold truth for scientists and engineers working to develop biotechnologies intended for clinical use is that few medical devices ever leave the laboratory. As Harvard chemist and Wyss faculty member George Whitesides notes, "the transition to a commercial device is a difficult and expensive task in most fields, but it can be Herculean in medicine" (Kumar et al. 2015, 5837). One governmental response to this challenge was the Catalyst, a federal program funded by Clinical and Translational Service Awards (CTSA) to medical research institutions throughout the United States. The Harvard Catalyst, for example, has as its goals (1) to reduce the time it takes for laboratory discoveries to become treatments for patients, (2) to engage communities in clinical research efforts, and (3) to train the next generation of clinical and translational researchers (see the CTSA website, http://ctsa.web.org). Over the past decade, the Harvard Catalyst has succeeded in promoting translational medical research, but its mission has not specifically been to identify and solve challenges in bioengineering and biotechnology.

In response to a challenge by the Harvard Provost to re-envision the future of bioengineering across the entire university, an interdisciplinary group of faculty proposed to leverage "how nature builds, controls, and manufactures" as a foundation for a new bioengineering institute. Initially seeded with Harvard funds in 2008, the Wyss Institute for Biologically Inspired Engineering at Harvard was established with a $125 million gift to the university by Hansjorg Wyss in 2009 (http://www.wyss.harvard.edu/). Its mission statement is as follows:

> The Wyss Institute aims to discover the engineering principles that Nature uses to build living things, and harnesses these insights to create biologically inspired materials and devices that will revolutionize healthcare and create a more sustainable world. In medicine, the institute is developing innovative materials, devices, and disease reprogramming technologies that emulate how living tissues and organs self-organize and naturally regulate themselves. Understanding of how living systems build, recycle, and control is also guiding efforts focused on development of entirely new approaches for constructing buildings, converting energy, controlling manufacturing, and improving our environment.

This book highlights work being done at the Wyss Institute, but not exclusively. As an Associate Faculty member at the Wyss, I am privileged to have an insider's view. But I take a much wider perspective by including work being done at other institutions collaborating with the Wyss, such as my home institution, the Boston Children's Hospital, as well as in the broader scientific community. The book takes a biologically inspired approach to materials and devices for remodeling injured nervous systems. At this juncture, it is crucial to note that not all biologically inspired approaches are alike. The approach at the Wyss explicitly (1) emulates principles from mechanical biology (for example, the effects of flows and forces on tissue) as a way of leveraging developmental processes at the multiple scales, (2) emulates nature's regulatory systems (for example, gene regulatory networks, and neuron-glia interactions) for maintaining dynamic homeostasis throughout the body, and (3) emulates self-organizing collective behavior (for example, "swarming" of local "agents") as the basis for building devices with emergent functionality.

Biological inspiration within the unique collaborative environment at the Wyss has accelerated the discovery of new, and even unexpected, technologies. One illustration is development of organs on chips, led by the Wyss Institute's founding director, Don Ingber (see Chapter 2). What distinguishes these microfluidic systems from other labs' chip technologies is that they emulate the effects of mechanical forces on cell function, in health and disease, a kind of reverse engineering (Ingber 2016). Here, the technology follows from first principles in mechanobiology: living cells are placed within microfluidic environments that emulate the mechanical forces characteristic of particular body organs, such as the healthy (Huh et al. 2010) or diseased lung (Benam et al. 2016).

Discoveries in the areas of biological networks and cellular devices by Wyss Core faculty members George Church, Jim Collins, Neel Joshi, and Pam Silver have led to breakthrough technologies, ranging from personal genomics (M. P. Ball et al. 2012) to programmable bacteria (Kotula et al. 2014). Breakthroughs in mathematics by Wyss Core faculty member L. Mahadevan, in collaboration with the engineering wizardry of Wyss professor Jennifer Lewis, has crossed the threshold to "four-dimensional" bioprinting of materials that, like plants, undergo fluid-controlled growth (Gladman et al. 2016). And advances in decentralized control systems by Wyss Core faculty member Radhika Nagpal, and in microrobotics by Wyss Core faculty member Rob Wood, have made possible my own work in wearable robots for young children (see, for example, Goldfield et al. 2012). A challenge in writing this book has been to place the problem of emulating nature's assembly and repair processes in the context of this exciting work at the Wyss and in the context of the dizzying pace of advances in laboratories around the world.

✦ 2 ✦

How Nature Builds: Physical Law, Morphogenesis, and Dynamical Systems

Forces are invisible. However, they impel particles that we are able to detect with our eyes, ears, and skin and with the aid of devices such as cloud chambers, microscopes, and motion capture cameras. With unaided organs for seeing, hearing, smelling, tasting, and touching, we encounter dust motes, compressions and rarefactions of air molecules, water droplets, chemical molecules, grains of sand, and moving bodies. Particles may move in probabilistic Brownian motions in still air, but thermal or chemical gradients, or lines of force, may stabilize their fluctuations into organized patterns. Their tendency to obtain electrical charge or chemical affinity also makes particles attract or repel each other. When particles travel in large numbers, those that attract each other tend to move together, and those that repel each other form layers and boundaries between them. When they encounter enclosures, trailing particles follow the lines of force created by leading ones and begin to whirl and flow in vortices.

When particles come to rest, they leave physical traces of the forces that have brought them there. For example, dunes form from velocity differences of sand particles being transported by fast moving air along a surface. As sand accumulates, dunes grow as accumulations create a "wind shadow," a region of suppressed wind velocity that slows sand particles so they settle onto the surface (Goehring and Grill 2013). Outside the limit of each wind shadow, the process begins again, creating the characteristic spacing between dunes. The physical traces of forces acting on particles may create even larger geologic patterns. For example, in his *Hebrides Overture,* the composer Felix Mendelssohn captures in sound both the echoes and the striking visual patterns of columnar jointing in Fingal's Cave, which he visited in 1828 as part of a journey to Scotland. Closer inspection of these columns in cross-section reveals a roughly hexagonal pattern. Such patterns are geometrically similar to crack formation and propagation in materials that initially flow and then solidify at the interface between an underlying moving substrate and an environmental source of thermal energy (Goehring, Mahadevan, and Morris 2009).

Living cells are also particles that are impelled by mechanical (for example, extracellular) forces but, significantly, have evolved receptor systems for detecting other particles that are attractive or repulsive, as well as the ability to propel themselves along attractive or repulsive gradients. Moreover, cells have evolved feedback loops and actuation to selectively modulate how they impel themselves along attractive or repulsive gradients. Cells that attract each other move together, and the mutual feedback loops of cells making contact with each other result in the emergence of a more macroscopic collective, biological tissue. The traces left behind when cells form a layer between underlying growing tissue and the environment exhibits the same hexagonal cracking pattern that Mendelssohn observed at Fingal's Cave. For example, patterns of crocodile head scales emerge from the cracking that occurs when the skin layer undergoes stresses due to the rapid growth of the underlying facial and jaw skeleton (Milinkovitch et al. 2013). How may we understand the process by which cells are both guided by forces that impel them along gradients and lines of force and, at the same time, use internal energy flows to select some behavioral pathways but not others? This question applies not only to individual cells at the microscopic scale, but also to networks of microscopic components that behave as goal-directed agents, including animals, humans, and machines.

D'Arcy Thompson: Growth and Form

At the turn of the twentieth century, in a classic volume entitled *On Growth and Form* (1942), biologist D'Arcy Thompson, proposed that the shapes of cells; the compartments of honeycombs, corals, and mollusc shells; and the striped markings of animals all were governed by physical law and could be expressed mathematically. Thompson offers some unique insights for understanding the relation between fields of energy and the forms assumed by living tissue and made the claim, revolutionary for its time, that biological tissue, like inanimate matter, reveals a "diagram of forces" acting on it. Thompson recognized the patterns common in the mechanical stresses on inanimate materials and in the growth of biological forms under the same physical forces. So, for example, he considered the evolution of bone morphology with reference to the tension and compression resulting from heavy loads on beams and girders.

D'Arcy Thompson's discovery of a fundamental mathematical scaling relationship between natural material structures and permissible forms hinted at an underlying heterogeneity in the structures self-assembled by nature at different scales. During the century since publication of his book *On Growth and Form* in 1917, the development of microscopes capable of revealing structure down to the nanoscale (and beyond) has confirmed many of D'Arcy Thompson's mathematical insights. For example, as revealed in the skeleton of the sponge *Euplectella*, self-assembly of biological materials involves joining constituent parts to form a distinctive geometry at a microscale and then joining together already-established geometries into new geometries at increasing scales in the hierarchy (Aizenberg et al. 2005). Under extreme loads, the structural stability that holds at a macroscale breaks down at more microscopic scales because the forces acting on the structure cascade via feedback loops through component geometries unable to resist those forces. The heterogeneity and scale-dependent vulnerability of geometric structure across a hierarchy of scales, providing both flexibility and stability, is one of the foremost challenges in fabricating biologically inspired materials.

The seahorse offers a tale that sounds more like Aesop than D'Arcy Thompson. The tail of the seahorse has become modified for grasping and holding onto seagrasses to maintain a vertical posture in a water column during feeding on small crustaceans (Neutens et al. 2014). While holding on to the substrate with the tail, the seahorse rotates the head, brings the mouth close to the prey, and uses suction feeding to draw them through the snout into the mouth (van Wassenbergh et al. 2013). In other words, the tail serves as an anchor to the substrate during feeding. The seahorse tail is a most curious structure.

Rather than being cylindrical, it is square: its jointed skeleton consists of a bony armor with L-shaped plates surrounding a central vertebra (Neutens et al. 2014; Porter et al. 2015). Is there a relationship between the seahorse tail architecture and its use for grasping onto seagrasses? To investigate this question, Porter et al. (2015) emulated the square geometry of the tail, and the three joints made possible by this geometry, in comparison with a cylindrical model. A ball-and-socket joint connects adjacent vertebra and constrains bending. A peg-and-socket joint connecting the plates of adjacent segments more effectively restricts twisting in square compared to a cylindrical model. Under experimental impacts and compression, the plates of the square shape slide past one another, making it stiffer, stronger, and more resilient than the cylindrical model. Physical modeling, thus, reveals that nature's square design is well suited to the functional requirements of seahorse suction feeding.

In *On Growth and Form,* D'Arcy Thompson also proposed a theory of transformations that related the morphology of one plant or animal to another. For example, to transform the body of one plant or animal into another, he drew its outline onto a sheet with a Cartesian grid and then geometrically changed the shape of the grid. With his grid method, the transformation of one type of fish revealed the morphology of another creature entirely. Perhaps the most important lesson to be learned from this approach to growth transformation is that, like the points on the grid, developmental processes occur in a coordinated, rather than piecemeal, fashion (Arthur 2006). It is noteworthy that this transformational method takes as a given an already mature form of a plant or animal, without addressing the nature of the influences on growth that produced that morphology at its origins.

Turing and Morphogenesis

Biologists have been slow to embrace mechanical forces as a major contributor, along with biochemical processes, to the emergence of complex systems from cell progenitors during evolution and ontogeny. This is evident in the history of the subdisciplines of biology that have studied this question since the 1917 publication of *On Growth and Form,* including mechanical biology, embryology, and evolutionary biology. With the dawning over the next two decades of the "modern evolutionary synthesis" of Mendelian genetics and Darwin's theory of natural selection, the focus in developmental biology shifted to population thinking—that is, how small genetic changes influenced phenotypes in their surrounding environment (Mayr 1982).

There were, however, some notable exceptions among scientists who nurtured the idea that mechanical forces in concert with biochemical processes influenced the morphology and behavior of cells. One was the biologist Paul Weiss (for example, Weiss and Garber 1952) who, in the 1940s and 1950s, conducted experiments that systematically manipulated the properties of the substrate on which cells were grown in order to induce strain forces and then observed the effects on cell shape. Like D'Arcy Thompson, Weiss was able to show a lawful relationship between the mechanical properties of the substrate and cell shape.

A second scientist with profound contributions to many fields, including mathematics and biology, was Alan Turing. He was, of course, best known for the introduction of what is now called the "Turing machine," a foundation of modern computation and cybernetics, as well as for leading the effort to crack the German "Enigma Code" during a pivotal period of World War II. In "The Chemical Basis of Morphogenesis," published in 1952, Turing considers how an embryo's development unfolds at each instant from its *molecular and mechanical state* (Reinitz 2012). To do so, Turing created the term "morphogen," an abstraction for a model of a molecule that could induce tissue differentiation (Reinitz 2012). He further proposed a mathematical model of an idealized embryo consisting of an initially uniform concentration of morphogens from which patterns emerge.

Turing's abstract morphogen concept became realized more than three decades later in the discovery in the *Drosophila melanogaster* zygote of the first morphogen, the transcription factor Bicoid (Frohnhofer and Nusslein-Volhard 1986), and reaction-diffusion mechanisms have become central to understanding pattern formation in what developmental biologists call morphogenesis (Guillot and Lecuit 2013). Morphogenesis is the shaping of an organism by cell movements, cell–cell interactions, collective cell behavior, cell shape changes, cell divisions, and cell death (Keller 2013). Morphogenesis occurs simultaneously at short temporal scales in subcellular rearrangements and at long temporal scales in tissue. The process of morphogenesis has been studied in the mechanics of epithelia—densely proliferating cells organized into layers and joined by strong cell-cell adhesive contacts (Guillot and Lecuit 2013). The function of epithelia in embryos is to provide structural integrity and act as a barrier to pathogens and as protection from dehydration (Purnell 2013). At the tissue level, epithelia exhibit distinctive hexagonal cell topology in which edges are junctions between two cells and vertices identify points of contact among three or more cells (Guillot and Lecuit 2013). Hexagonal topology appears to emerge from both rapid cell proliferation and maintaining structural integrity (Gibson et al. 2006). A seemingly paradoxical effect of mechanical forces on cell shape is apparent in experiments in which epithelial tissue is stretched for either short time periods (less than a few minutes) or for tens of minutes. In the former case, cells change their shape, resulting in a change in tissue geometry, but in the latter, cells begin to exhibit fluid behavior and tissue flows (Guillot and Lecuit 2013). How may we understand these phenomena?

In Vivo Imaging of the Process of Morphogenesis

A fundamental technical challenge for *in vivo* imaging of morphogenesis is that cell movements, cell–cell interactions, collective cell behavior, and cell death are processes that occur at spatial scales from hundreds of nanometers to several millimeters and at temporal scales from fractions of a second to days (Keller 2013). Figure 2.1 presents images of a zebrafish embryo at multiple spatial and temporal scales using light-sheet microscopy, a technology that allows entire multi-mega-pixel images to be acquired by using a laser to illuminate a micrometer-thin volume, and collecting the fluorescence emitted by reporter molecules with an objective lens (Keller 2013). Once these

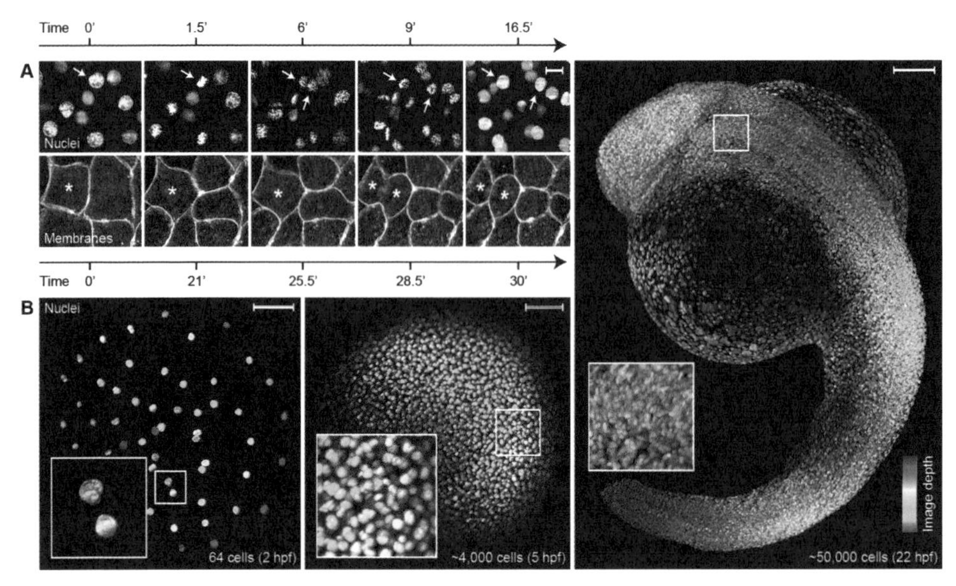

Figure 2.1 Spatiotemporal scales of morphogenesis. Morphogenesis, illustrated here for the zebrafish embryo, involves multiple spatial and temporal scales: from fast, local cell shape changes and cell movements (a) to slow but large-scale structural changes on the tissue- and whole-embryo level (b). Images were recorded with digital scanned laser light-sheet microscopy (DSLM) and simultaneous multiview light-sheet microscopy (SiMView). (a) Nuclei (top) and membrane (bottom) dynamics in two different locations of a two-hour old zebrafish embryo. Arrows and asterisks highlight examples of dividing cells. (b) Whole-embryo projections of nuclei-labeled zebrafish embryos at different developmental stages, showing the rapid increase in cell count and morphological complexity in early embryogenesis. Insets: enlarged views of highlighted areas. Scale bars: 20 mm (a), 100 mm (b). Keller, P. J. (2013). Images reprinted with permission from AAAS.

images are obtained, visualization tools and physical modeling are used to interpret the resulting data. Here, we can actually see how the process of development at the microscopic scale of cells translates to the emergence of macroscopic forms.

More recently, it has been possible to significantly improve spatial imaging resolution of large multicellular organisms, such as *Drosophila* and zebrafish, with a super resolution imaging technique called IsoView microscopy (Chhetri et al. 2015). Through the use of scanned spatially offset Gaussian beams for sample illumination, IsoView microscopy simultaneously acquires four orthogonal views of the specimen (Liu and Keller 2016). The spectacular images obtained with this new technology raise the question of how morphogenesis emerges from interaction of genetic, cellular, and mechanical inputs. Gilmour, Rembold, and Leptin (2017) propose that tissue shaping proceeds *from morphogen to morphogenesis and back*: gene regulatory networks control when and where a specific tissue-shaping process will occur, but interact with proteins and mechanical influences to dynamically feed back and modulate their activity. What is the nature of this dynamical feedback on gene regulatory networks?

Tensegrity and the Emergence of Mechanical Biology

Scientific advances may occur when one or more individuals bring together critical observations from disparate disciplines, leading to nonobvious hypotheses and, in a process that may take decades, creating entirely new fields. Such is the case for the hypothesis that cells are tensegrity structures proposed by Wyss Institute Founding Director Don Ingber (see, for example, Ingber and Jamieson 1985). The application of tensegrity to the cell was based upon an insight he had as a student in the 1970s and has contributed to the emergence of the fields of mechanical biology and mechano-therapeutics in the twenty-first century (Shin and Mooney 2016). I showed earlier that D'Arcy Thompson promoted mechanical forces as a fundamental influence on growth. Most notably for relics of extinct animals, such as dinosaurs, D'Arcy Thompson attempted to deduce how forces shaped the skeletal anatomy. However, it is not until we put soft tissue back on those bones (for example, through modern computer simulations and finite element modeling) that it becomes clear how hard structures are mechanically stabilized through tensile linkages with the muscles and tendons. Indeed, unless museum pieces are carefully wired together, they will collapse [as in one of my favorite films with Cary Grant and Katharine Hepburn, *Bringing Up Baby* (1938)]. I also discussed earlier Alan Turing's 1952 proposal of a reaction-diffusion mechanism for morphogenesis: a slowly diffusing local activator and a rapid long range inhibitor, through their mutual interactions, created spatial patterns. We now know that mechanical forces are able to travel at the speed of sound through biological tissue (a million times faster than diffusion), so long-range mechanical stress may be involved in generating biochemical patterns with length scales exceeding that of diffusion alone (Howard, Grill, and Bois 2011).

Cells. Ingber's tensegrity hypothesis of cell morphogenesis (Ingber 1998, 2006) provides a nonobvious critical link between the biochemical processes of Turing morphogens and the lines of force of D'Arcy Thompson. According to Mammoto and Ingber (2010), the form of living cells is controlled via a tensional integrity, or *tensegrity*, in which shape emerges from the balance between tensile and compressive forces, establishing a state of isometric tension or prestress (see Figure 2.2). Tensegrity offers a mechanism by which cytoskeletal

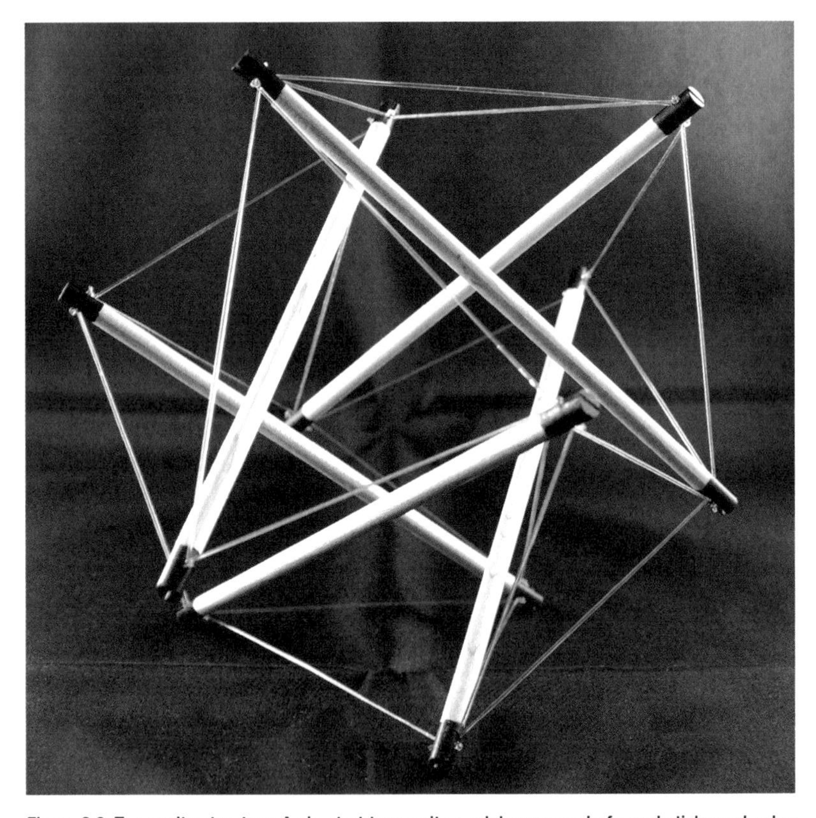

Figure 2.2 Tensegrity structure. A six-strut tensegrity model composed of wood sticks and nylon strings. Note that the struts do not come in direct contact, but rather are suspended open and stabilized through connection with the continuous series of tension elements. Tensegrity structure by Donald E. Ingber. Photo courtesy of Wyss Institute for Biologically Inspired Engineering at Harvard University.

structures may provide mechanical pathways for molecular signaling within cells: soluble morphogens may exert their effects, in part, by inducing changes in cell and tissue mechanics that, in turn, feed back to modulate chemical signals governing collective morphogenetic cell movements. To unpack this hypothesis—and its implications for inspiring the engineering of devices for promoting cellular remodeling—I turn to key developmental steps during embryogenesis and how mechanical forces may influence three fundamental processes: cell division, gastrulation, and morphogenesis (see Table 2.1). As summarized by Wozniak and Chen (2009), embryogenesis begins as zygotic

Table 2.1 Roles of mechanical forces during embryogenesis

Developmental process	Role of mechanical forces
Asymmetric cell division (morula)	Mechanical forces generated in the cytoskeleton modulate spindle position through establishing a cytoskeletal force balance with resisting microtubules, responsible for asymmetric cell divisions of the morula.
Blastocyst formation	Local changes in physical forces and in mechanical properties (stiffness) of the ECM contribute to control of gene transcription that drives cell fate switching.
Cell sorting	Sorting of progenitor cells into three distinct germ layers (ectoderm, mesoderm, and endoderm). The three germ layers display different surface tension properties, which prevents random mixing.
Axis formation	There is a simultaneous narrowing and elongation of tissue, called convergent extension, that transforms the entire embryo from a round to an elongated shape.
Tissue folding and invagination	Coupling of cell-generated mechanical forces through cell–cell adhesions results in apical constriction of epithelial cells, producing tissue invagination.
Dorsal closure to form hollow tube	Actin cables at the leading edge of the dorsal epidermis generate traction forces that pull the edges toward the midline. At the same time, cell shape changes generate additional tugging forces toward the midline to produce closure.

Source: T. Mammoto, A. Mammoto, & D. Ingber (2013), Mechanobiology and developmental control, *Annual Review of Cellular and Developmental Biology, 29,* 27–61.

proliferation generates a blastula, which then forms a blastocyst. Gastrulation transforms the blastocyst into a gastrula, with its characteristic (usually three) germ layers: mesoderm, ectoderm, and endoderm. This transformation involves the sorting of progenitor cells and positioning of the mesoderm and endoderm beneath the prospective ectoderm by means of apical constriction and internal movements. Events called epiboly, including intercalation, then expand and thin these nascent germ layers. Finally, the embryo narrows medio-laterally, and lengthens antero-posteriorally by means of the process of convergence and extension.

Tensegrity may provide a link between cell behavior at an individual level and their collective organization as tissue (Blanchard and Adams 2011). By what means do cells engage in collective migration? One possibility is that there is a mechanism of self-organization by which the dynamics of cell interactions themselves result in a transition from rearrangements to one of collective flowing behavior. One requirement for a transition from individual to collective behavior by means of self-organization is cell–cell communication: cells must act as sensors. Prestress identifies a mechanism by which mechanical forces imposed on cells may spread through tissue over long distances and thus allow cells to behave as mechanosensors (Ingber 2006; Kollmannsberger et al. 2011). A second requirement for self-organization of collective behavior is a process that connects the output signals of each cell back as an input to the cell—that is, a source of positive feedback (Brandman and Meyer 2008). The way that mechanical signals are amplified to emerge as order at a more macroscopic scale is summarized by Kollmannsberger et al. as follows:

> When many cells are mechanically linked directly, or via the extracellular matrix, the response of individual cells to mechanical signals is transmitted to other cells and leads to a feedback loop, extending the interaction range. In multicellular tissues, cooperative behaviors emerge such as mechanically patterned proliferation and differentiation, extracellular matrix alignment, or curvature controlled tissue growth. This gives rise to order over distances much

larger than the range of interaction between individual cells. (2011, 9557)

Before considering the mechanical forces acting during embryogenesis and organogenesis, I turn briefly to the question of how it is even feasible to measure forces at the scale of a cell. Some of the techniques currently used to measure cell-level forces include two specialized types of microscopy: atomic force microscopy that deforms the surface of individual cells and records force-modulation curves (Krieg, Heisenberg, and Muller, 2008) and traction force microscopy that measures traction forces of single cells cultured on a substrate of known elasticity (Young's modulus) (Li et al. 2009). A novel technique developed at the Wyss Institute uses time-lapse fluorescent microscopy to reconstruct *in situ* the shape of three-dimensional, cell-sized microdroplets deformed by contact with cells within living tissue. Each microdroplet functions as a kind of force transducer because it has known mechanical properties, and deformation from its equilibrium spherical shape reveals the forces acting on it, as determined by confocal microscopy and computerized image analysis (Campas et al. 2014). For example, using this technique, Serwane et al. (2017) have discovered that zebrafish body elongation during embryogenesis involves spatial variation of tissue mechanics along the anteroposterior axis.

Tissues and organs. All of these remarkable techniques are revealing how mechanical forces exert control of tissue and organ development. During early embryonic development, physical interactions between cytoskeletal microtubules and actin myofilaments modulate spindle position to control asymmetric versus symmetric cell divisions of the morula (Grill and Hyman 2005). In the formation of the blastocyst, local changes in physical forces and mechanical properties of the extracellular matrix (ECM), such as stiffness, appear to actively contribute to the control of gene transcription that drives cell fate switching (Mammoto, Mammoto, and Ingber 2013). In the process of progenitor cell sorting into three germ layers, each layer displays different surface tension properties, which prevents random mixing (Halder, Dupont, and Piccolo 2012). A contractile cytoskeletal network generates inward-directed tensional forces on the cell's surface adhesions to the underlying ECM and neighboring cells as the basis for tissue folding and invagination (Mammoto et al. 2013). Elongation of embryonic tissue is propelled by traction forces exerted at cell–cell junctions, which elongate the tissue by inducing shortening in the mediolateral direction (convergence) and extension in the antero-posterior direction (Mammoto et al. 2013). Finally, tissue growth regulation, as in the *Drosophila* wing, involves gene-regulated growth inhibition of centrally located cells as they become physically compressed. At the same time, cells in the periphery slow their proliferation because they have grown beyond the influence of morphogen gradients. The result is uniform growth throughout the entire wing.

A further illustration of the mutual influences between cellular responses and the mechano-chemical environment is the branching architecture of tubular forms during organogenesis. Branching morphogenesis is evident in biological organs that include the gut and gastrointestinal tract, the heart and vessels of the circulatory system, and the lungs. Nelson and Gleghorn (2012) demonstrate that in order to build tubes, nature uses a process of wrapping and budding of epithelial sheets. Branching morphogenesis occurs when an epithelial sheet is induced to form a tube: a thickened region of the epithelium induces its cells into a wedge-like shape that locally bends the epithelial tissue. Local reiteration of this process results in the fractal-like branching patterns of organs, such as the lungs. As the lungs mature during the fetal stage of development, specific mechanical environments promote integrated function. Breathing movements, secretions, peristalsis, flow, actomyosin contractility, and differential material properties are all involved in lung development of the fetus (Nelson and Gleghorn 2012).

Emergent Forms

Our evolutionary heritage in animals that lived in the oceans, and then moved to land, raises the question of the possible influences that mechanical forces in the flow of water may have played for the formation of new functions in tissue. One way to investigate this possibility is to look for extant animals that continue to live in environments at the transitional boundary of water and land. Consider the mudskipper *Periopthalmus barbarous*, a member of the most successful group of fishes capable of extended terrestrial foraging (Michel et al. 2015). A most curious characteristic of the mudskipper is that it comes out onto land with its mouth cavity filled with water. The mudskipper does not have a tongue, but during the initial capture and subsequent intra-oral transport of food, it uses the water in its mouth as a "hydrodynamic tongue" (Michel et al. 2015). To use water to function as a tongue, the mudskipper propels a meniscus of water from the mouth to engulf its prey and then suck the engulfed prey back into its mouth. This "protrusion and retraction" of water exhibits kinematic and functional resemblance to tongue movements in lower tetrapods. Moreover, the kinematics of the hyoid during use of the hydrodynamic tongue bears striking resemblance to that of the hyobranchial structures supporting the true tongue of closely related land animals, the salamandrids (Michel et al. 2015).

Mechanical forces also induce a characteristic vertebrate growth pattern in the embryonic gut tube. For example, Savin et al. (2011) demonstrate experimentally and mathematically that tissue-scale forces play a role in determining the characteristic number and amplitude of gut loops in different vertebrate species. Their experiments include different types of dissections of the gut tube and mesentery, with the idea that specific mechanical forces generated on the gut tube by the mesentery are at least partially responsible for gut looping. The critical comparison was dissections that either left the two organs attached to each other or completely separated them from each other. The latter dissection, but not the former, resulted in an uncoiling of the gut into a straight tube and a relaxation of the mesentery into a thin, uniform sheet. Thus, the two organs generate complementary mechanical influences on each other during growth, creating the distinctive geometry of each. As a further test, Savin et al. (2011) constructed a physical model of the gut (a rubber tube) and mesentery (an elastic sheet), with dimensions appropriate to particular vertebrate species, and made predictions from each animal of the number of loops that the model would form. The model accurately predicted species-specific number and amplitude of gut loops, demonstrating the fundamental contribution of mechanical forces to the process of organogenesis.

The crucial role of mechanical forces for promoting patterning of flows during organogenesis is dramatically demonstrated in the role that motile cilia dysfunction may play in idiopathic scoliosis, a deformation in vertebrate spinal curvature (Grimes et al. 2016). Long, motile cilia are present on the surface of specialized cells that project into the extracellular space, where they generate directional flow of extracellular fluids, including cerebrospinal fluid (CSF). During formation of the vertebrate spinal cord, the polarized beating of ependymal cell cilia lining the ventricles breaks left-right symmetry of flow. However, in a mutant zebrafish model of idiopathic scoliosis, in which a particular zebrafish genetic mutation exhibits all of the characteristics of the human disease, disruption of cilia motility during embryogenesis appears to have profound effects on the curvature of the developing spine. Grimes et al. (2016) tracked CSF flow by means of fluorescent microsphere movement across the ventricles in mutants and normal zebrafish. In the normal zebrafish, there was a robust anterior to posterior flow, but in mutants, the microspheres exhibited both irregular trajectories and significantly reduced speeds (Grimes et al. 2016). These results support the hypothesis that cilia-driven CSF flow plays a critical role in spinal development and that irregularities of flow are an underlying biological cause of idiopathic scoliosis.

Mechanical Constraints on Energy Flow and the Emergence of Cellular Networks

Living cells are "excitable media" that aggregate into complex organizations and have the capability of channeling energy flows for growth differentiation and the emergence of new forms (Newman 2012). During evolution and development, new functionality emerges in cell aggregates by channeling energy flows in complex ways, such as via the formation of networks (Nicosia et al. 2013). The formation of large brains with many distinct functional regions has high survival value. However, there is a limiting constraint of spatial scale on wiring costs (that is, the distance between regions at a large distance from each other in large brains). Nature's evolutionary solution to the high survival value of new network architectures, and the concomitant constraints of wiring costs, is a performance trade-off between local aggregates and some long-distance connections (Bullmore and Sporns 2012). In humans, the wiring of these "small-world" brain networks is not only expensive to build, operate, and maintain, but as I will show in Chapter 6, *the developmental processes that build the human brain result in particular structural vulnerabilities and tendencies toward failure.*

How does nature channel energy flows to build complex networks? D'Arcy Thompson demonstrated mathematically that growth of biological forms could be understood as geometric transformations of one shape into another. While he recognized that different material structures imposed constraints on growth transformations, neurobiologists have only recently considered the potential influences of mechanical forces to impose geometric transformations on cell growth (Hilgetag and Barbas 2006; Taber 2014). These reciprocal influences include how path-finding axons exert mechanical strain on neural tissue and how transformation of the neural tube into more complex solid geometries imposes bending and other shape changes on fiber pathways.

A suggestion from recent findings in neuroembryology is that circuit pathways may follow morphogenetic (mechanical and chemical) growth gradients. One illustration of the principle that neural circuitry follows morphogenetic growth gradients is the discovery by Wedeen et al. (2012) that cerebral fiber pathways of nonhuman primates and humans form a rectilinear three-dimensional grid based upon the rostro-caudal, medio-lateral, and dorso-ventral principle axes of embryonic development. Wedeen et al. (2012) used diffusion MRI to map the path crossing and path adjacency of fiber pathways. When they computed paths within regional neighborhoods—for example, in the frontal lobe of the rhesus monkey—they found transverse callosal paths and longitudinal paths of the cingulum crossing like a woven fabric. Wedeen et al. propose that this architecture is the fiber substrate of a pervasive sheet structure, found throughout the cerebral white matter and in all orientations and curvatures in homologous structures of all species examined (that is, galago, marmoset, owl monkey, rhesus, and human). They hypothesize that a grid structure could provide a means for axonal pathfinding throughout the entire brain, not just a few limited pathways. More recent work by Wedeen and colleagues (Mortazavi et al. 2017) provides additional support for this hypothesis by combining dMRI tractography, axon tract tracing, and axon immunohistochemistry. Through use of these methods, Mortazavi et al. (2017) demonstrate that axon fibers navigate deep white matter via microscopic sharp turns, that is, approximately right angles, between the three orthogonal primary axes of a geometric grid.

These findings suggest that neuroembryological changes in sheet geometry may create opportunities for the emergence of new circuit architectures. One way to examine this possibility is to compare circuit architecture during transitions in the three-dimensional shapes of the developing nervous system. Douglas and Martin (2004, 2012) compare the circuits extant during the period when the nervous system takes the form of a segmented, axially symmetric neuroepithelial tube with the circuits that emerge during formation of the forebrain. The forebrain has two major subdivisions: the more rostral telencephalon and the more caudal diencephalon. The dorsal telencephalic pallium forms a cortical sheet of neurons that, in humans, has a characteristic six-layered cortex. Douglas and Martin propose that the modified coordinate framework of the forebrain induces interactions between the circuit organizers. However, the opportunity for the emergence of new neocortical circuit functionality is constrained by its geometry as fundamentally a two-dimensional sheet. The result is a patchwork of heterogeneous regions localized to some region of the sheet. For example, motor cortex consists of regions that are "patches, overlaps, and fractures" separated by "fuzzy borders" (Graziano and Aflalo 2007).

Organizers

The waxing and waning of subdisciplines in biology during the early twentieth century is exemplified by work on "organizers" in the field of experimental embryology (see, for example, de Robertis 2006). The most famous experiment in embryology, conducted by Spemann and Mangold in 1924, demonstrated that animal development results from successive cell–cell inductions in which groups of cells, or organizing centers, signal differentiation of their neighbors (de Robertis 2006). The discovery that development was induced by organizing centers earned Hans Spemann a Nobel Prize in 1935. However, the idea of organizer—and the entire field of experimental embryology—fell into disfavor in 1941, when it was discovered that saline solutions could also induce neural tissue in explants. This set the field back for forty years.

Spurred, in part, by biologist Viktor Hamburger's 1988 memoir describing Spemann's earlier work, laboratories armed with the late twentieth-century cloning technologies of molecular biology cloned the endogenous factors of the Spemann organizer. It is now known that Spemann's organizer produces signals by secreting a "cocktail" of bone morphogenetic protein (Bmp) and antagonists in the Wnt family (de Robertis 2006). Embryonic self-regulation now appears to result from multiple signaling centers—for example, at opposite poles of the embryo—to form a network of interacting secreted proteins: the dorsal center secretes chordin and anti-dorsalizing morphogenic protein, while the ventral center secretes Bmp (de Robertis 2006). Similarly, in the developing nervous system, two major signaling centers organize dorso-ventral patterning of the neural tube: the (dorsal) roof plate secretes Bmps and Wnts, and the (ventral) floor plate secretes sonic hedgehog (Shh) (Kiecker and Lumsden 2012). Multiple organizers regulate neural development by inductive signaling events between neighboring tissues but, significantly, are constrained by cell lineage restriction boundaries (Kiecker and Lumsden 2012).

The Epigenetic Landscape

Required reading for developmentalists within several subdisciplines of biology—including developmental biology, developmental neurobiology, developmental psychobiology, and developmental psychology—is a book by Conrad Waddington (1957), *The Strategy of the Genes.* It contains a powerful visual metaphor of cell development: the epigenetic landscape. The epigenetic landscape is depicted as a ball rolling down a surface of undulating hills and valleys, eventually settling in a valley. In the book, Waddington explains that the ball is a visual representation of a cell in the developing embryo and that the particular topology of the landscape characterizes branching points at which the ball is nudged down one path or the other by the action of embryonic inducing factors and / or homeotic genes (Slack 2002). Waddington was influenced by the work of French dynamicist Henri Poincare, and biologists' subsequent use of a dynamical systems toolbox has led to a characterization of the epigenetic landscape as a layout of attractors (Ferrell 2012).

During embryogenesis, many different cell types are formed, spatial patterns emerge, and there are major changes in cell shape. Each cell within a region, such as the future germ layers (mesoderm, ectoderm, and endoderm), has a particular cell fate—that is, what it will develop into. One cell becomes functionally different from another through a process of interacting with other cells, called induction (Wolpert et al. 2007). During cell fate induction, a cell or group of cells produces an inductive stimulus that causes another cell to adopt a new function—for example, during mesoderm induction in the early *Xenopus laevis* embryo (Dale, Smith, and Slack 1985). The fate of a cell may be modeled by the way that potential changes in the topology of "sticky" hills and valleys influence the behavior of a rolling ball (Ferrell 2012). With cell differentiation, the landscape changes from one valley, with a single steady state, to additional valleys separated by ridges. According to this model, each alternative cell fate corresponds to a stable state. When one valley turns into two valleys separated by an intervening ridge, the process is called a *bifurcation.* Bifurcations arise because the system includes positive feedback: a differentiation regulator promotes its own synthesis via a feedback loop. The stable states of the system may be depicted within a potential framework where, by analogy to a physical landscape, the slope steepness gives impetus to the ball. For a system that has bifurcated, the two stable steady states sit at the bottoms of two valleys, with one stable steady state being the global minimum of the potential and the unstable steady state at the top of a ridge (Ferrell 2012).

How Biological Systems Leverage Physical Instabilities to Build Devices

Television rebroadcasts of the cult film classic *Little Shop of Horrors* (1960) motivated me to purchase my first houseplant, a Venus flytrap. I hoped, but did not expect, that my plant would grow to the human-eating proportions of "Audrey Junior," as it did in the film. Nevertheless, I did manage to get the plant to snap shut and ingest a sacrificial fly. That was my plant's last meal. In high school, I discovered Charles Darwin and, to my great satisfaction, found that he also was fascinated by the "wonderful" device by which a plant devoured an insect, as in his 1875 book on "insectivorous plants." However, it was not until I started to read articles on the physical mechanisms harnessed by biological systems that the means by which the Venus flytrap snaps shut was revealed to me (Forterre et al. 2005).

Mathematician and Wyss faculty member L. Mahadevan has conducted many ingenious experiments with plants to demonstrate how nature harnesses the nonlinear mechanical response of curved elastic surfaces to perform work (see, for example, Vaziri and Mahadevan 2008 for a mathematical description of Gauss curvatures). One source of energy for doing work occurs when some parts of a surface grow more than others, leading to "growth strains" (Cerda and Mahadevan 2003). For example, in the process of blooming, the edges of the petals of the Asiatic lily *Lilium Casablanca* wrinkle due to differential rates of growth at the margins relative to the center of the petals (Liang and Mahadevan 2011). A particular kind of growth strain occurs when the strains are not homogeneous and are sufficiently large to result in buckling and bending out of plane, known as "snap buckling" (Koehl et al. 2008).

Snap buckling is also how the Venus flytrap snaps shut in about 100 msec. By painting sub-millimeter ultraviolet fluorescent dots on the external faces of leaves and recording closure with high-speed video, Forterre et al. (2005) measured and modeled the leaf curvature of Venus flytrap plants and found a double curvature in which bending and stretching modes of deforming are coupled. They propose that when the trigger hairs are touched, the plant regulates the natural curvature of the leaf to release stored elastic energy for rapid closure. More generally, the nonmuscular hydraulic movements of plants and fungi (swelling / shrinking, snap buckling, and explosive fracture) may be classified by the relation between the duration of movement and the smallest dimension of the moving part (Skotheim and Mahadevan 2005).

By using a set of tools that included high-speed video to measure strain, dissection experiments to modify leaf curvature, and physical and mathematical modeling, Forterre et al. (2005) were able to identify an instability due to leaf geometry itself as the mechanism by which the leaves snap shut to trap an insect within 100 msec of contact. Each leaf has a doubly curved surface—that is, it is curved in two orthogonal directions. As the leaf shape is deformed, its middle surface simultaneously bends and stretches. The plant is able to actively change one of the leaf curvatures when stimulated, thus releasing the energy required to generate a rapid snap closing. Once shut, the leaf's water content induces a rapid, dashpot-like damping that keeps the prey trapped.

Physical instabilities may provide nature with other opportunities for building devices that require a rapid response for the function of predation. For example, work on worms by Mahadevan and his colleagues (Concha et al. 2015) demonstrates that a physical instability may drive geometric amplification of a syringe-like system used by the velvet worm to eject a jet of slime that entangles and immobilizes prey in a disordered web. Through high-speed videography, dissection experiments, and physical and mathematical modeling, Concha et al. (2015) discovered that the characteristic oscillating motion of the jet was due to the interplay among a gradually contracting reservoir of slime, the inertial effects of the exiting fluid jet, and the elasticity of the oral papilla. A physical model of the oral papilla—an elastic tube—demonstrates how instabilities arise as fluid flowing through it increases its speed. Thus, velvet worms may entangle their prey by harnessing the instability associated with rapid flow through a long, soft nozzle, which causes oscillation of the exiting jet.

Dynamical Systems and Criticality

Dynamical models of behavior are based upon mathematical equations, called differential equations, that express the temporal dynamics of a system's state variables according to the physical laws governing the system (Breakspear 2017). In mechanical systems, these differential equations derive from Newton's second law, and in neural systems, the differential equations are derived from the biophysical properties of neurons (Breakspear 2017). Dynamical systems are characterized by a set of "signatures." One of the signature properties of dynamical systems is that transitions in macroscopic behavior—such as the changes from a liquid to a gaseous state, the polarity of a magnetic field, or local field potential of the brain—occur *when the fluctuations of the individual components (molecules, magnetic spins, neurons) suddenly become correlated at all length scales.* What governs such behavior? The criticality hypothesis states that far-from-equilibrium dynamical systems function at the boundary between order and chaos (Chialvo 2010). Transitions in the macroscopic behavior of dynamical systems occur as the parameters governing the system's behavior find a "critical point." Consider the *exploration* of the parameter space of temperature and pressure for the physical states of matter. Here, "exploration" means that fluctuations move the system through a space defined by temperature and pressure. At certain parameter values of temperature and pressure, there are phase transitions, where small changes in parameters produce qualitative changes in macroscopic behavior (for example, between liquid and gas). At certain "critical points," fluctuations (for example, in the density of the liquid) become correlated on all length scales, from molecular to macroscopic (Bialek et al. 2014).

Nature has harnessed criticality for a range of biological devices, in plants and animals, that are poised at the edge of chaos to perform

Table 2.2 Signatures of dynamical systems

Signature	Description
Stability and adaptive flexibility	Intrinsic fluctuations shift a system away from its attractors. The function of perceiving and acting is to stabilize behavior around an attractor, while maintaining adaptive flexibility.
Symmetry breaking	There are temporal and spatial imperfections that detune a dynamical system and drive the system away from its attractor(s). When symmetry is broken, the result is a bifurcation in the form of a pitchfork.
Embodiment	Due to the hybrid nature of neuromechanical systems interacting in the environment, there may be forces generated that require a jump from one continuous vector field to a new condition for a new vector field.
Collapse of dimensionality	The neuronal and mechanical architectures of animals effectively collapse high-dimensional dynamical systems into much lower-dimensional ones.
Goals and control	Behavioral goals constrain agents to act within the workspace of their control architecture by tuning controls to adapt a dynamical system to different operating regimes within a given environment.

Source: P. Holmes, R. Full, D. Koditscheck, & J. Guckenheimer (2006), The dynamics of legged locomotion: Models, analyses, and challenges, *SIAM Review, 48,* 207–304.

behaviors that involve switching between stable states. Chaotic systems provide useful randomness and the ability to anticipate the uncertain future (Crutchfield 2012). Nature leverages chaos by selectively stabilizing it. One example of nature's chaos control is the use of time-delayed feedback (Schöll 2010): a control signal chosen so that it is proportional to the difference in value of some output variable. As I discuss in Chapter 5, nature may harness neural circuits poised at the edge of chaos to build nervous system networks with emergent functions. Other signatures of dynamical systems are presented in Table 2.2.

Attractors (Primitives)

A finger of one hand extends to touch a door handle and exhibits a trajectory that tends toward a stable equilibrium. An infant excitedly kicks her legs, waves her arms, and squeals at the sight of a rotating overhead mobile. A child is pushed on a swing, which may remain in a bounded region of state space, but does not settle into a stable orbit. A hand is used to turn the handle of a crank or hold a paddle to repeatedly bounce a ball tethered by an elastic band. All of these behaviors, revealed in the topological trajectories of state space, constitute a limited bestiary of "attractors" or "primitives" (Dominici et al. 2011; Flash and Hochner 2005; Hogan and Sternad 2012, 2013; Ijspeert et al. 2013). The trajectory of a finger to touch a door handle exhibits the behavior of a point attractor, a pendulum is a periodic attractor, and forced oscillation of heart pumping is a chaotic attractor. Mechanical impedance is additionally proposed as an attractor for managing contact and physical interaction with objects and surfaces (Hogan and Sternad 2012). These signatures of unique attractor dynamics have been proposed as elementary building blocks, or "*primitives*" for action (Flash and Hochner 2005; Hogan and Sternad 2013; Ijspeert et al. 2013). Primitives may be assembled into combinations through operations or transformations such as vector summation of spinal force fields (Mussa-Ivaldi and Bizzi 2000) or by means of more complicated coarticulation between consecutive elements (Sosnik et al. 2004).

Since at least the 1980s, behavioral scientists have proposed primitives to characterize emergent behavior in the coordination dynamics of rhythmic motions of the mouth during speech (Saltzman and Kelso 1987), the hands of musicians playing a string quartet (Winold, Thelen, and Ulrich 1994), and the arms and legs of an excited infant during looking at an overhead mobile (Hsu et al. 2014). Three groups were especially influential in introducing this approach: Elliot Saltzman and Scott Kelso published their theoretical paper on task dynamics (Saltzman and Kelso 1987; and see Kelso 1995 for a review of early work), Peter Kugler and Michael Turvey (1987) published their influential book on the self-assembly of rhythmic behavior (see Turvey 2007 for a review of later work on coordination), and Esther Thelen (reviewed in Thelen and Smith 1994; see also Thelen and Ulrich 1991; Thelen 2000; Thelen et al. 2001) began her landmark research into the dynamics and development of infant kicking and reaching behavior. In all three cases, behavioral organization could be captured by the mathematics of a small set of interacting coupled non-linear dynamical systems (differential or difference equations). The power of nonlinear dynamics for developing models of the underlying laws of behavior has now become a standard tool in fields such as biomechanics, motor control, neuroethology, and neuromechanics (see, for example, Holmes et al. 2006).

From a dynamical systems perspective, nature may build the ubiquitous human rhythmic behaviors of finger tapping, breathing, speaking, and walking by *coupling together self-sustaining oscillators*. Turvey (1990) identifies four components of self-sustaining oscillators: an oscillatory component, which both guarantees overshoot and returns a system to equilibrium; an energy source that makes up for losses through friction; a gate that admits energy to the oscillatory component in the right amounts and at the correct instant; and a feedback component that controls the gate. As discovered in musical instruments and electrical circuits by the English physicist Rayleigh, the feedback component controlling the gate was a sum of velocity and cubed velocity. The Dutch radio engineer van der Pol found that resistance was related to the square of current (Turvey 1990). Each of these oscillators may be mathematically characterized as an *equation of motion* and may be coupled as part of a single function. The coupled hybrid oscillator is a powerful tool for modeling the particular behaviors of complex systems, such as the space–time behavior of bimanual rhythmical movements (Kelso et al. 1981; Kay et al. 1987). Another set of important components in dynamical system equations of motion is parameters. A parameter is a term that changes on a slower time scale than does a state variable (Warren 2006). The influences on a parameter may be organismic, such as when stiffness and damping properties of muscle are modulated by spinal systems, or environmental, as when a metronome is used to drive the frequency of an oscillatory behavior.

Kay et al. (1987), for example, were interested in revealing the underlying attractor dynamics that could account for observed behavior on a task in which a human subject rotated two manipulanda that allowed flexion and extension about the wrist (radio-carpal) joint in the horizontal plane. In their experiments, they first established a subject's "comfortable rate" and then used a metronome to pace the frequency at which they cyclically moved their wrists. In the unimanual condition, the experimenters increased frequency (in steps of 1 Hz) from 1 to 6 Hz and found that amplitude dropped inversely. To model the underlying dynamics of this behavior, Kay et al. (1987) combined the van der Pol and Rayleigh oscillators to form a *hybrid oscillator* and then compared the data from subjects with that generated in simulation by the model equations. The identical changes that occur in actual behavior, as well as in simulations driven by the hybrid oscillator model, confirm the underlying dynamics of behavior on this task. However, to understand how changing goals and environmental opportunities influence the layout of attractors requires that we go beyond primitives—that is, to address what Saltzman and Kelso (1987) call "task dynamics."

Task Dynamics

The scientists at Haskins Laboratories in New Haven are most celebrated, perhaps, for introducing the motor theory of speech perception (Liberman et al. 1967; Liberman and Mattingly 1985; Galantucci, Fowler, and Turvey 2006), the view that human speech *perception* is influenced by the peculiar manner by which human speech is *produced*. Haskins scientists also cracked the code that made it possible to create synthetic speech on a computer. By revealing one of nature's secrets, the speech code, Liberman et al. (1967) helped set the stage for the current revolution in machines that can communicate with us. Even after fifty years, the motor theory continues to generate scientific interest, garnering new support from studies of human infants (see, for example, Bruderer et al. 2015).

Haskins Labs also remains a nexus for a dynamical systems approach to speech and gesture production (see, for example, Byrd and Saltzman 2003; Saltzman et al. 2008) and for modeling motor control in human behavior, more generally. In the late 1990s, I had the good fortune of spending a summer at Haskins and also sat in on a graduate-level course on dynamical systems in behavior taught by

Elliot Saltzman at the University of Connecticut. One assignment in the course was to learn how to juggle so we could appreciate its underlying dynamics (see Beek 1989). Alas, I am hopelessly uncoordinated. However, my efforts were not entirely wasted because my daughter, Anna, picked up the course materials and rapidly became a highly proficient juggler at around age 10. She soon was juggling soft cubes, balls, and plastic chickens. Tutorials in the mathematics of dynamical systems with Elliot over the years have been instrumental in my own scientific work, and I am forever grateful that he introduced me to the jazz piano of Bill Evans, initially by directing my attention to a CD of *Waltz for Debby* in a bin at the Yale Co-Op (when it was still possible to physically browse CDs). Elliot's approach, called *task dynamics*, holds a central place in this book, and I offer a brief introduction here (interested readers may find more technical presentations on task dynamics in Byrd and Saltzman 2003; Saltzman and Byrd 2000; Saltzman et al. 2008). Then, in Chapter 8, I leverage task dynamics to address one of the challenges in neuroprosthetics—coupling the dynamics of the body with the sensory fields of the environment—and with robotic devices (see Warren 2006). Task dynamics characterizes how behavior emerges from the relation among goals in a task space consisting of a layout of attractors, the biomechanical properties of the body, and information actively obtained by organs transported through the environment (Saltzman and Kelso 1987; Warren 2006). More generally, a task space includes the minimal number of relevant dimensions (task variables or task-space axes) and a minimal set of task-dynamic equations of motion (Saltzman and Kelso 1987).

The Developmental Origin of Primitives

At Harvard in the 1960s, Peter Wolff—drawn to veterinary medicine but trained as a psychiatrist, fascinated with rhythms in music for insights into neuromotor disorders, and guided by the earlier work of von Holst (Holst 1973) on oscillations in bony fish—turned his attention to rhythms in human infants and animals. Wolff had the important insight that the ubiquitous rhythmic behaviors seen in animals and infant humans were a window into the system generating their underlying dynamics (Dreier et al. 1979). Von Holst had found in fish that the oscillatory behavior of one fin influenced the others, and he formulated mathematical principles to capture the different types of coordination influences. Inspired by this work and the behavior of human infants, Wolff conducted a set of classic observations of his own and other infant children (see, for example, Wolff 1960), discovering the emergence of behavioral states of sleep and wakefulness in respiratory and movement patterns (Wolff 1987). He also made some of the earliest measurements of pacifier (non-nutritive sucking) in human infants and other mammals, such as baby goats (Wolff 1973), providing a comparative perspective on "central" rhythms.

When I returned to Boston Children's Hospital in the mid-1990s to join Peter in the Psychiatry department, he encouraged me to go back to von Holst's work as I developed testing procedures to study the role of underlying dynamical systems for the developmental emergence of coordination in the behavior of typically developing infants and those at risk due to complications at birth, such as prematurity (see Goldfield and Wolff 2002). An initial study with Richard Schmidt, an expert in dynamics now at College of the Holy Cross, and Paula Fitzpatrick, a psychologist at Assumption College, specifically used von Holst as a starting point for understanding respiratory movements of the infant's chest and abdomen as a coupled oscillatory system (Goldfield, Schmidt, and Fitzpatrick 1999). We began with von Holst's discovery of fundamental entrainment phenomena of biological oscillators in the rhythmically moving fins of the fish *Labrus*. Von Holst noticed that each fin attempted to move the other to its inherent frequency, the frequency exhibited when oscillating alone, and called this phenomenon a "magnet effect." If the coupling process between the fins was sufficiently strong, it caused the oscillation of the fins to (1) settle at a coupled frequency different from the inherent frequencies of the fins and (2) become phase locked, such that the phases of their cycles maintained a constant relation. In further experiments, von Holst found that each oscillator attempted to maintain its own identity, what he called a "maintenance tendency": each oscillator attempts to pull, or attract, the other to its frequency. Von Holst deduced that the biological coordination of oscillatory behavior had two reciprocal properties: a cooperative coupling process and a competitive detuning process (Goldfield et al. 1999). If the cooperative and competitive processes *balanced,* both oscillators had a constant phase—1:1 phase locking—called *absolute coordination.* However, if the *competitive process dominated,* the two oscillators would continue to influence each other, for example, by one oscillator lapping the other in its cycle. Von Holst called this *relative coordination* (Goldfield et al. 1999).

Our goal was to characterize the coordination of chest and abdomen respiratory rhythms in full-term infants compared to infants born prematurely. To do so, we measured movements of chest-abdomen during quiet breathing and used a technique called cross-spectral analysis to calculate their relative phase: a phase angle of 0 degree indicated that the two rhythms were in perfect synchrony, while a 180-degree phase angle indicated that the two rhythms were in perfect anti-phase. Mean phase ranged between –30 and 71 degrees, and for most infants, the abdomen led the chest in the cycle. A group of high-risk preterm infants, who had a history of respiratory complications, *exhibited the greatest chest-abdomen phase lag,* compared to healthy term and preterm infants (Goldfield et al. 1999). We then examined whether chest and abdomen breathing rhythms could be modeled by the cooperative and competitive influences between two interacting oscillators, as indexed by their relative phase. To do so, we followed the modeling strategy developed in studies of adult finger and limb oscillations, introduced earlier.

We used an equation of motion with terms for (1) competing influences indexed by the difference between the oscillators' inherent uncoupled frequency; (2) cooperative influences, indexed by a coupling term; and (3) a Gaussian white-noise process characterizing the inherent fluctuations in the system (Haken, Kelso, and Bunz 1985; Schmidt, Shaw, and Turvey 1993). We also determined whether frequency of respiration might function as a control parameter of coordination, decreasing its stability. Of particular interest for understanding the differences in coordination in typically developing and premature infants was the relationship between the competitive and cooperative processes of chest and abdomen oscillations. The model identifies two ways that the respiratory phase lag of the high risk preterm infants could be produced: (1) by means of greater frequency detuning or (2) by a weaker coupling of chest and abdomen. We found that respiratory phase lag increased as coordination strength decreased, evident graphically in the systematic change in phase lag with slope (coherence) of the zero crossings (Goldfield et al. 1999).

The confirmation by Goldfield et al. (1999) of coupled oscillator dynamics in infant breathing raises the question of the origins of primitives; how primitives apparent in utero and at birth are related to subsequent patterns; and whether primitives underlie other functionally specific "action systems" that develop during the infancy period, including locomotion and reaching (see Goldfield 1995). These questions have a very long history in the fields of developmental neurobiology (Jacobson and Rau 2005; Preyer 1885), developmental psychobiology (Michel and Moore 1995), and developmental psychology (Adolph and Robinson 2013), as well as in studies of animal and human fetal behavior (for example, de Vries, Visser, and Prechtl 1982) and in a dynamical systems approach to locomotion (Thelen and Fisher 1983) and reaching behavior (Thelen et al. 1993).

In a more recent comparative study of the locomotor primitives underlying stepping behavior, Dominici et al. (2011) used EMG recordings to examine muscle activation patterns in human newborns and compared these to older children as well as to existing data on adult animals. To identify patterns of muscle activation, Dominici et al. (2011) applied a technique called *non-negative matrix factorization* to the EMGs, pooled across all steps. They found several dominant patterns of muscle activations—weighted combinations of three basic sinusoidal waveforms—in newborn stepping, and that two of these patterns were also present in toddlers. Further, toddlers exhibited two other patterns that were not apparent in newborns but, instead, were transitional shapes similar to that of adults. The newborn human patterns were also apparent in comparisons with the rat, cat, macaque, and guinea fowl. From these data, Dominici et al. (2011) conclude that locomotion is built starting from common primitives. But are the newborn patterns hardwired in the spinal cord, without the need for experience? What can we learn from animal studies of fetal behavior that may address the nature of the primitives from which all mammalian behaviors are built and, perhaps, may form the basis for machine primitives that are used for shared control of neuroprosthetic devices?

Peter Wolff recognized that the earliest forms of organized behavior could be observed not only when newborns were awake, but also when they were asleep, falling asleep, and waking from sleep. He called these organizational changes behavioral states (Wolff 1987). This prescient insight was not only a forerunner of the discovery of resting states in brain function (see Chapter 5), but also has inspired decades of research on the origins of spontaneous behaviors in sleep and wakefulness in fetuses and newborns. One surprising finding, for example, is the role of spontaneous myoclonic twitches during the active sleep of rat fetuses in providing sensory feedback to refine sensorimotor circuits (Blumberg, Marques, and Iida 2013). Like the retinal waves seen in the developing visual system, spinally generated twitching appears to prepare the developing motor cortex for the sensory input that becomes available during postnatal life (Blumberg et al. 2015).

Like Wolff, Esther Thelen (see, for example, Thelen 1989, 2000) appreciated that the spontaneous self-generated activities of newborns and older infants provided a unique window on the origins of behavior and led a series of studies of how exploratory behavior transformed the earliest newborn activities into functional capabilities. Reminiscent of current work in neuromechanics, Thelen and her students famously demonstrated, through studies of infants partially immersed in water, that infant stepping was influenced by the biomechanical properties of the body, as well as by the effects of gravity (Thelen, Fisher, and Ridley-Johnson 2002). Her studies of reaching behavior reintroduced into the study of behavioral development an appreciation of the role of variability for transforming the intrinsic dynamics of spontaneously produced behaviors into functionally specific, goal-directed behaviors (Thelen et al. 1993). The "primitives" of reaching development, in this view, were not simply a pattern of muscle activations, but rather a heterogeneous assembly of parts

(Thelen et al. 1993). Infants' exploration of this assembly of parts allows them to discover how to control the behavior of their body in order to achieve a goal, such as reaching for a toy.

Dynamics of Breathing and Device Development

Respiratory rhythms emerge from networks of synaptically coupled microcircuits in the ventral respiratory column of the brainstem (Smith et al. 2013). The ring-like recurrent architecture of these circuits makes possible the use of modulatory influences to generate multiple rhythmic breathing patterns, and adaptive switching between them (Smith et al. 2013). Cortical midbrain and cerebellar sensorimotor inputs to the pons, as well as peripheral sensory afferent inputs to the nucleus tractus solarius of the medulla, provide excitatory drives to these circuits (Smith et al. 2013). The same circuits are reconfigured via neuromodulation to generate distinct types of respiratory activity, including eupneic normal breathing, sighs as part of an arousal mechanism, and gasping during severe hypoxia (Koch, Garcia, and Ramirez 2011). Is there a dynamical basis for the variety of rhythms and of the context dependency of switching between them? Is network excitability a parameter that is used to control progression from stable periodic activity to disorganized aperiodic activity?

To examine the dynamical foundations of respiratory rhythm generation, Del Negro et al. (2002) turned to the Poincare map, a tool from the toolkit of dynamicists, for geometrically diagnosing the underlying dynamics of inspiratory rhythms recorded in neonatal rats and humans. The Poincare map is a scatter plot in which each data point $x(n)$ in a time series is plotted against an adjacent point $x(n+1)$ (Kantz and Schreiber 1997). For example, a Poincare map may be used to distinguish among (1) a simple periodic system with noise, consisting of a normally distributed single-point cluster; (2) a system that is periodically modulated (a "mixed-mode" oscillator), producing several distinct points; and (3) a quasi-periodic system, exhibiting ring-like structures. Del Negro et al. (2002) found that human neonatal breathing is characterized by both stable limit-cycle patterns of inspiration-expiration with a normally distributed cluster of points and a transition to quasi-periodic breathing with its characteristic ring structure. Thus, the neonatal human respiratory network is periodically modulated and destabilized and then returns again to stable breathing.

The generation of emergent rhythms by the circuits of the respiratory network is but one part of a complex, integrated system for ventilatory behavior. Integrated respiratory output is additionally dependent upon (1) the synaptic connectivity between premotor respiratory neurons and the respiratory motor neurons innervating the diaphragm, chest wall, and muscles of the upper airway and (2) afferent activity from mechanoreceptors in the lung and peripheral chemoreceptors (Gauda and Martin 2012). Therefore, apnea of prematurity may result from the central respiratory network, peripheral and central chemoreceptors and mechanoreceptors, and the compliance of the chest wall and soft tissues of the upper airway (Di Fiore, Martin, and Gauda 2013).

David Paydarfar, now at the University of Texas, is a neonatal neurologist and an expert in the dynamics of respiratory control (see, for example, Forger and Paydarfar 2004). The characteristics of nonlinear networks in the emergence of different forms of respiratory behavior inspired him to develop a dynamically based intervention strategy for promoting healthy breathing in preterm neonates exhibiting apnea of prematurity. This strategy takes into account both the role of intrinsic (endogenous) fluctuations for transitions between neonatal respiratory patterns and the role of mechanosensory input for stabilizing nonlinear systems. Bloch-Salisbury et al. (2009) hypothesized that small-amplitude, noisy inputs from an experimental mattress making physical contact with the dorsal skin surface of a supine neonate might transform the state of the central respiratory system from subthreshold arrhythmic activity to rhythmic breathing. In their experiment, Block-Salisbury et al. placed preterm neonates on a mattress embedded with actuators generating noisy mechanosensory stimulation (that is, stochastic resonance) during multiple 10-minute intervals. During the experimental session, these infants exhibited both a reduction in the variability of inter-breath intervals, and a 65 percent reduction in the duration of O_2 desaturation. In a similarly designed study of the SR mattress, Smith et al. (2015) conducted a randomized crossover study with preterm infants (mean 30.5 weeks gestational age and birthweight of 1409 grams) with documented apnea, bradycardia, and / or oxygen desaturation event. Smith et al. (2015) report that during up to two, three, or four intervention periods in which they received mattress SR stimulation, there was a 20 percent to 35 percent decrease in oxygen desaturation events.

In his earlier work with adults, David examined the role of deglutition on respiration using another important tool from the dynamicist's toolkit: phase resetting (Paydarfar et al. 1995). Phase resetting reveals a critical feature of endogenous rhythms: when a perturbing stimulus is delivered during an ongoing oscillation, the response to perturbation will be different, depending on the phase of the oscillation in its cycle (Glass 1998). The differential nature of the response to a stimulus delivered at a particular location in the cycle may, thus, be another means for revealing the underlying dynamics of a system. Paydarfar et al. (1995) found with adults that the respiratory cycle exhibits different responses to swallow perturbations, depending upon the location of the swallow with respect to the transition between expiration and inspiration. Their conclusion was that the underlying oscillatory system generating the respiratory pattern has a cycle-dependent response to the discrete perturbation that occurs during deglutition.

My own clinical and translational work with prematurely born neonates has been similarly inspired by the dynamics of interactions between deglutition and respiration. An example of some earlier work is a device developed for assisting feeding of premature infants, called the active bottle (see, for example, Goldfield 2007). A core idea behind the active bottle is that the infant's observed feeding pattern is comprised of multiple subsystem components, each with its own rate of development. The appearance of a particular coordination pattern is dependent not only on the ongoing development of each of the individual components (for example, sucking burst lengths), but also on the compatibility of the parts with each other. The development of individual components may be asynchronous or even regress. Moreover, any one of the individual parts may hinder the development of cooperation with other parts. The possibility of one or more rate-limiting factors in development is particularly noteworthy for understanding oral feeding by preterm infants, who have several comorbidities that may limit the organization of stable feeding patterns, including respiratory distress and gastroesophageal reflux.

The development of stable feeding patterns is believed to reflect the organization of the individual components into a single functional entity. This is apparent when the components compensate together in order to maintain the functional integrity of the feeding process. For example, oral feeding involves the coordination of sucking, swallowing, and breathing into a functional system for withdrawing milk from the nipple and moving it through the pharynx to a sphincter between the esophagus and stomach. When no milk flows from the nipple, mandibular oscillations during non-nutritive sucking may couple with the respiratory system in a stable pattern of three sucks for every two breaths. When milk flow is initiated, the functional relation between sucking and breathing is maintained, even though their pattern of coordination changes to two sucks for every breath.

This adaptive process of maintaining the organization of a stable feeding pattern despite perturbations is also evident as the pharynx rapidly changes its shape and its contact with other anatomical surfaces during the transition from its function in respiration to its function for swallowing. The pharynx is the shared anatomical pathway for both air to the lungs and nutrients to the esophagus and gastrointestinal tract. Respiration and deglutition are incompatible: swallowing when the airway is open results in material entering the lungs. Therefore, at each swallow, the pharynx must be reconfigured from its respiratory function to seal the airway and allow nutrients to pass through the pharynx. Because of the brief duration of each breath during the respiratory cycle, there are only limited "windows of opportunity" for swallowing.

The active bottle was designed to control the flow of milk from a bottle nipple—based upon ongoing sensor input from a respiratory sensor below the nostrils and from a pressure sensor in the nipple—

to a computer-controlled micro-pump inside the bottle. Active bottle control of milk flow is personalized to the current capabilities of each infant, based upon clinical diagnosis of feeding problems, including weak sucking and poorly coordinated swallowing and breathing. From recordings of the infant's sucking, swallowing, and breathing, active bottle diagnostic software is used by clinicians to set thresholds for sucking and breathing parameters that determine rate of milk flow. So, for example, consider the case of an infant making the transition from nasogastric tube feeding (thus entirely bypassing sucking and swallowing) to complete oral feedings. Initially, the clinician provides the infant with an opportunity to receive small amounts of milk from the active bottle, while continuing to be fed by nasogastric tube, in preparation for the initiation of bottle feedings. Following a review of graphic and numeric data derived from the active bottle recordings, the clinician may conclude that there has been a clinically significant increase in the organization of sucking relative to breathing, with high and stable levels of oxygenation, and that the infant is ready to begin oral feedings, with any volume remaining after 30 minutes fed by nasogastric tube. Thus, the active bottle diagnostics provide an objective basis for helping to decide when to begin oral feedings.

• 3 •

What Nature Builds: Materials and Devices

Complementarity, the balancing of opposing tendencies, is a central motif throughout nature (Kelso and Engstrom 2006; Marder 2012). Complementarity in *what* nature builds may reflect *how* nature builds by coupling the internal energy flows of metastable far-from-equilibrium systems (for example, nervous systems) with flows of obtained environmental information (Haken 1983; Kugler and Turvey 1987; Warren 2006). This central motif of the coupling of opposing tendencies is evident in the prestress of tensegrities (Ingber 2006); in the homeostatic balance between excitation and inhibition achieved in nervous system circuits interconnected via interneurons (Marder 2011); in neural control systems that leverage the body's mechanical properties in performing sensorimotor functions, such as locomoting on land, in water, and in the air (Cowan et al. 2014); in homeostatic plasticity (Hensch 2014; Turrigiano 2011); and in the initial exuberance of synaptic connectivity (Innocenti and Price 2005) and subsequent apoptosis, or programmed cell death (Buss, Sun, and Oppenheim 2006), that occurs during human cortical development (all discussed in the chapters of Part II).

The central motif of complementarity between how and what nature builds addresses a fundamental paradox of open systems continuously exchanging energy and materials with the environment: how to maintain robust and stable functioning in spite of material turnover and perturbations of activity (Marder 2012). In neurons, for example, regulatory systems maintain robust electrical signaling despite ongoing protein turnover and perturbations of function (Marder 2012). At the level of behavior, the use of available body metabolic resources to temporarily assemble and dissolve functional systems for locomoting, eating and drinking, and communicating may be stabilized by the flow of information obtained by perceptual systems

(Goldfield 1995; Turvey and Fonseca 2014; Warren 2006). The functional relation between agent and environment remains stable despite variability in the composition of elements, such as which muscles comprise a synergy (Goldfield 1995).

Complementary functional properties may emerge from the self-assembly processes that built them. For example, the glass sponge *Euplectella* introduced in Chapter 2 is a beautiful paradox (Aizenberg et al. 2005). Glass is inherently brittle, but its structure at progressively larger spatial scales reveals how nature is able to tame brittleness in order to use glass as a building material. At the nanoscale, silica spheres are embedded within layers glued together by an organic matrix to form flexurally rigid composite beams at the micron scale. At the most macroscopic scale, *Euplectella* is a cylindrical, square-lattice, cage-like structure reinforced by diagonal ridges (Aizenberg et al. 2005). Thus, nature builds hierarchical structures that combine physical behavior at multiple scales into a material with emergent macroscopic properties. In addition to hierarchical structure, biological materials exhibit fracture resistance, multifunctionality, adaptive behavior, and self-healing (see Table 3.1).

At the macroscopic scale of behavior dynamics (Warren 2006), acting and perceiving are complementary: critical points in the attractor dynamics of action have a complementary form in the flow of obtained information from sensory systems, creating a seamless set of loops for the flow of energy and information. In discussing behaviors such as locomoting, reaching and grasping, gesturing and speaking, and eating and drinking, this chapter focuses on the complementary relation between *special-purpose devices* for goal-directed adaptive behaviors and the obtaining of information with *smart perceptual instruments* (Runeson 1977; Turvey and Carello 2011). Here, I address questions that include:

- What are interfaces in the hierarchical structure of materials?
- What is the nature of information that joins together the physical and biological worlds?

Table 3.1 Properties of biological materials

Property	Description	Examples
Hierarchical structure	Structural hierarchy in *Euplectella* form silica nanospheres arranged in concentric layers to yield fibers and bundling of fibers into beams within a silica matrix forms a mechanically resistant glass cage	Sponge *Euplectella,* wood, bone, tendon, ligaments, skin, cornea
Fracture resistance	Biological mineralized composites generate fracture toughness (resistance to initiation and growth of a crack)	Mollusc, e.g., abalone shell (nacre)
Multifunctionality	Rapidly and reversibly alter stiffness of connective tissue	Invertebrate dermis (e.g., sea cucumber *Cucumeria frondosa*)
Adaptive behavior	Remodeling	Trabecular bone
Self-healing	Ability of a material to heal or repair damage automatically and autonomously	Plant microvascular network (a centralized network for distribution of healing agents)

Source: P. Fratzl & F. Barth (2009), Biomaterial systems for mechanosensing and actuation, *Nature, 462,* 442–448.

- What is the nature of the perceptual organs and task-specific devices that have evolved for actively exploring and performing work in the environment?

Nature's Living Materials

The processes and products of growth result in biological solid materials that emulate the physical properties of solids in the environment. Tensile materials—such as spider silk, cellulose (plants), chitin (insect exoskeleton), collagen (animal tendon), and composites (animal bone)—all resist tensile stress (Vogel 2003). Pliant materials—including resilin (tendons in insect wing hinges)—and pliant composites (the gel matrix of jellyfish bodies and animal cartilage and skin) all regulate the extent to which, and speed at which, a material recovers its form after removal of a stress) (Vogel 2003). And rigid materials—such as arthropod cuticle, animal bone, keratin, wood, sponge, coral, eggshell, and tooth enamel—all resist stress with very little deformation, and are highly anisotropic (that is, their mechanical response depends on the direction of loading) (Vogel 2003).

Materials such as bone and wood are adaptive to the natural environmental forces of their niches, including gravity, wind, and surrounding material substrates. They are characterized by particular shapes, internal architectures, fiber orientations, and composite structures (see Table 3.2). For example, during development, long bones grow to form the shape of a hollow cylinder by removing material from the inner surface and depositing it on the outer surface, while wood starts with a hollow tube that already has its final diameter (as in bamboo) (Weinkamer and Fratzl 2011). By placing all its material along the main loading directions (see, for example, Wolff's law), the growth of trabecular bone acts as a truss to avoid bending. Wood tissue reorients in response to load-bearing changes by generating internal stresses (resisting compression in soft woods and tension in hard woods) (Weinkamer and Fratzl 2011). The fiber geometry of bone consists of parallel alignment where loading is unidirectional and helical winding improves extensibility, while growing trees use control of the microfibril angle for adaptation in different ways, depending upon their age: young trees allow the stem to bend in response to external forces, while in old trees, the stem must prevent failure by buckling and so require stiffness (Weinkamer and Fratzl 2011).

Nature also leverages fibers for adaptive structures. Nanofibers on the surface of the lotus leaf make the leaves super-hydrophobic so that droplets of water containing dust and collected insects roll off to maintain a clean surface (Pokroy, Epstein, et al. 2009). In *Euplectella,* high aspect-ratio silica fibers not only have superior mechanical properties, but also effectively channel ambient light (Aizenberg et al. 2005). The control of transmission and focusing of light is used for signaling marine life. Gecko feet are each comprised of half a million setae fibers, and each seta is tipped with about 1000 nanometer-sized spatula (Autumn et al. 2002). This multiscale fibrous assembly makes possible a reversible adhesion mechanism, holding to, and selectively releasing from, a surface

Table 3.2 Mechanical adaptation strategies in bone and wood

Strategy	Bone	Wood
Shape	Grow a hollow cylinder by removing material from the inner surface and depositing it on the outer surface (e.g., long bones during development).	Start with a hollow tube that already has its final diameter (e.g., as in bamboo).
Internal architecture	Trabecular bone acts as a truss to avoid bending: its growth places all material along the main loading directions (cf. Wolff's law).	Wood tissue reorients in response to load bearing changes by generating internal stresses (e.g., resisting compression in soft woods and resisting tension in hard woods).
Fiber orientation	Bone is a composite material that uses particular fiber geometries, e.g., parallel alignment where loading is unidirectional and helical winding to improve extensibility.	Growing trees use control of microfibril angle (MFA) for adaptation as the tree grows in size. In young trees, allowing the stem to bend to in response to external forces (e.g., animals) requires flexible (high MFA) materials. In old trees, by contrast, stem must prevent failure by buckling, and so requires stiffness (low MFA).
Nano-composite structure	Bone consists of a composite of two materials—one stiff but brittle and the other tough but soft—and achieves a composite at multiple hierarchical levels that is both stiff and tough.	In wood, stiff cellulose fibrils are embedded in a soft matrix of hemicellulose and lignin, allowing for release of stress.

Source: R. Weinkamer & P. Fratzl (2011), Mechanical adaptation of biological materials—the examples of bone and wood, *Materials Science and Engineering C, 31,* 1164–1173.

(Autumn et al. 2006). The body surfaces of fish and amphibians have fibrous cilia connected to a haircell at their base that detects water flow (Coombs, Görner, and Münz 1989; Sane and McHenry 2009).

A most remarkable characteristic of nature's adaptive processes in forming materials is its use of composites (see Tables 3.3 and 3.4). Nature's materials achieve the seemingly contradictory requirements of being sufficiently stiff to support a load yet tough enough to resist crack propagation, by the self-assembly of *composite* materials into

Table 3.3 How composites are used in nature for interacting with and controlling patterned energy

Property	Illustration
Electrical	Fish sensory organs
Magnetic	Bacteria
Light scattering	Incandescent structures of bird feathers and butterfly wing scales
Light transmitting and focusing	Basalia spicule of *Euplectella*

Source: J. Dunlop & P. Fratzl (2010), Biological composites, *Annual Review of Materials Research, 40,* 1–24.

Table 3.4 How composites are used in nature for controlled actuation

Type of controlled actuation	Illustration
Swelling/shrinking	Cellulose angle in plants.
Bending	Spruce branches combine tissue with small microfibril angles (for tensile stress) on the upper side with small microfibril angles (for compression) on the lower side.
Snap buckling	Venus flytrap uses an elastic buckling instability to catch insects in 0.1 sec. This is based on storage and sudden release of elastic energy.
Explosive fracture	Rapid geometric change of a thin shell due to tissue tearing.

Sources: J. Dunlop & P. Fratzl (2010), Biological composites, *Annual Review of Materials Research, 40,* 1–24;. and L. Mahadevan (2005), Physical limits and design principles for plant and fungal movements, *Science, 308,* 1308–1310.

hierarchical structures with interfaces (Barthelat, Yin, and Buehler 2016; Dunlop and Fratzl 2010; Dunlop, Weinkamer, and Fratzl 2011). Nature's interfaces improve material toughness, bridge different materials, allow materials to plastically deform, and allow materials to serve as actuators of motion or stress (Barthelat et al. 2016; Dunlop et al. 2011) (see Table 3.5). For example, the protein layers found in the skeleton of *Euplectella,* as well as in nacre, are interfaces that improve the fracture resistance of inherently brittle materials by introducing soft interfaces. The gradient in the mechanical properties found along the byssus connecting the soft body of the mussel to a hard, rocky substrate, as well as the dentine-enamel junction of teeth, are interfaces that act as bridges, or joints, between materials with different properties (Dunlop et al. 2011). The suture of the turtle shell and the noncollagenous protein layers found in bone are interfaces that easily

Table 3.5 Internal interfaces in biological materials

Category	Description	Examples
Improving material toughness	Improve the fracture resistance of inherently brittle materials by introducing soft interfaces	Protein layers found in skeleton of *Euplectella,* as well as in nacre
Bridging different materials	Act as bridges or joints between materials having different material properties	The gradient in the mechanical properties found along the mussel byssus connecting the soft body of the mussel to a hard, rocky substrate, as well as in the dentine-enamel junction of teeth
Allowing materials to plastically deform	Easily deform in response to applied forces	The suture of the turtle shell, or the non-collagenous protein layers found in bone
Allowing materials to act as actuators of motion or stress	Develop forces that change the relative orientation of their components	Motion of pine cone scales in response to drying and wetting

Source: J. Dunlop, R. Weinkamer, & P. Fratzl (2011), Artful interfaces within biological materials, *Materials Today, 14,* 70–78.

deform in response to applied forces (Barthelat et al. 2016; Dunlop et al. 2011). And the motion of pinecone scales in response to drying and wetting are interfaces that develop forces changing the relative orientation of their components (Dunlop et al. 2011).

Life's Devices

In his wonderful 1988 book, *Life's Devices,* as well as in a more recent comprehensive textbook on *Comparative Biomechanics* (2013), the late biologist Steven Vogel introduces a new generation of scholars to the physical laws that govern the structural forms revealed by D'Arcy Thompson. One focus of the work in comparative biomechanics is on materials. For example, Vogel (2013) classifies the properties of solid materials and provides examples of how nature uses these materials for particular biological functions. The degree to which a material exhibits tensile strength is an indication of the way it behaves in response to imposed forces that may disrupt its structural integrity (as in spider silk, exoskeleton, animal tendon, and bone). Pliant materials (for example, tendons of insect wing hinges, the gel matrix of jellyfish bodies, and animal cartilage and skin) are able to recover with a certain degree of resilience from imposed stresses, and rigid materials (for example, arthropod cuticle, coral, eggshell, and tooth enamel) resist stress with very little deformation and are highly anisotropic (that is, their mechanical response depends upon the direction of loading). Thus, Vogel describes solids as they are rather than as they self-assemble during growth and development.

Comparative biomechanics joins together the physical and biological worlds—for example, by explaining how plants and animals adapt to flow patterns, harness forces and media for propulsion, and use a plethora of body forms to perform work. However, as a subdiscipline of biology, comparative biomechanics has not ventured into the nature of the information available to animals as they interact with these flow patterns, how animals use these flow patterns to guide behavior, or how animals learn about the possibilities for action in unpredictable environments. A starting point for understanding the specialized devices that nature builds with available materials, at all spatial and temporal scales, is the behavior of creatures that lived on the young earth. There are clues about this behavior in the compressed Middle Cambrian animal remains of the Burgess Shale, located in British Columbia (Conway Morris 1998; Knoll 2003). A discovery in these 505 million-year-old rocks is that early animals, or "acorn worms," lived individually in tubes. The shape and position of an individual's remains in each tube, and the shape of its proboscis, indicate that they were motile, constructed the tubes, and were able to leave their tube habitat in times of stress (Caron, Morris, and Cameron 2013; Gee 2013). Thus, even 500 million years ago, living creatures were freely moving and were building shelters. The Burgess Shale has yielded other clues about the lifestyles of the earliest molluscs, and, in particular, the feeding apparatus of swimming individuals who grazed the cyanobacteria biomat for food about 40 to 50 million years later. *Odontogriphus omalus* was a soft-bodied animal that had a molluscan feeding structure, called the radula, and a broad foot for locomotion (Caron et al. 2006). Large swimming anomalocarids, such as *Tamisiocaris borealis*, had a frontal appendage specialized for filter feeding (Vinther et al. 2014). So, within a relatively short period of time after acorn worms, animals were locomoting in water, exploring for food, and using specialized organs for feeding on available resources.

The body plans of these animals, as well as their exploration, locomotion, and feeding behaviors have, thus, been conserved for hundreds of millions of years (Kirschner and Gerhart 2005). And yet, over the subsequent hundreds of millions of years of evolution, nature has invented innumerable changes in shape, size, and configuration of parts that achieve the same behaviors in different ways. Consider the dinosaurs. How did they disappear? How are we and other animals related to these (sometimes) giant creatures that lived tens of millions of years ago? Fiction books and film reveal the endless fascination that we have for these animals. As one illustration, the popular book *Dinotopia* imagines an island inhabited by shipwrecked humans and sentient dinosaurs that peacefully live together (Gurney 1992). Some films imagine that dinosaurs did not go completely extinct and, for example, somehow survive deep within the earth's mantle (*Journey to the Center of the Earth*) or that through alt-history, an asteroid did not smash into the earth initiating the end-Cretaceous (K-T) extinction so that dinosaurs survive to the present (*The Good Dinosaur*). Other films allude to the evolutionary transformation of dinosaurs into birds, as in the opening scenes of *Jurassic Park*. These are all fun. However, the reality of modern science tells a tale that is, perhaps, even more engaging (see X. Xu et al. 2014 for an integrative review). The skeletal architecture of bipedal carnivorous dinosaurs (Triassic theropods) was remodeled for new functions as its body size was miniaturized from 163 kg to 0.8 kg over a period of 50 million years (Lee et al. 2014); fossils from the early Cretaceous show that theropod dinosaurs had feathers (Norell and Xu 2005; Zhou 2014), and these feathers likely had colored patterns (Li et al. 2014). The early Cretaceous five-winged *Microraptor* had feathers that first had nonaerodynamic functions and were later adapted to form lifting functions for flight (Dyke et al. 2013). What is the process by which adaptations that originated in dinosaurs are still evident in birds? Scientists in the field of evolutionary developmental biology, or "evo devo," have discovered that conserved gene networks control embryonic development and that adaptive morphological variations at the time scale of evolution are made possible by developmental mechanisms (Carroll et al. 2005; Carroll 2008).

Remodeling at Multiple Time Scales: Evo Devo

In nature, there are dramatic ontogenetic transformations that achieve species-specific changes in form and function at the time scale of evolution. There are three factors contributing to transformations of animal body parts for different functions (Carroll et al. 2005; Coyne 2005). The first is that most animals share a similar set of "toolkit genes" that regulate the development of different gene modules (Carroll 2008). These genes, which produce regulatory proteins, called transcription factors, have been highly conserved for more than 500 million years. The homeotic, or *Hox* genes, for example, are a family of regulatory genes expressed along the anterior–posterior axis of most metazoans and may have played a role in the evolution of specific axial variation by modifying the number of vertebral segments (Burke et al. 1995). The second factor in the emergence of body form is modularity of organization: the body plan of bilateral animals involves repeated segments that can evolve independently. For example, cis-regulatory modules are short stretches of (usually) noncoding genomic DNA that include clustered transcription binding sites (with enhancers, insulators, and silencers) that control gene expression (Lemaire 2011). The modules are organized into networks with some changes having larger effects than others (Erwin and Davidson 2009). Modularity may explain how modification of an ancestral body part—the limb—results in diverse sets of special-purpose devices of the lobster—antennae, mouthparts, claws, walking legs, swimming legs, and tail (Coyne 2005). A third factor in the evolution of form is a means of diversification that conserves vital functions. One account of diversification suggests that regulatory mutations in key developmental control genes, such as *Pitx1* in the stickleback, may selectively alter expression in specific structures and yet preserve expression at other sites required for the animal's viability (Shapiro et al. 2004).

Evo devo research on segmental animals provides insights into how nature remodels existing forms for new functions by (1) conserving some segments while transforming others and (2) using trade-offs for building robust systems. During axial growth, repeated segments provide a high degree of robustness because subsets of segments may be transformed for new functions, without perturbing the original (conserved) segments. For example, in crustaceans, there are morphological differences across taxonomic orders in whether appendages are used for feeding or locomotion (Mallarino and Abzhanov 2012). The origins of the differences can be observed during embryonic development: in some orders, an appendage called T1 starts out very leg-like but undergoes a series of transformations into mouthparts. At the molecular level, these changes are due to *Hox* genes, a set of transcription factors that regulate identities of structures and functions along the main anterior–posterior axis (Mallarino and Abzhanov 2012). Experimental manipulation of *Hox* expression demonstrates how gene regulatory networks dramatically influence whether appendages are used for locomotion or eating (Liubicich et al. 2009).

The ten appendages of crustaceans are like a "Swiss Army knife" for performing multiple functions: the appendages of the posterior head segment are part of the jaw apparatus that crushes food and moves it to the mouth. In lobsters, but not all crustaceans, modified thoracic appendages, called maxillipeds ("jaw-feet"), are morphologically similar to the mouth parts, and there is now regulatory genetic evidence that thoracic appendages were transformed into jaw-feet during evolution, perhaps to give lobsters competitive advantage in obtaining food (Liubicich et al. 2009). The neural control system for using maxillipeds for multiple functions, and for switching between functions, must also have evolved through a process in which existing circuits were remodeled (Nishikawa et al. 2007). Marder and Bucher (2007) have demonstrated, for example, that neuromodulatory substances reconfigure circuit dynamics and that individual neurons can switch among different functional circuits.

New methodological synergies have now made it possible to examine how changes at the level of developmental genetics are related to morphological diversity in the evolution of the vertebrate limb (Moore et al. 2015). One such change is digit reduction, a decrease in the number of digits from a base pentadactyl (five-digit) morphology during the evolution of the four-toed pig, three-toed rodent, two-toed camel, and single-toed horse (Cooper et al. 2014). These changes enabled the limbs of these animals to be used for specialized functions adapted to their ecological niches, such as the surface differences that sand, turf, or rock afford for locomoting. The studies by Cooper et al. (2014) and by Lopez-Rios et al. (2014) identify two developmental processes involved in digit reduction: early limb-bud patterning and later remodeling and growth of digit precursors. Early limb-bud patterning involves polarity in the expression of the signaling molecule sonic hedgehog (SHH), as well as the influence of fibroblast growth factors in the apical ectodermal ridge (AER), and remodeling is due to cell death and proliferation that cause the AER to regress (Huang and Mackem 2014). One illustration of the relation between developmental processes and the evolution of a new functionality is the remodeling of the skeleton and digit loss in the evolution of bipedality in the rodent jerboa (Moore et al. 2015). During the transformation to bipedalism, the hindlimbs of the jerboa became elongated, and the three central metatarsals were fused into a single bone, increasing metatarsal resistance to bending loads as the hindlimbs support the entire body weight (Moore et al. 2015). A functional consequence of hindlimb elongation was to increase stride length, promoting digit reduction that decreases the limb moment of inertia and, perhaps, providing an energetic advantage by decreasing the energy required to propel and redirect limb motion (Moore et al. 2015).

Bodies and Nervous Systems: Scaling Relationships

There is a complex scaling relationship among brain, body, environment, and behavior evident in the influence of physical constraints on the evolution of body morphologies and nervous systems. Nervous systems have evolved to transform the elastic mechanical properties of the body into special-purpose devices for performing the adaptive functions that ensure survival in an unpredictable environment. Even the smallest animals are capable of orienting the body (taxis), locomoting, ingesting foodstuffs, forming social groups, and communicating. With small body size, large social networks evolved for collective functions, including constructing shelters and farms (Hölldobler and Wilson 2009). Large-bodied animals evolved internal networks as well as multifunctional appendages to perform these same functions (Hall 2006). This suggests that multifunctionality may be an adaptive solution to the numerous possibilities for acting in an environment with rich energy fluctuations.

Despite the diversity and complexity of life, key biological processes, such as basal metabolic rate and heart rate, exhibit an elegant simplicity in how they scale as a function of body size. The scaling takes the form of a power law, $Y = Y_0 M^b$, over many orders of magnitude in body mass (West and Brown 2005). The exponent b is often a multiple of $\frac{1}{4}$, and such scaling appears to underlie and constrain many organismal time scales (growth rates, gestation times, and lifespans). The relation between basal metabolic rate and body mass may reflect fundamental properties of branching networks that have evolved to minimize power loss when delivering resources to body cells (West 2012). This relation may be determined by the scaling relation between the volume of a hierarchical space-filling (fractal) vascular network and the number of its endpoints.

There are several findings on the relation among body size, biomechanics, and neural control of movement that may be illustrated by comparing large animals with small insects. One is that nervous systems in animals with a large limb mass may take advantage of momentum, but small insects do not. For example, in horse and human locomotion, muscles accelerate the leg at swing onset with a brief action potential burst and then go silent, relying on momentum to sustain the limb motion (Hooper 2012). By contrast, stick insect (*Carausius morosus*) swing motor neurons fire throughout the entirety of swing (Berg et al. 2015). This is because muscle passive force varies with muscle cross-sectional area, but limb mass varies with limb volume. Thus, in animals with large limbs, muscle passive forces are so small relative to limb mass that momentum and gravity dominate limb mechanics. In small-limbed animals, muscle passive force is so great relative to limb mass that gravity and momentum are inconsequential.

A second difference owing to scaling of neural control with body size is that in large animals, regardless of the direction of gravity, body parts (for example, the fingers) assume the equilibrium posture at which the antagonist muscles exert equal and opposite force. In a standing posture (with head up), relaxed human shoulders remain next to the sides but are fully extended if the body is rotated 180 degrees with the head down. However, in the stick insect, the limbs assume a constant gravity-independent posture in the (experimental) absence of muscle nerve activity (Hooper et al. 2009). One consequence of this difference for motor control is that for large animals, nervous systems must monitor gravity direction to calculate the muscle contractions required to achieve a movement in that posture. No gravity monitoring is required for insects.

Even the smallest brains in nature, such as nematodes and spiders, are capable of behaviors adapted to their ecological niches (Eberhard and Wcislo 2011), but there is a more restricted repertoire of behaviors in response to unpredictable events. By contrast, animals with large brains and bodies are characterized by behavior flexibility and ability to anticipate events not apparent in individual insects. It requires "swarms" of insects, informationally and physically connected to each other by chemical and other gradients, for their collective behavior to self-assemble and reorganize in the face of changes in the ecological niche (Hölldobler and Wilson 2009). In both cases, low dimensional behaviors emerge through the cooperative rearrangement of microscopic parts.

According to an allometric scaling relationship called Haller's rule (Eberhard and Wcislo 2011), smaller animals have relatively larger brains for their size. Miniature animals—such as spiders, insects, and other invertebrates—sacrifice some morphological aspects of body size to accommodate their disproportionately large nervous system (Quesada et al. 2011). Spiders are reliant on mechanical sources of information transmitted by the size limitations of their orb in order to catch prey. Each orb meets the metabolic needs of a single insect.

For the individual spider, a precise orb requires a nervous system that is so large, relative to its head, that it flows over into other parts of its body. The size limitations of a constructed mechanical sensor (the orb) have been overcome by social insects that work together to construct devices that do many different types of work, including thermoregulation and food storage. With the availability of niches requiring different kinds of work, social insects have evolved enlarged sensory organs for transducing chemically diffused gradients or optical flow, as well as the ability to informationally link their individual nervous systems (Camazine et al. 2001; Turner 2000). Structured gradients carrying the same information are available to guide different kinds of work and make it possible for an insect at a microscopic scale to contribute its individual resources to the behavior of its social group at a more macroscopic scale. For example, honeybees have a rich behavioral repertoire of behaviors at a microscopic scale, whose function emerges at the scale of the collective behavior of the hive: individuals build hexagonal honeycombs, manipulate pollen to glue it onto special places on the body, perform "dances" to communicate to other bees the location of a food source, and warm the brood by shivering (Chittka and Niven 2009).

Nature Builds Organs Capable of Detecting Structured Energy

Gibson and ecological physics. Sensor arrays have evolved across species for detecting diffusion of chemicals (for example, in ants), flow of patterned light (for example, during bird flight), and cascades of mechanical vibrations (for example, in spiders detecting prey in a web). An insight of psychologist (and first ecological physicist) James Gibson (1966, 1986) was that despite differences in the form of the organs that have evolved in different animals, the same information is available in these arrays. All that is required for revealing information is for animals to actively move their body organs through the array in order to detect patterns useful for guiding adaptive behavior. At the heart of Gibson's vision of ecological reality is that the relation between animal and environment constitutes a fundamental unit of analysis for understanding behavioral functions, such as locomoting and communicating (Gibson 1966, 1986). This means, for example, that to understand how animals use information to guide behavior requires an analysis that includes the way in which the sensory organs are moved through structurally patterned energy fields that envelop them (for example, optical flow for the organs of vision). The information revealed by the animal's motion through these fields is specific to each kind of displacement made by the animal. Consider the optical information of a bird during its flight through the medium of air. When the bird flies in a straight line, there is a flow of structure of the field past the head relative to a moving point of observation. From the bird's observation point, the center of outflow specifies approach and direction of approach, and with each change in direction, there is a shift in the center of outflow. The energy fields vary in the relative density of their structure, or gradients, and may attract the bird toward or away from the source of the gradient.

Gibson's description of the structuring of energy fields by the substances and surfaces of the environment, as the animal actively moves its sensory organs through these fields, is the basis for an ecology of relationships of living things with the world and the things in it. The temporal flow of information during active exploratory behavior progressively reveals a world of surface topologies, attachments, openings, enclosures, embeddings, and overlaps. Each animal's particular ecological niche consists of the experiences of particular events that distinguish living from inanimate things in motion (translation, looming, zooming, radial flow, perspective transformation), interaction between things (for example, deformation, accretion–deletion, common fate, occlusion), events unique to living things (for example, growth / decay, biological motion), and changes of state (for example, melting of a solid into a liquid, construction, and destruction). The underlying link between all animals on our planet, in Gibson's view, is their mobility. Regardless of the difference in their forms and kinds of receptors, all animals are able to reveal certain unchanging transformations resulting from motion, such as surfaces. However, some animals, by virtue of the differentially evolved or developed state of their nervous systems and their size, form, and capabilities, are able to reveal information that others cannot. While they may be able to obtain the same information, some animals may be able to select or use that information differently than others.

The eyes have it. The diversity of devices nature uses to transduce light for vision is testimony to the information-carrying potential of optical structure, as well as to the trade-offs that emerge from body-size constraints and the demands of particular ecological niches for the evolution of body organs. Relatively large animals that swim, walk, and fly use two mobile eyes in a mobile head as well as the entire wavelength spectrum of light to achieve high visual spatial resolution in a range of habitats. Giant deep-sea squid (for example, *Architeuthis*) have the largest eyes of all animals, giving them long visual range in deep water, useful for avoiding predatory sperm whales (Nilsson et al. 2012). Fiddler crabs have eyes on long, vertical stalks, providing a panoramic visual field attuned to their ecological niche, the flat terrain of intertidal mud flats (Zeil and Hemmi 2006). Avian visual systems are particularly sensitive to ultraviolet wavelength light, giving them enhanced ability to detect rapid motion during flight, pursue prey, and escape predators (Rubene et al. 2010). By contrast, insect compound eyes consist of a mosaic of tiny ommatidia that offer a panoramic field of view with negligible distortion but trade off high spatial resolution (Floreano et al. 2013).

Despite the dramatically different design of vertebrate and insect eyes, there are at least two remarkable similarities in the organization of neural circuits underlying their visual capabilities (Masland 2012; Sanes and Zipursky 2010). First, in both insects and vertebrates, photoreceptors transduce light into relatively simple electrical signals: in mammals, these are conveyed by cells that tile the retinal surface with apertures of varying sizes for diverse encoding (Masland 2012), and in flies, cells are organized into a repeated crystal-like columnar structure (Sanes and Zipursky 2010). Second, distinct functions are mediated by parallel pathways that transform visual information in multiple ways before transmission to the brain. These similarities support the view that nature has conserved the architecture of neuronal circuits across species, even as the organs for transducing patterned energy have adapted to the specific needs of different animals (Sanes and Zipursky 2010).

Measurements of tetrapod eye socket size and eye location in the skull relative to the visual ecology of fish and tetrapods, respectively, reveal how eyes that function in both water and air may have given some animals an adaptive advantage. MacIver et al. (2017) propose that the vantage point of a partially submerged body allowed certain animals with larger eyes on raised brows elevated above the water surface (as in extant crocodilians) to see clearly enough in air to visually explore distal dry land for tasty invertebrate prey. According to their "buena vista" hypothesis, exploring with the eyes in this way promoted transformation from fins to digited limbs, eventually enabling legged tetrapods to leave the water to approach, capture, and eat their terrestrial prey. MacIver et al. (2017) support the hypothesis by showing a close correlation between measured eye socket size and skull length with a time-calibrated tetrapod transition from water to land.

Affordances. The concept of affordance is a central part of the ecological approach to perceiving and acting (Gibson 1986; Goldfield 1995; Warren 2006). *Affordances are opportunities for action that depend upon the fit between the environment and an individual's action capabilities at a given moment* (Fajen, Riley, and Turvey 2008; Gibson 1986; Warren 1988). There are three key features of affordances that distinguish the concept from the view of information in approaches such as information theory. First, affordances are relational properties defined relative to action capabilities. They are not inherent in objects or environments themselves. Second, affordances are prospective, revealing opportunities for active exploration. And finally, affordances are dynamic, arising and dissolving with movements of the actor in a structured environment (Fajen et al. 2008). An illustration of the relational nature of affordances is the classic study by Warren (1984), in which the same units were used to measure a bodily property, leg length (L) and riser height (R). Warren (1984) found that the resulting dimensionless ratio (R / L) predicted whether subjects perceived a stair riser as climbable. This dimensionless number is called a body- or action-scaled ratio, or *pi ratio*. In a more general form, the pi ratio environmental property E, and an action-relevant property A, of an agent, has been used to investigate a range of affordances, including reach and graspability of objects (Carello et al. 1989; Choi and Mark 2004) in adults as well as children (Soska and Adolph 2014).

Research on affordances for reaching reveals that an individual's perception of whether an object can be reached by extending an arm, by bending at the waist and extending the arm, or by standing (see Figure 3.1) is determined by a scaling of the distance and height of the object to be reached (Choi and Mark 2004; Gardner et al. 2001). Affordances also reveal the boundaries within which a particular manner of reaching is possible. When the ratio between the relevant properties of the environment and an individual's action system reach an affordance boundary, other more stable affordances emerge, and behavior is spontaneously reconfigured for different manners of reaching. Affordances not only guide perception, but also promote selection of the manner of action actually performed. For example, Franchak, Celano, and Adolph (2012) asked participants to judge whether they could walk through horizontal openings without shoulder rotation and through vertical openings without ducking. Participants then actually walked through the openings. Results indicate that they turned their shoulders with more space available than the space they left themselves for ducking. Shoulder turning may have left more room for the body's lateral sway compared to the room required for the body's vertical bounce, suggesting that affordances are scaled to action, not just body dimensions (Franchak et al. 2012).

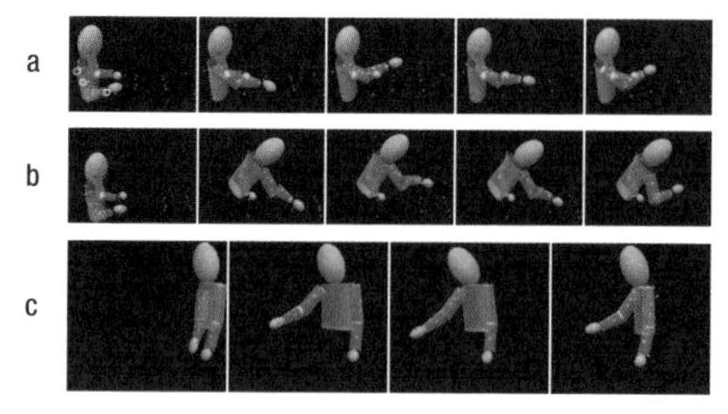

Figure 3.1 Affordances for reaching. Whether an object can be reached by (a) extending an arm, (b) bending at the waist and extending the arm, or (c) standing, is determined by a scaling of the distance and height of the object to be reached. From unpublished data of the author.

Smart Instruments for Mechanotransduction

The lessons of D'Arcy Thompson remind us that there is a limit to increasing the size of animal organs, such as the eyes for low light conditions. However, nature has arrived at another solution for organs guiding animal behavior in addition to size changes: the use of other types of receptor systems to transduce information equivalent to the structure available in patterned light (see Tables 3.6 and 3.7). For example, fish guide themselves through their watery econiches not only with their eyes, but also by means of a mechanosensory system called the peripheral lateral line (Coombs, Görner, and Münz 1989). The lateral line system consists of superficial neuromasts, occurring in pits or pedestals raised above the skin surface (or as free-standing receptors), and a second type called canal neuromasts. Both types of neuromasts are mechanosensory organs, with hair cells much like those in the vestibular system of other vertebrates (Mogdans and Bleckmann 2012). An extreme example of the adoption of an available mechanosensory organ in place of vision, when the eyes become no longer useful for guiding behaviors such as navigation and predation, is the teleost *Astyanax mexicanus* (Yoshizawa et al. 2010). Two forms of this teleost have been identified: one that is surface dwelling, and a blind cave-dwelling form, called the cavefish. Yoshizawa et al. (2010) conducted a set of experiments with blind cavefish to demonstrate

Table 3.6 Smart perceptual instruments of animals that measure complex variables

Animal	Instrument
Ant	Odometer
Honeybee	Personal and community odometers
Crab	Stride integrator
Ants, bees, wasps	Toolbox of guidance systems
Fly	Flight and altitude controllers
Migrating butterflies and birds	Time-compensated sun compass
Birds, seals, humans, dung beetles	Celestial (Milky Way) navigation

Sources: M. H. Dickinson (2014), Death Valley, Drosophila, and the Devonian toolkit, *Annual Review of Entomology, 59,* 51–72; B. Holldobler & E. O. Wilson (2009), *The superorganism,* New York: Norton; M. Wittlinger, R. Wehner, & H. Wolf (Eds.), The ant odometer: Stepping on stilts and stumps, *Science, 312,* 1965–1967.

Table 3.7 Smart perceptual instruments for mechanotransduction

Information detected	Smart perceptual instrument
Gentle touch	In *C. elegans,* there are specialized receptors at the nose for gentle touch.
Strain	In spiders, an organ for strain detection consists of cuticle, pad, and lyriform (compound) slit organ. The cuticle is a composite material with fiber-reinforced laminations in a protein matrix, giving it specific mechanical properties. The spider leg tarsus pushes against the cuticular pad, which compresses and deforms the slit organ. The combined mechanical properties of these components act as a high pass filter for detecting mechanical vibrations
Air flow	In spiders, each tactile hair bends away from an applied force and exhibits elastic restoring forces at its coupling with the body exoskeleton.
Fluid flow	In fish, the lateral line system is a biomechanical filter consisting of a cluster of neuromasts, or mechanosensory hair cells on the body surface. These detect flow because of their linkage to a gelationous structure, the cupola.
Insect prey	Whiskers are a sensory specialization of mammalian hair. Rodents (e.g., mice, shrews) use active whisking to detect, pursue, and capture insect prey.
Soft-bodied prey	The star organ of the star-nosed mole is like a "tactile eye": it has the highest density of mechanoreceptors and nerve endings of any mammal.
Fish prey	Tentacled snakes have a pair of appendages that protrude from their face. These snakes are entirely aquatic and feed exclusively on fish. Each tentacle is an analog of a whisker, in the form of a scaled appendage. The tentacles are densely innervated and extremely sensitive to vibration in water.

Sources: P. Fratzl & F. Barth (2009), Biomaterial systems for mechanosensing and actuation, *Nature, 462,* 442–448; K. Catania (2012), Tactile sensing in specialized predators—from behavior to the brain, *Current Opinion in Neurobiology,* 22(2): 251–258.

how the mechanosensory function of superficial neuromasts has evolved from the lateral line of the original surface dwelling form for enhanced detection of vibratory water disturbances that afford finding food by the cavefish form.

Lowly Worm is a beloved character appearing in many of the late Richard Scarry's picture and word books for children. Lowly has an inviting smile, jauntily sports a trademark green hat, is attired in shirt and pants leg, and wears a single shoe. In *What Do People Do All Day?* (1979), Lowly is seen peering over the plans for a new road, perhaps as a consultant to the Busytown workers (pigs, foxes, rabbits, dogs, and bears) using machines for digging and moving earth. This is an inspired depiction for the anthropomorphized worm, because Lowly's biological counterparts are masters of leveraging physical principles for boring, burrowing, burying, and excavating various granular media (Dorgan et al. 2005; Che and Dorgan 2010; Dorgan 2015). Consider the process of extending a burrow by the marine annelid worm *Nereis virens*. By simulating mud sediment with gelatin having similar mechanical properties and using polarized light to visualize the burrow around *N. virens*, Dorgan et al. (2005) were able to see evidence of a discoidal crack, held open by the dorsal and ventral surfaces of the animal.

An astounding and unexpected discovery was that the worm's burrowing behavior is directed toward propagating this crack! In this species, the crack is propagated as the worm everts its pharynx, producing dorsoventral forces against the crack walls, and then moving forward with a combination of undulation and peristalsis. In an anatomically different species, the peristaltic burrower *Cirriformia moorei*, the worm moves forward to the tip of the crack, extends anteriorly, and thickens the body to widen the crack laterally (Che and Dorgan 2010). A peristaltic wave then travels along the body, moving the body wall forward. In other words, this worm uses its anterior body as a wedge to drive the crack (Dorgan 2015). Apparently, as in other worms, such as the nematode *C. elegans*, mechanosensory organs are indeed smart devices for detecting the relation between the body and the mechanical properties of the granular media (Schafer 2016). Alas, neither Lowly nor these biological worms are legged creatures, and while they can burrow and rearrange localized grains, their anatomical toolkit does not include a device for pulling or scraping grains, such as a leg, as is apparent in burrowing by legged crustaceans and land-dwelling burrowing animals (Dorgan 2015).

Land-dwelling, burrowing animals have retained their visual systems for above-ground forays but, for underground predatory behavior, heavily depend on tactile organs that use specializations of mammalian hairs (whiskers) or of the epidermis (nasal rays) (Catania and Henry 2006). For example, the twenty-two nasal rays (eleven per side) that ring the nostrils of the star-nosed mole are a tactile organ

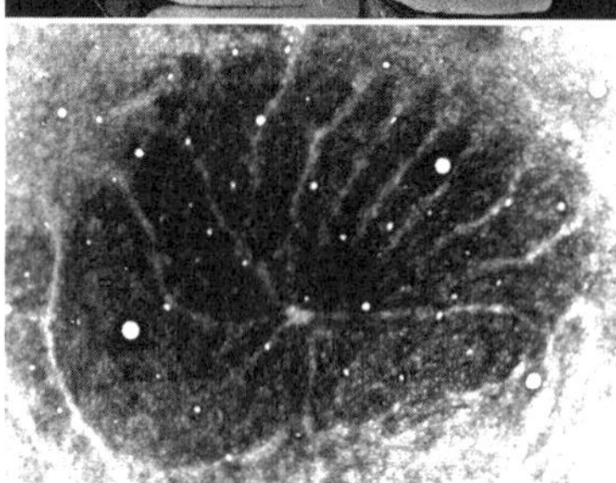

Figure 3.2 The remarkable sensory system of the star-nosed mole. *Above:* A mole emerges from its tunnel showing the twenty-two rays (eleven per side) that ring the nostrils. The eleventh ray is the tactile fovea. *Left:* One half of the nose (half star). *Below left:* The striking representation of the half star in primary somatosensory cortex. Images courtesy of Kenneth C. Catania.

with the remarkable appearance of a star in the middle of the mole's face (see Figure 3.2) (Catania 2012). A great adaptive advantage for this predator that eats small, soft-bodied animals (each providing a small amount of energy), is to minimize the metabolic cost of capturing and eating each one, by doing so very rapidly. The star organ is exquisitely adapted for high precision detection: it is only a centimeter in diameter, but is covered with 25,000 epidermal touch domes (papillae) called Eimer's organs, innervated by more than 100,000 myelinated fibers. Another striking characteristic of the star organ is that its functioning makes it seem more like a retinal surface scanning the environment, than like a nose. The analogy to a visual organ is an apt one for another reason: the eleventh ray on each side is a tactual fovea. But the most compelling case that the star functions like a "tactile eye" may be observed in the way that the rays of each half star and the tactile foveas are mapped in the principal trigeminal sensory nucleus (PrV), in the thalamus, and in layer 4 of the somatosensory cortex. The PrV contains eleven wedge-shaped regions, a trigeminal mapping of each ray, and an over-representation of the tactile foveas in the primary somatosensory cortex (Catania, Leitch, and Gauthier 2011).

How Smart Are Insect Perceptual Instruments?

Insects occupy every imaginable niche of our planet: they live and work underground and forage on blades of grass and in trees (ants and beetles), fly through the air (butterflies and dragonflies), and even walk on the surface of ponds and lakes (water striders) (Hu, Chan, and Bush 2003; Bush and Hu 2006). They use their nests as bases for journeys that may extend for tens or thousands of miles and across multiple seasons of the year (Merlin, Heinze, and Reppert 2012). To find their way, insects use available information from polarization of the daytime sky (Weir and Dickinson 2012), according to the position of the moon and Milky Way at night (Dacke et al. 2013), and from the position of their nests relative to local topography (Buehlmann, Hansson, and Knaden 2012). What kinds of biological instruments do insects use to find their way?

One illustration of the range of instruments used for way-finding by a flying insect, *Drosophila,* comes from Michael Dickinson's reconsideration of classic mark and recapture experiments in the open desert of Death Valley (Dickinson 2014). In these experiments, tens of thousands of flies were released to directly test whether they could fly long distances to one of two habitable oases (as far as 14.6 km) with inhospitable terrain intervening (Coyne et al. 1982). Only a very small number reached either of the two oases. Dickinson (2014) proposes that the small number of flies that were able to cross the desert in one night did so without any flight plan, or autopilot, stored in the brain. Instead, their flight may have consisted of *iterative applications of actions, each elicited by the local environment, and guided by the repertoire of instruments common in many insect species.* This iterative, information-guided process, bound together by the intrinsic dynamics of the fly's behavior and the availability of information-rich arrays encountered in local niches, is called *stigmergy*. How might a sequence of actions, bound together by locally available information over time, promote the use of a set of instruments to guide a long journey?

To begin their journey across Death Valley, the flies began with a take-off sequence, which, Dickinson (2014) notes, is highlighted by the way they position their middle set of legs to propel them away from the supporting surface. Their initial heading was likely characterized by a stereotyped zigzag pattern, which includes directed straight flights interspersed with collision avoidance reflexes (Frye, Tarsitano, and Dickinson 2003). To maintain a constant heading while flying in this zigzag pattern, the flies may have used a sky compass to read the polarization of the sky (Weir and Dickinson 2012). In continuing their flight, the flies may have adjusted their altitude by actively tracking horizontal edges in their visual surround (Straw, Lee, and Dickinson 2010). As they approached their destination, a set of banana-baited traps, they were likely guided by the odor plume of the traps and followed its concentration gradient to a landing spot. To make a safe landing, each fly oriented to a landing spot; decelerated, extended legs forward; and made contact with the surface. Thus, on their long journey, the flies were likely to have used a sky compass, optical flow detector, and odor plume tracker—all bound together by the requirements of navigating through a sequence of local environments.

Among the thousands of flies that never reached the banana traps were some that may have encountered a species of hunting dragonfly, *Libellula cyanea*. How does the dragonfly track and capture a fly engaging in zigzag flight patterns? Hunting dragonflies exhibit foraging modes that vary along a continuum from ambush to active predation (Combes et al. 2012). Ambush predators typically hide and wait for the approach of an unsuspecting fly, then attack by accelerating to capture their prey, while active predators engage in prolonged searches and / or chases to find and subdue their prey (Combes et al. 2012). An unresolved question is how flying insects such as the dragonfly are able to distinguish information in the visual field due to their own motor output in tracking a target from information due to external perturbations, such as air turbulence. One possibility is that dragonfly interception steering is guided by a neural mechanism that predicts the future state of a system given the current state and control signals, what is called a *forward model* (Webb 2004).

At an indoor dragonfly flight arena on the Janelia Farm campus, experiments are being conducted to examine the predatory behaviors of dragonflies and their prey (Mischiati et al. 2015). The experiments used an eighteen-camera motion capture system for free-flight measurement of dragonfly head and body orientation during computer-controlled translation of synthetic prey (retroflective beads) along a wire suspended between two height-adjustable pulleys. The logic behind the experiments follows from the head predictively leading the orientation of the body. If neural mechanisms for prediction guide dragonfly behavior, then head rotation used to track the prey should guide rotation of the body to align it with the prey's flightpath in preparation for attack. From their experiments, Mischiati et al. (2015) discovered that the dragonfly rotates the head to keep the target foveated and orients the body to the prey's direction of motion, while remaining directly below the prey. The finding that the head of the dragonfly uses foveation of its target relative to the orientation of the body in anticipation of attack has important implications for what makes a perceptual system "smart": animals use their perceptual systems prospectively to obtain information for attuning the body for action and, in complementary fashion, use their actions to orient the perceptual systems. The behavior of land-dwelling insect species further illustrates this fundamental complementarity of perceiving and acting.

Ants widely use smart perceptual instruments for navigation, including chemoreceptor organs for detecting gradients of chemical pheromones (Hölldobler and Wilson 2009); compasses for constructing navigational vectors (Wehner 1997); and odometers, or distance gauges, for measuring distance (see Wolf 2011 for a review). There are few landmarks in the desert, so the most reliable reference to determine distance traveled is measurement of the body in relation to the ground surface, which is the number and length of strides. A series of experiments demonstrate that desert ants use odometry by stride integration to measure the distance they have traveled to a source of food (Wittlinger, Wehner, and Wolf 2006; Steck, Wittlinger, and Wolf 2009; Bolek, Wittlinger, and Wolf 2012). Ants were trained to collect food at a feeder located 10 meters from the nest and were then captured. The experimental treatment involved manipulating leg length, either by severing the tibiae to "stumps" to reduce stride length by about 30 percent or by gluing "stilts" to the tibiae to increase stride length by more than 30 percent (Wittlinger et al. 2006). If the ants were relying on a stride counter, then when given a food item and released at a distance from the nest, those with reduced stumps would be expected to underestimate distance to the nest by 30 percent, and those with stilts to overestimate homing distance by more than 30 percent. The results indicate that it is not the number of strides counted because both load carrying and speed influence stride number. Instead, the number of ant strides is summed considering both their length and number—that is, they are integrated. The conclusion, therefore, is that to measure distance, ants use a stride integrator (Wolf 2011).

The way that cockroaches use their antennae to measure distances between their body and environment further illustrates how insects use body properties to scale perception to action (Mongeau et al. 2013). In their natural habitat of caves, cockroaches use their antennae, a passive mechanosensory organ, to track relative distance to a rock wall, in order to avoid collisions during high-speed escape running. The antennae passively switch from a "forward" orientation at rest to a "backward" orientation during high-speed running. Information from mechanosensory hairs in contact with the wall during running promotes a switch to the backward orientation to provide near-zero time lag detection of distance to the wall surface.

Prestress: Mechanotransduction Across Scales in Hearing and Haptics

Body organs are systems that are functionally integrated across scales ranging from the molecular nanometer (10^{-9} meter) scale to

10^{-2} meter macroscale. Consider the human auditory system (Gillespie and Muller 2009; Hudspeth 2014; Vollrath, Kwan, and Corey 2007). As a functional organ, the mammalian ear is an acoustical amplifier and frequency analyzer, as well as a three-dimensional inertial guidance system (Hudspeth 1989, 2014). Ingber (2006) proposes that mechanotransduction is accomplished across these spatial scales by means of keeping systems under mechanical tension, or prestress: at the macroscopic scale, the ear drum and ossicles (tiny bones) are stiffened by muscles that keep the system under mechanical tension; at a mesoscopic scale within the cochlea, the oval window induces fluid pressure waves within the cochlea that induce corresponding vibrations in the extracellular matrix (ECM) of the basilar membrane. At the micro- and nanoscales, the sensory epithelia in the organ of Corti of the cochlea feature bundled rows of outer hair cells that amplify sound signals and receive efferent innervation and inner hair cells innervated by afferent neurons that carry sound information to the nervous system (Zhao and Muller 2015). There are mechanically gated ion channels located in stereocilia on the hair cells, and the stereocilia are connected by tip links, or extracellular filaments composed of protein complexes that may gate the transduction channels (Zhao and Muller 2015). The tip links may serve as springs that mechanically gate ion channels. The spring-like architecture of the tip links suggest that they too may be prestressed—that is, they experience a resting tension—to immediately respond to physical signals. Ingber proposes that

> Nature has developed an ingenious strategy for mechanotransduction that involves use of structural hierarchies (systems within systems) that span several size scales and are composed of tensed networks of muscles, bones, ECMs, cells, and cytoskeletal filaments that focus stresses on specific mechanotransducer molecules. (2006, 815)

Turvey and Fonseca (2014) hypothesize that prestress in muscle, connective tissue, and bone across spatial scales may provide a transmission medium for mechanotransduction, as evident in haptic perception. At the heart of their argument is that the muscular, connective net, and skeletal (MCS) organ systems of the body have the characteristics of a tensegrity system (see the discussion of tensegrity in Chapter 2). The functional morphological unit of the MCS, in their view, is

> an in-series organization of muscular, connective and skeletal tissues, with the receptive structures concentrated in the tissue-transition zones. This responsive viscoelastic architecture is both active (its muscular tissue generates contraction forces) and passive (its connective tissue conveys tensile forces, and its skeletal tissue sustains compressive forces). Further, it is seemingly repeated at all articulated segments. (2014, 163)

The spine is one example of how the interconnectivity of the muscles and connective tissue are organized in a triangular pattern to provide mechanical support for the lumbar area: when bones of the spine are experimentally removed, the connective net maintains its integrity. Thus, like the ear as an organ for hearing, the *prestressed architecture of body tissues* make the entire body an organ for haptic perception.

Muscular Hydrostats

A range of soft-bodied animals, or their body parts, possess a hydrostatic skeleton, a liquid-filled internal cavity surrounded by a muscular body wall (Kier 2012; Kurth and Kier 2014). The liquid inside the body cavity resists changes in volume so that muscular contraction does not significantly compress the fluid. The resulting internal pressure increase provides possibilities for multiple functions, including support, muscular antagonism, mechanical amplification, and force transmission (Kier 2012). Differentiation of function is achieved, in part, through the arrangement of muscle fibers into circular and longitudinal geometries and by division of the body into septa, or segments, as in annelids, such as the earthworm.

Many of the structures of soft-bodied animals, such as the arms of the octopus, *Octopus vulgaris*, lack the fluid-filled cavities of annelids. Instead, these structures, called "muscular hydrostats" (Kier and Smith 1985; Smith and Kier 1989), consist of a densely packed, three-dimensional array of muscle and connective tissue fibers. A characteristic of muscular hydrostats is that the selective muscle contraction that decreases one dimension of the structure must result in an increase in another dimension (Kier 2012). The selectivity of the change of shape of the organ is achieved by the effect of contraction on the muscle fibers comprising its particular architecture. Elongation results from the contraction of muscle fibers arranged transversely, radially, or circumferentially: shortening decreases diameter and, hence, increases length (Kier 2012). Bending, by contrast, is caused by selective contraction of longitudinal muscle fibers on one side of the structure and simultaneous contraction in the transverse, circular, or radial musculature (Kier 2012).

The capability for elongation, shortening, bending, torsion, and variable stiffness provides cephalopods—such as the octopus, an intelligent hunter and tool user—with the ability to use its arms for multiple functions. Each of its eight arms exhibits a characteristic row of sophisticated active suckers for standing, locomoting, capturing prey, opening bivalves, reaching for a target, grasping, fetching, grooming, digging to build a shelter, and more (Hochner 2008). The flexibility and multifunctionality of the muscular hydrostatic body organs of the octopus raise the question of how its many muscles are coordinated to perform specific tasks. The octopus has a highly distributed nervous system, with the majority of its 500 million neurons located in the periphery, close to the sites of information flow between its body and the environment (Hochner 2012). A distributed control system may allow the octopus to simplify using its eight arms for complex tasks, such as visually guided reaching, by leveraging the properties of its muscular hydrostatic actuators.

The arms of the octopus do not have joints in any fixed location. Instead, following the principles for muscular hydrostats identified earlier, any arm may form a bend wherever it is needed. This functionality drastically reduces the high dimensionality of the muscular architecture to just three degrees of freedom: two for the direction of the base of the arm and a third for scaling the propagation velocity profile of the bend along the arm (Gutfreund et al. 2006). To form a special-purpose, quasi-articulated feeding structure, the octopus nervous system divides the section from the base of the arm to the target into proximal, medial, and distal segments. The distal segment acts like a hand by grasping food with its suckers, while the proximal and medial segments act like a forelimb to bring food to the mouth (Sumbre et al. 2005). This example of one of nature's solutions to controlling a body part with a soft architecture has implications for building neuroprosthetic devices with a rich capability for dynamics built into the arm itself and a simple activation signal from networks generating intentions, a point further discussed in Chapter 8.

Nature's Pumps

Nature builds highly specialized pumps for obtaining nutrients to meet the particular metabolic requirements of an animal species. Hummingbirds, for example, exhibit uniquely high metabolic activity, fueling an astounding rate of wing beats for hovering, as well as for nimble and rapid aerial maneuvering. To satisfy their fuel needs, hummingbirds preferentially seek out the nectar of flowers, a nutrient source with a high energy density. The beak of hummingbirds is particularly well suited for dipping deeply into the corolla of a flower and for rapidly extracting the nectar. However, the manner by which the nectar is extracted with the tongue has been a source of some controversy. Elegant studies using high-speed photography, mimesis of flower forms, and precise measurement of nectar flow from flower to beak have now come to the extraordinary conclusion that what moves nectar inside hummingbird tongues is not capillarity, as once believed. Instead, hummingbird tongues are elastic micropumps (Rico-Guevara, Fan, and Rubega 2015). Other animal species leverage their particular anatomy to rapidly assemble the anatomy into pumps for drinking.

Dogs are notoriously messy drinkers because their incomplete cheek anatomy prevents the use of suction for drinking (Crompton and Musinsky 2011). Instead, dogs use an "open pumping" mechanism for lapping: the tongue forms a "ladle shape," and splashes when it enters the water surface, and as water adheres to the rapidly retracting tongue due to inertia, a water column is produced providing a brief window of time for the dog to bite into the column (Gart et al. 2015). To examine how a dog controls the properties of the water column for drinking, Gart et al. (2015) conducted physical simulations of the tongue with a glass rod dipped into, and then extracted from, a water bath at particular accelerations. Like the dog's tongue, when the rod traveled upward, a water column and pinch-off was created, and the extracted volume was positively correlated with rod acceleration. Dogs, therefore, appear to control acceleration of tongue retraction, and must time the bite so that the jaws are closing at the moment of water column pinch-off. Those of us who have shared our home with a dog know that the contents of the water bowl are likely to end up all over the floor. The mess is apparently because of this open pumping process in all dogs, not just ours.

The tongue of dogs, like other mammals, is a hydrostatic skeleton attached to bone (the hyoid). It is able to rapidly change shape and precisely appose its surfaces to multiple intraoral organ surfaces, or extend outside the mouth (Bramble and Wake 1985; Hiiemae and Crompton 1985). Several elegant imaging and modeling studies have directly measured how orthogonally aligned muscle fibers contribute to this volume-conserving tissue deformation (Mijailovich et al. 2010; Gaige et al. 2007; Gilbert et al. 2007). Mijailovich et al. (2010) used MR imaging and diffusion-based tractography to display and measure the orientational coherence (alignment of local strain vectors) of myofiber populations. Lingual myoarchitecture consists of a core region of orthogonally aligned and crossing fiber populations encased within longitudinally aligned fibers, all of which merge with externally connected extrinsic fiber populations. Thus, the tongue is a unique form of muscular hydrostat, in that it displays actuation that is both untethered to any external skeletal structure (intrinsic muscles) and tethered by mechanical relationships with the extrinsic musculature and extrinsic bony structure.

The soft tissue and muscular architecture of the hydrostatic tongue reveals how its structure provides hybrid functionality at the boundary between the inside and outside of the body. The tongue is sufficiently soft to apply pressure to other soft tissues inside the oral cavity without causing injury, but is able to change its shape and stiffness to extend outside the mouth in order to obtain nutrients. For food in a solid state, the tongue moves the bolus inside the oral cavity so that it may be chewed and processed for swallowing. Liquid nutrients are swallowed directly through the propulsive muscular forces of the tongue in contact with palatal and pharyngeal surfaces (Goldfield et al. 2006, 2017). Anatomically, the tongue is also at the interface between airway, vocal cords, and our gaseous atmosphere. It has evolved to use the same set of muscles for shaping and stiffening to make contact with internal surfaces of the oral cavity which modulate the flow of expelled air molecules, so that they carry highly structured arrays of acoustic energy (Goldfield, Perez, and Engstler 2017).

Human swallowing involves muscle groups and organs that may serve other functions, including respiration and communication (Crompton et al. 2008). Swallowing requires that these muscle groups be flexibly recruited, so that the organs for swallowing may *rapidly change their shapes and the timing and force of their contact with each other* for different functions. In air-breathing mammals, ingesting food requires that the tongue and respiratory organs be dynamically coupled in such a way that it is possible to rapidly switch between the *vegetative (life-supporting) mode* and the *appetitive mode.* As an illustration, consider the challenge faced by newborn humans learning to ingest milk during breast or bottle feeding (Goldfield 2007; Goldfield et al. 2017). The pharynx is a shared anatomical pathway to the lungs and digestive tract, guided by a complex network of mechanoreceptors, thermoreceptors, and chemoreceptors. The pharyngeal muscles create and release constrictions to select the direction taken by a flow of air or food (Miller 2002). In the vegetative mode, the attractor dynamics of the brain stem respiratory control networks generate a cycle of inspiration-expiration (Feldman, Del Negro, and Gray 2013), bind the breathing rhythm to orofacial perception (Kleinfeld et al. 2014), and regulate coupling to the attractor dynamics of the tongue and pharynx (Smith et al. 2013). However, in the transition to swallowing, perceptually guided brain stem networks trigger a reorganization of muscle groups during which the pharynx is reconfigured (Jean 2001).

How do the tongue and pharynx of the human infant act together during breast and bottle feeding to move milk into the mouth? The familiar pattern of suckling may be thought of as the soft assembly of multifunctional organs into a positive displacement pump (Goldfield 2007; Goldfield et al. 2017; Vogel 2003). Such pumps accumulate successive bits of fluid into a confined space, reduce the space enclosing a volume of fluid to force it to leave by a specific outlet, and create a pressure difference that moves a volume. During infant suckling, cooperation between oral, labial, mandibular, and lingual-pharyngeal muscle groups draws milk into the mouth, seals the lips around the breast nipple and areola, and forces accumulated milk boluses into a small space so that the elastic forces of the tissue surrounding the forming bolus builds to a certain pressure for a particular volume. During deglutition, anatomical constrictions and openings arise to define a fluid pathway that excludes the bolus from the airway, as it is forcefully pumped through the pharynx on its journey to the upper esophageal sphincter (Sawczuk and Mosier 2001).

From Attractor Dynamics to Function

Modes of behavior. Locomoting, finding food, avoiding predators, eating and drinking, and communicating are among the preferred, stable *modes of behavior* of animals on this planet (Goldfield 1995; Kugler and Turvey 1987; Reed 1988; Warren 2006). *Each mode of behavior emerges from a rapid and temporary assembly of bodily resources for generating particular attractor dynamics as the animal is attracted to or repelled by particular environmental resources.* The same underlying attractor dynamics of each mode may be implemented in different ways, depending upon the particular anatomical resources of the animal, such as the size, form, and musculature of its appendages. Animals that transport the body in contact with solid surfaces have evolved slithering, crawling, and walking modes of locomotion. Animals that live in water use swimming as their mode of transport. And animals that have wings are able to glide or flap their wings to fly. Organizational transitions *within* modes—for example, from walking to running on land—or hovering rather than changing spatial location during flying, as well as transitions between functions of the same appendage—for example, using a fin for propulsion or for sieve feeding in water (Koehl 2004)—have also evolved. The frog, an amphibian that makes a living at the physical boundary between water and dry land, uses a tail appendage for swimming in an exclusively aqueous environment when in its tadpole form. But during the developmental process of metamorphosis in many frogs, encompassing an individual lifespan, the tail is resorbed and replaced in function by four limbs useful for both swimming and walking (see Chapter 7). And during the developmental process of embryogenesis in other vertebrates, appendages are transformed in ontogenetic time from a form characteristic of earlier vertebrates to one specialized in various ways for locomotion on land; in water; and, in the case of gliding mammals, in the air.

A personal perspective. As a postdoctoral fellow, my major research project was to study how infants leverage the intrinsic dynamics of their body motion for locomotion (see, for example, Goldfield 1989). Freshly armed with the dynamical systems toolbox, I was attracted to two intriguing rhythmic phenomena in prewalking infants, pointed out to me by Peter Wolff and reinforced by my literature search on the classic work on infant crawling by Myrtle McGraw (1943; see also Gottlieb 1998) and Arnold Gesell (1946; see also Thelen and Adolph 1992). First, Wolff had clinically observed that typically developing infants often rocked to and fro on all fours before they began to crawl, one of many rhythmic infant "stereotypies" (and see Thelen 1979, 1981, 1996). Second, infants supported under the arms with the soles of their feet on the ground excitedly flex and extend their legs, as if "bouncing." Indeed, when placed vertically in a harness attached to a spring, infants will joyfully bounce for sustained periods (Goldfield, Kay, and Warren 1993). From a dynamical systems perspective, both of these rhythmic motions phenomenologically seemed to exhibit attractor dynamics and so might be the kernel from which functional behaviors, such as locomotion, might emerge. My major hypothesis was that infants may discover how to assemble their body resources into a functional device for locomotion by exploring the dynamics of their body motions and systematically varying them to satisfy their intentions.

At that point, I had no motion capture system to measure the dynamics of rocking, but I was able to conduct longitudinal video observations in the homes of Boston-area infants to better understand the potential significance of rocking. I also was influenced by some of the ideas on the role of lateral asymmetries of hand use from George Michel, a developmental psychobiologist, also working with Peter Wolff at Harvard, and now at the University of North Carolina. During that period Michel and I had done some research together on infant reaching (Goldfield and Michel 1986a,b), and so, during my longitudinal video recordings of the dynamics of learning to crawl, I also assessed each infant's hand preference. This was a good decision because it helped solve a puzzle about infant rocking. When infants first independently support themselves on their hands and knees, the four points of contact of the body with the floor form a symmetrical based of support. So, when they try to move forward by planting their feet and extending their legs, gradual application of equal and opposite forces from their hands pushing against the floor generate a to-and-fro rocking motion. The puzzle for infants, and for me, was to figure out how the body breaks free from this balance of forces. The answer, it turned out, involved hand preference. Infants gradually learn to shift their body mass so that their preferred hand for reaching is free to lift off the floor surface and reach forward. This leaves a three-point support adequate to keep infants from falling; at the same time, it allows them to make forward progression. In other words, the intrinsic lateral asymmetry of hand use may break the biomechanical symmetry seen in rocking in order to create a sequential cycle of hand placement driven by oscillatory kicking motions of the legs, the emergent behavior that we call "crawling" (Goldfield 1989).

In order to probe the nature of the underlying dynamics of a behavior, such as kicking while supported upright, dynamicists use models to make and test predictions about a system's behavior. These may take the form of task dynamic modeling using equations of motion (Saltzman et al. 2006) or may be based on robot physical and algorithmic architectures (Berthouze and Goldfield 2008). As part of the longitudinal recordings made of prewalking infants in their homes, I provided each family with a commercially available infant bouncer, called a "Jolly Jumper." This allowed me to make weekly recordings of infant progress in learning to bounce. Bill Warren at Brown University and Bruce Kay, now at the University of Connecticut, and I developed a model of the infant in the bouncer as a tunable forced mass-spring system (Goldfield, Kay, and Warren 1993). An equation for a forced mass spring is one that includes, on one side of the equal sign, some regular periodic source of energy input (such as the force produced by infant kicking) and, on the other side, the parameters of the system, its mass, stiffness, and damping, representing the properties of the physical spring. We then tested predictions of the model and found good agreement with the observed behavior of the infants. Robotic modeling of infant bouncing extended this work by determining whether the equations governing the way a robot explores its own behavior were able to generate the same patterns of learning that we observed in longitudinal measurements of infants (Berthouze and Goldfield 2008; see also Chapter 9).

Biomechanics and dynamics. Learning to bounce may not be directly related to learning to walk, and indeed, bouncing is more like running than walking. Classical biomechanical models of walking, based upon some version of inverted pendulum systems (for a classical model see, for example, McMahon 1984), demonstrate that once the limbs begin to swing, they move ballistically—that is, entirely under the action of gravity. But how do neuromuscular forces initiate the swing and sustain its motion throughout the gait cycle? Another University of Connecticut–trained dynamicist, expert in biomechanics, and physical therapist now at Boston University, Ken Holt, and his former student Sergio Fonseca, now at Universidade Federal de Minas Gerais, in Brazil, have proposed an escapement-driven, damped, hybrid-inverted pendulum system (see, for example, Holt, Obusek, and Fonseca 1996; Holt, Fonseca, and LaFiandra 2000). In this model, a clock-like escapement mechanism replaces energy lost during the cycle (for example, due to friction) with a pulse of energy released at an appropriate phase of the pendular oscillation (Holt et al. 2000). Young children learning to walk gradually discover the pendulum and spring dynamics of their body (Holt et al. 2006). The model has also made successful predictions about the consequences of excessive stiffness on the gait cycle of children with cerebral palsy (Fonseca et al. 2001, 2004).

Navigation: Neural Attractors and Behavioral Dynamics

Careful observers of the natural world, including Charles Darwin, have noticed that most animals keep track of their location relative to a "home base," an integration of self-motion called *path integration* (McNaughton et al. 2006). Head direction (HD) cells, abundant throughout the limbic system, use the animal's head direction in the horizontal plane, independent of the animal's location and ongoing behavior, for path integration (Taube 2007). The HD cells are hypothesized to be arranged in a circle, exhibiting ring attractor dynamics, and through a process of interaction between excitation and inhibition, a localized, self-sustaining "bump" of activity emerges on the ring (Knierim and Zhang 2012).

Navigation based solely on path integration accumulates errors over time, so nature's navigation systems use sensory information to update the intrinsic dynamics of path integration. Two studies illustrate this relation between environment-driven and intrinsically generated contributions to navigation. The first is a demonstration in mammals that the ring attractor remains temporally organized even in the absence of vestibular or other sensory input. Peyrache et al. (2015) determined whether a temporally correlated group of HD neurons (called an activity packet) moved on a virtual ring as a mouse turned its head, as well as when the animal was asleep. Peyrache et al. (2015) found that during rapid eye movement sleep, movement of the activity packet was similar to that during waking, evidence for ring attractor dynamics.

A second approach to identifying the attractor dynamics of the brain networks involved in navigation is to measure activity of a simpler nervous system, such as the fly. It is well documented that to navigate, insects use a combination of path integration and sensory information from polarized light relative to the sun (Heinze and Homberg 2007). Seelig and Jayaraman (2015) have now also identified the actual neural dynamics of the ring attractor during landmark orientation and path integration of the fly. In the fly, the central complex (CX), especially the ellipsoid body (EB), is involved in navigation. Seelig and Jayaraman (2015) used two-photon imaging and a genetically encoded calcium indicator (see Chapter 5) to monitor neural responses in the CX while a head-fixed fly walked on an air-supported ball within a light-emitting diode (LED) arena, also at Janelia Farms. There are several remarkable findings in support of a ring attractor model during navigation, including organization of neural activity into a localized bump, movement of the bump based on self-motion, and drift in bump location during darkness. So, together, studies of mammals and insects support a view that brain networks generate attractor dynamics for navigation devices.

The evidence for neural path integration in insect and animal navigation raises the fundamental question of how behavioral functions, such as navigating and feeding, may emerge from interactions between the particular neural attractor dynamics of a different species and information available in its ecological niche (Goldfield 1995). This question has important implications not only for an understanding of brain and behavior (see Chapter 5) but also for control of autonomous robots (see Chapter 4) and for design of robotic assistive devices that are a seamless part of biological systems capable of emergent functions (see Chapters 8 and 9; and see, for example, Krakauer et al. 2017). I discussed earlier Gibson's discovery of the informational gradients, invariants, and affordances that characterize a structured environment. How might such information be the means for mapping brain and environment?

There is some evidence that in animals endowed with even the simplest neural circuits, such as the worm, *C. elegans*, basic smart perceptual instruments for navigation and feeding are specialized to detect time-varying fluctuations in chemical gradients. For example, Kato, Xu, Cho, Abbott and Bargmann (2014) recorded calcium responses of sensory neurons as well as exploratory olfactory behavior, and found that adult worms modified their head-swing and turning frequency behavior in order to follow experimentally manipulated odor concentration changes. Remarkably, their behavior was modified at two time scales: head orientation within a few seconds of chemical gradient change, and turning frequency over about a minute. A further evolutionary advance of smart perceptual instruments guiding feeding behavior may have involved increased robustness to sensory perturbations. For example, the feeding behaviors of the mollusc *Aplysia* appear to emerge from neural circuit architectures that make possible a range of responsiveness to sensory input (Lyttle et al. 2017).

But while worms and molluscs may be "smart" in how they use available sensory information to explore their ecological niches for food, larger ensembles of brain circuits may have evolved for assembly of action-guided perceptual instruments robust to a wider range of sensory information. In the mouse, there is evidence that cortical gamma oscillations may coalesce ensembles of subcortical circuits to form particular networks for regulating food-seeking navigation (Carus-Cadavieco et al. 2017; see also Chapter 5). Behavioral navigation by mice during feeding may, therefore, emerge from cortical synchronization of certain subcortical circuits driving cycles of behavior, guided by online sensory information (Goldfield 1995; Goldfield, Kay, and Warren 1993). I further explore the possible role of cortical synchronization for behavioral dynamics in Chapter 5. In closing this chapter, though, I turn to the contribution of the attractor layout of the environment to the behavioral dynamics of human navigation.

Neuroimaging studies of human navigation in virtual environments demonstrate that particular brain regions track self-motion (translation and rotation) to update knowledge of position and orientation (Chrastil et al. 2016). Other work has indicated that during animal exploratory behavior, place cells in the medial temporal lobe are modulated by a strong theta rhythm (see Chapter 5), such that within each theta cycle, place cells are activated so that they are able to "look ahead," exploring multiple paths at a choice point (Pfeiffer and Foster 2015). What are these choice points in the environment? One possibility is the layout of attractors and repellors.

Bill Warren and his students at Brown University have been conducting a research program on human exploration of, and behavior within, the attractor layout of structured environments (see, for example, Warren 2006). To illustrate, in a series of studies by Warren and former student Brett Fajen, now at Rensselaer Polytechnic Institute, participants wearing a head-mounted display walked through computer-generated virtual environments consisting of floor surfaces, goalposts, obstacle posts, and moving targets and obstacles (see, for example, Fajen and Warren 2003; Warren and Fajen 2008). A motion-tracking system was used to reveal how each participant steered toward environmental attractors (the goalposts) and away from repellors (the obstacle posts). Then participant behavior was compared to a simulation based on a set of ordinary differential equations defining attractors and repellors in the direction of locomotion (heading). From this dynamical systems perspective, global paths in navigating environments *emerge* from local interactions with these attractors and repellors. It was possible, in these studies, to successfully model how an individual steers to stationary goals, avoids obstacles, intercepts moving targets, and avoids moving obstacles. In another set of studies, groups of subjects were instructed to walk behind a confederate who randomly increased or decreased walking speed (Rio, Bonneaud, and Warren 2012; Rio, Rhea, and Warren 2014). Here, a speed matching model based on nulling of optical expansion captured the behavioral dynamics of one person following another. Taken together, these studies support the view that while navigating structured environments, humans steer toward attractors and away from repellors. An implication of these studies is that there is a natural affinity between the attractors emerging in neural activity and the layout of environmental attractors and repellors, a crucial relationship discussed further in Chapter 5.

✦ 4 ✦

Building Devices the Way That Nature Does

In the short, clay animation film, *A Grand Day Out,* Wallace and his silent, but brilliant, canine companion Gromit build a homemade rocket in order to journey to the moon, which, of course, is made of cheese. Wallace, you see, is a cheese fanatic. When they land, Wallace and Gromit lay out a picnic blanket, china, and cutlery and puzzle over which variety of cheese they are eating. They soon discover that they are not alone. On the moon with them is a wheeled robot, a 1950s-looking, mustard-colored, British oven. We never find out how this machine arrived on the moon, but its behavior is a humorous take on how to build an adaptive (albeit coin-operated) robot. Wallace inserts ten pence into a coin slot on the robot, but to no avail. It is not until he and Gromit walk away that the coin registers on a gauge, and the robot awakens from an apparently long nap. The robot's telescoping arms emerge from side compartments to open and rummage around a bottom drawer, eventually finding a viewscope that the robot then screws into a socket. The robot comically gestures with chagrin when Wallace and Gromit's picnic blanket comes into focus.

With our heroes nowhere in sight, the robot rolls itself over to the blanket and unceremoniously dumps all of their cups and saucers into its bottom drawer. The prize of this haul, we discover, is a pamphlet on "ski touring." The pictures provoke an idea, in the form of a video "thought bubble" generated by its top-mounted antennae: an imagined future in which the robot effortlessly skis down a slope, emulating the human skier in the pamphlet. The robot's arms droop as it apparently realizes that its current wheeled locomotion does not afford skiing. What to do (indicated by a head scratching gesture)? First, the robot repairs the cheese landscape that the interlopers cut from the moon by gluing back a remaining piece. Then, eyeing the rocket ship in the distance, the robot steers around craters and, not knowing the nature of the ship's substances, bumps into it. It is quite hard, not like the moon's cheese. Again retrieving and installing the viewscope, the robot sees Wallace and Gromit as well as planet Earth in the distance. We see the robot's "thought process" again, this time making the connection between Wallace and Gromit and this place where there are opportunities for skiing. As Wallace and Gromit grab some cheese and attempt to escape, the robot accidentally yanks two strips of metal from the ship before it lifts off. Now in orbit, Wallace and Gromit are safe. Back on the surface, gestures indicate that the robot is forlorn and angry. But wait! The confiscated strips of metal are hard, yet pliable, and they can be bent on one end. A solution is at hand: the short film ends as we see the robot using the metal strips and other leftover parts for skis and poles. The robot is now able to ski up and down the hills of the moon, whirling its arms in windmill fashion to help climb up one side and then glide down the other. Finally, in a gesture of amity, the robot on the surface and Wallace and Gromit in orbit wave to each other.

This comic gem presents a late twentieth-century vision of an autonomous robot, following its programming by stubbornly assuring the integrity of the moon's cheese, and yet capable of inventing ways to use available resources to solve problems. But this robot seems to go beyond its programming for preserving the moon's cheese or for solving other problems: it is able to satisfy its own desires. This chapter considers how far we have come in being inspired by biology to develop new adaptive materials and devices, including soft robots. We are still far from building machines that are able to go beyond their programming. But we have made a beginning.

Emulating Life's Devices

Tinkering in a basement workshop may be fine for Wallace and Gromit, but the challenges of building devices that emulate nature require advanced technologies, as well as a deep appreciation for nature's construction motifs and how they are orchestrated over time. At the cellular level, nature's construction milieu is an open system with constant flows of matter and energy to power reaction networks, maintain concentration gradients, and enable active transport (Grzybowski and Huck 2016). For example, biological construction of organs takes place within cellular "factories" that use compartments—for example, cell vesicles—to maintain nonequilibrium conditions for synthesis (Grzybowski and Huck 2016). Nature uses feedback mechanisms interacting with the physical dynamics of animals to maintain stability in response to perturbations (Cowan et al. 2014). Nature additionally leverages feedback loops to build structures with characteristic length scales as evident in pattern formation in animal skin stripes or the clocking of circadian rhythms (Novak and Tyson 2008). A grand challenge for bioinspired materials, micro-machines, and factories, then, is to integrate these and other motifs into functional, adaptive devices that transport materials and themselves and that exchange energy and matter with the environment. One bioinspired embodiment of this challenge, for example, would be a soft robot powered autonomously by a chemical "factory" with interconnecting organ-like systems as well as feedback connectivity. Are we there yet?

The Wyss Institute Bioinspired Robotics lab in Cambridge hummed with activity, like a hive of bees. The laser darted rapidly on its computer-guided stage to cut precise microchannels in a sheet of material. Just outside the doorway, a high-resolution 3D printer was additively sculpting a computer-drawn machine component. Around the corner, an insect-scale robotic fly was tracked by motion capture cameras as it rose off the ground, dipped, hovered, and returned for a four-point landing (Ma et al. 2013). In a nearby building, somewhat larger robots each carried a brick up an inclined ramp formed by previously deposited bricks. Like mound-building termites, each TERMES micro-robot was using its whegs (specialized wheels well-adapted for climbing), guided by onboard sensing, to identify the location for its current load (Werfel, Petersen, and Nagpal 2014).

In the cases of both fly and termite, engineers and computer scientists collaborated with biologists to design and build specialized micro-robots guided by principles of evolution and physical scaling laws applied to biological systems. These principles include building at scale and using existing parts for new functions (the dipteran thorax for attachment of the synthetic muscles of an insectoid robot) and using local sources of information, or *stigmergy*, to guide behavior (the use of locally available information by TERMES). The technologies of the coming era in bioinspired devices—including advanced biomaterials and microrobots—may provide new solutions from evolution and developmental processes for addressing the enormous engineering challenges of remodeling injured nervous systems.

Being a Wyss Institute faculty member has given me the opportunity to learn from, and collaborate with, these and other leading innovators in biologically inspired engineering. For example, Wyss faculty are working together to use advanced materials, soft sensors, actuators that leverage material compliance, and soft robotics. The translation of biological principles into bioengineering practice has been possible because of groundbreaking ideas by colleagues, such as the concept of "self-assembly" by George Whitesides (Whitesides and Grzybowski 2002). Self-assembly is the autonomous organization of components into patterns or structures (Whitesides and Grzybowski 2002). In nature, the cooperative behavior of insects, especially ants, illustrates how self-assembly is used for *cooperatively coupling available body parts* to form structures, such as rafts. Why would ants build a raft? In the Brazilian rain forest, the native habitat of the fire ant *Solenopsis invicta*, there is regular flooding. An individual ant will struggle in water, but when a large group of ants link their bodies together, the air pockets formed when they are submerged allow the ants to float together effortlessly for days (Mlot, Tovey, and Hu 2011; Tennenbaum et al. 2016). The collective formation is cohesive, buoyant, and water-repellent. How do ants build rafts and other structures with their own body parts? Scanning electron microscopy (SEM) images and micro CAT scan reconstructions illustrate that (1) ants use their mandibles to connect to the legs of another ant, (2) increase the strength of a structure by increasing the number of connections, (3) use their legs to push against their neighbors to control their orientation, and (4) allow small ants to fit between the legs of larger ones to regulate packing density (Foster et al. 2014). Thus, ant rafts and other constructions made from their own body parts are not only self-assembled, but also informationally regulated.

Self-Assembly of Bioinspired Adaptive Materials

Contemporary bioinspired, or "smart," materials constitute a fourth generation in the evolution of materials science, with lineages that may be traced to the classic biomaterials of the 1960s, bioactive materials of the 1990s, and biodegradable materials of the late twentieth century (Holzapfel et al. 2013) (see Table 4.1). As discussed in Chapter 3, the use of advanced imaging tools, such as SEMs, has revealed principles for building the way that nature does, including materials characterized by hierarchical structure. For example, one of the most striking images adorning the walls of the Wyss Institute is what appears to be a ball nested within many pudgy fingers (see Figure 4.1). However, further investigation with the aid of SEM reveals that those fingers are microscale bristles from the labs of Wyss colleagues Joanna Aizenberg and L. Mahadevan.

Coiled and spiral forms, such as the ones in Figure 4.1, are examples of mesoscale structures formed from helical fibers and then further self-assembled into hierarchical helical assemblies (Pokroy, Kang, et al. 2009). By using a physico-chemical process, called elastocapillary coalescence, Pokroy, Kang, et al. (2009) were able to emulate

Table 4.1 Evolving goals of biomaterials over the decades

Era	Goals
Classic biomaterials	Early biomaterials from the 1960s sought an appropriate combination of chemical and physical properties to match those of the replaced tissue with minimal foreign body response in the host.
Bioactive materials	The goal of bioactive materials, since the 1990s, has been to elicit a specific biologic response at the interface of the material, such as implants with coatings and structures that facilitate growth.
Biodegradable materials	Biodegradable, or bio-resorbable, materials are initially incorporated into the surrounding tissue and, eventually, completely dissolve.
Bioinspired "smart" materials	Smart biomaterials are adaptive and modify their structural properties in response to their environment.

Source: B. Holzapfel, J. Reichert, J.-T. Schantz, U. Gbureck, L. Rackwitz, U. Noth, F. Jakob, M. Rudert, J. Groll, & D. Hutmacher (2013), How smart do biomaterials need to be? A translational science and clinical point of view, *Advanced Drug Delivery Reviews, 65,* 581–603.

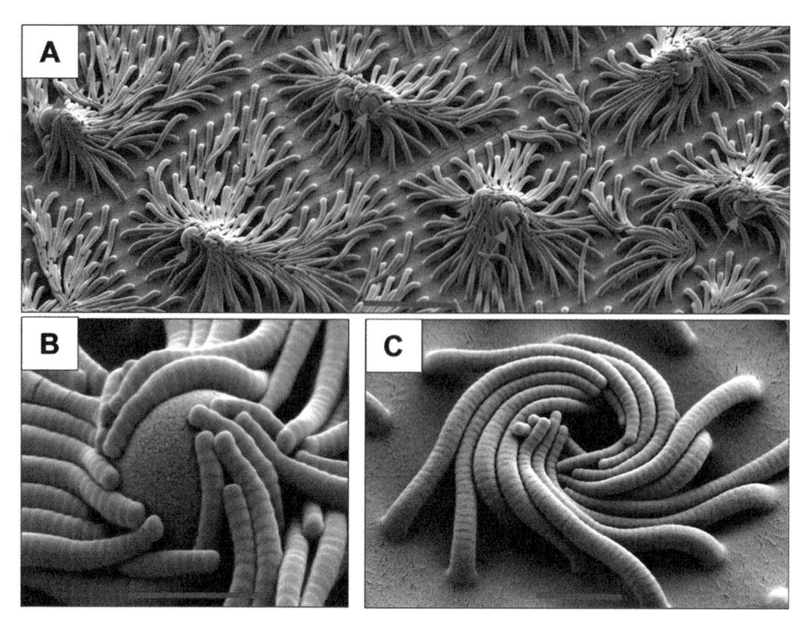

Figure 4.1 Mesoscale coiled bristle. Illustration of the adhesive and particle trapping potential of the helically assembling bristle. (A) Low-magnification SEM showing the capture of the 2.5-mm polystyrene spheres (indicated by arrows). Scale bar, 10 mm. (B) Magnified view depicting a single sphere trapped through the conformal wrapping of the nanobristle. Scale bar, 2 mm. (C) Coiled whirlpools remain after the removal of the spheres. Scale bar, 2 mm. Boaz Pokroy, Sung H. Kang, L. Mahadevan, and Joanna Aizenberg (2009). Images reprinted with permission from AAAS.

biological morphogenesis. The coalescence occurs when an array of bristles is wetted and then allowed to evaporate: the bristles self-assemble into helical forms. But why should a seemingly simple evaporation process result in the emergence of helical patterns? Microscopic images reveal that evaporation of the bristle array results in the formation of a meniscus connecting neighboring pillars. A mathematical model based on earlier work by Mahadevan (Cohen and Mahadevan 2003) proposes that imperfections in the geometry of the bristle array, competition between bristle bending elasticity and adhesion, and local differences in evaporation rate result in local self-assembly. The model is able to account for the formation of lower-order braids, and then the propagation of these local effects throughout the array forms long-range ordered domains and, hence, hierarchical assembly into large, coiled clusters. Once assembled into clusters, the

Figure 4.2 Controlled synthesis of hierarchical complex structures. False-colored SEMs showing a field of purple $SrCO_3$-SiO_2 vases containing $SrCO_3$-SiO_2 stems (green) that were subsequently opened with a CO_2 pulse (blue). Image courtesy of Wim L. Noorduin.

array becomes a device with emergent functions, such as a particle-trapping system or a device similar to the adhesion mechanism on the tarsi of beetles (Eisner and Aneshansley 2000).

Another example of the morphogenesis-like process of fabricating bioinspired adaptive materials by self-assembly is the formation of complex microarchitectures, which appear to be beautifully colored delicate flowers (see Figure 4.2). However, this photo is, again, a SEM image at the microscale from the collaborative work of Mahadevan and Aizenberg and their coworkers (Grinthal, Noorduin, and Aizenberg 2016; Noorduin et al. 2013). The key to fabricating micro-architectures in the Noorduin et al. (2013) work is to actively modulate the fabrication process throughout their formation. For example, by placing minerals and organic molecules on a glass plate in an aqueous solution and then changing the "growing conditions," the solution pH, and the tilt angle of the glass plate, it is possible to control the structure of the grown forms. The particular forms grown—hemispheres, cones, stem shapes, vases, and corals—depend on the chemical and physical interactions that occur in three distinct "growth regimes." Because the morphology of the growing structures is responsive to the solution conditions, many different shapes can be stacked on top of each other by controlling the position, pH, temperature, and salt concentration from the elementary building blocks (Noorduin et al. 2013, 836).

The Aizenberg lab has drawn other insights from the relation between plants and insects (see Table 4.2). Consider the poor ants facing unknown peril while walking on the rim surrounding the maw of the *Nepenthes* pitcher plant. The rim is often highly conspicuous in color, and its ridges are "ant-sized" and angled toward the bottom part of the plant. Under humid conditions, the rim becomes extremely slippery, and ants lose their footing and fall into the bottom part of the pitcher, filled with digestive juices (Bauer and Federle 2009). The slippery surface results from its wettability, a combination of forces in the interaction between water, the solid surface, and air (Bauer and Federle 2009). Inspired by *Nepenthes,* Wyss founding core faculty member Joanna Aizenberg has developed self-healing, slippery, liquid-infused, porous surfaces, or SLIPS materials (Wong et al. 2011). These materials use nano- / microstructured substrates to embed an infused lubricating fluid. The liquid film is intrinsically smooth and defect-free down to the molecular scale, provides immediate self-repair by wicking into damaged sites in the underlying substrate, is largely incompressible, and may be designed to repel immiscible liquids of virtually any surface tension (Wong et al. 2011).

Soft lithography. Bioengineering and allied disciplines have developed remarkable new technologies to emulate nature's processes of self-assembly. For example, the Whitesides laboratory has been a leader in the use of soft lithography for nano- and microfabrication (see, for example, Qin, Xia, and Whitesides 2010). Soft lithography is the process of replica molding: fabricating a topographically patterned master with feature sizes from 30 nM to 100 μm on its surface, molding the master to generate a patterned stamp, and generating a replica of the original template in a functional material (Qin et al. 2010; Xia and Whitesides 1998; Gates et al. 2004). Mi-

Table 4.2 Bioinspired adaptive materials

Material	Characteristics
Actuated nanostructures	Finely tunable multifunctional, responsive, nanostructured material. Has the following properties: (a) self-cleaning (like lotus leaf), (b) capable of movement and reversible actuation (like echinoderm spines), and (c) sensing the force field (like the skin).
Self-healing, slippery, liquid-infused, porous surfaces (SLIPS)	Inspired by *Nepenthes* pitcher plants, uses nano/microstructured substrates to lock in place an infused lubricating fluid. The liquid film is intrinsically smooth and defect-free down to the molecular scale, provides immediate self-repair by wicking into damaged sites in the underlying substrate, is largely incompressible, and may be designed to repel immiscible liquids of virtually any surface tension.
Homeostatic materials	Bristle-like microstructures swell or contract in response to a chemical stimulus. When a catalyst is affixed to the microstructure tips, and immersed in a liquid bilayer, the mechanical action of straightening turns on a chemical reaction, and the act of bending turns it off.
Stretch-responsive liquid film supported by nanoporous elastic substrate	A liquid film overlays a nanoporous substrate. Stretch-induced changes in the substrate pores results in the liquid film flowing within the pores, changing the surface topography. Thus, the stretch provides a graded means of controlling properties, such as optical transparency and wettability.
Complex hierarchical microstructures	Minerals and organic molecules form by different processes. When these processes are coupled together by placing them together on a glass plate in an aqueous solution, controlling the growth conditions (e.g., pH of the solution, tilt angle of the glass plate) results in the self-organization of flower-like hierarchical microstructures.

Sources: B. Pokroy, A. Epstein, M. Persson-Gulda, & J. Aizenberg (2009), Fabrication of bioinspired actuated nanostructures with arbitrary geometry and stiffness, *Advanced Materials, 21,* 463–469; X. He, M. Aizenberg, O. Kuksenok, L. Zarzar, A. Shastri, A. Balazs, & J. Aizenberg. (2012), Synthetic homeostatic materials with chemo-mechano-chemical self-regulation, *Nature, 487,* 214–218; X. Yao, Y. Hu, A. Grinthal, T.-S. Wong, L. Mahadevan, & J. Aizenberg (2013), Adaptive fluid-infused porous films with tunable transparency and wettability, *Nature Materials, 12,* 529–534; and W. Noorduin, A. Grinthal, L. Mahadevan, & J. Aizenberg (2013), Rationally designed complex, hierarchical microarchitectures, *Science, 340,* 832–837.

crofabrication by means of replica molding uses an elastomeric material called poly(dimethyl siloxane), or PDMS (Ren, Chen, and Wu 2014). It is nontoxic, optically transparent, hydrophobic, gas permeable, and conformable. Recent advances in soft lithography, such as patterning on topography (PoT) printing, even make it possible to directly transfer proteins onto a complex surface (Sun et al. 2015; see also Chapter 10).

Microfluidics. According to Whitesides (2006), microfluidics "is the science and technology of systems that process or manipulate small (10^{-9} to 10^{-18} liters) amounts of fluids, using channels with dimensions of tens or hundreds of micrometers" (2006, 368). Put another way, microfluidics is a field dedicated to "miniaturized plumbing and fluidic manipulation" (Squires and Quake 2005), at a scale close to the sizes of biological cells (Ren et al. 2014). In microfluidic devices, the basic unit of fluid handling, the micromechanical valve, is fabricated by means of multilayer soft lithography (Melin and Quake 2007). Moreover, in the "bottom-up" fabrication process, this fundamental structural unit is transformed into complex integrated functional networks in which valves are used for pumping, mixing, metering, latching, and multiplexing (Melin and Quake 2007). A major advance in creating microchannels for microfluidic devices was the introduction of silicone-based elastomers, primarily polydimethylsiloxane (PDMS). The introduction of hydrogels has been a further advance in materials for microfluidic devices. Hydrogels contain water up to 90 percent of the total mass and are highly porous, allowing diffusion of molecules through the bulk. The combination of inherent compatibility with biological cells and high permeability make hydrogels outstanding material for encapsulating living cells (Ren et al. 2014).

Microfluidic devices use fluidic control-logic devices, valves with a pressure gain that enable sequential stages of logic without energy loss, to regulate flow (Weaver et al. 2010). Analogous to electronic resistor-transistor circuits, a nonelectric fluidic resistor is placed in series with the valve's source line, an orifice. By blocking the orifice, there is no flow through the valve, and the pressure at the source line input to the valve rises; when fluid is able to flow through the valve, the pressure loss drops pressure below a threshold level (Weaver et al. 2010). Two valves in parallel create an AND logic gate, while two valves in series serve as a fluidic OR logic gate (Weaver et al. 2010). Three-layer microfluidic architectures make it possible to convert a constant infusion of fluid into a transient outflow by means of through holes that serve as check valves and switch valves (Mosadegh, Mazzeo, et al. 2014). Interconnected physical gaps and cavities form fluidic logic gates that spontaneously generate oscillatory flow (Mosadegh, Mazzeo, et al. 2014). These control components serve as logical primitives from which have emerged the complex, nonelectronic network architectures of organ-on-chip technologies.

Reverse Engineering of Organs: Microfluidics as Enabling Technology

Microfluidic devices provide an enabling technology for reverse engineering organ systems. An organ combines tissues into a functional multiscale hierarchy in which intrinsic and extrinsic forces act on constituent cells via feedback loops (Blanchard and Adams 2011). Moreover, organs and organ systems generate or are influenced by mechanical forces such as cyclic strain, compression, and shear, as well as by neuroelectrical signals. To reverse engineer an organ is a nontrivial process, requiring means for establishing tissue–tissue interfaces, vascular perfusion, and a microenvironment that captures distinctive functionality (Ingber 2016). Consider organ-on-chip devices.

An organ-on-chip microfluidic device is used for "culturing living cells in continuously perfused, micrometer-sized chambers in order to model physiological functions of organs and tissues" . . . with the goal of "synthesizing minimal functional units that recapitulate tissue- and organ-level functions" (Bhatia and Ingber 2014, 760). These imaging-friendly (that is, optically clear) micro-engineered devices make it possible to recreate aspects of the gaseous, fluid, and mechanochemical microenvironments that cells experience when they function as part of organ systems, such as beating hearts and breathing lungs. The emphasis in organ-on-chip devices being developed by Don

Ingber's group at the Wyss Institute is (1) *control over fluid flow at microscale* and (2) *the effects of mechanical forces on cell function* (see also Chapter 2). For example, by using microfluidic channels less than approximately 1 mm in diameter, laminar flow dominates, such that when two streams meet, they flow in parallel with no eddies or currents (Vogel 2013). Control of laminar flow in organ-on-chip devices is a means for producing physical and chemical gradients. The Wyss-engineered organs on chips are distinctive for also emulating the mechanical strain and shear forces experienced by an organ's constituent cells—for example, by controllably deforming the flexible chamber walls to which cells are adhered (Bhatia and Ingber 2014).

A human "breathing" lung-on-chip incorporates laminar flow as well as strain and shear forces in order to recapitulate the microenvironment experienced by cells of a functioning lung (Huh et al. 2010). The architecture of this microfluidic device consists of a vascular channel lined with endothelial cells and an air-filled channel lined by human lung alveolar epithelial cells (Huh et al. 2010). To emulate a functioning lung, cyclic suction rhythmically distorts the flexible channel sides. As described by Bhatia and Ingber:

> The combination of fluid flow in the vascular channel, generation of an air-liquid interface in the alveolar channel, and application of cyclic mechanical strain strongly promoted differentiation of the epithelial and endothelial cells lining the channels, as indicated by enhanced surfactant production and vascular barrier function. (2014, 765)

In other words, this microfluidic device functions like a human lung!

Paper-Based Biotechnologies

Paper has, for millennia, been at the center of both cultural change and the ecological transformation of our planet's resources (Kurlansky 2016). The use of paper to carry our cultural heritage by words and images, first with inks applied by hand tools and then imprinted by machine, has literally been a rags to riches story. Now, in the twenty-first century, paper is being used in novel ways to carry biological information and, in the process, may herald a new transformation in our cultural response to the threat of disease in resource-poor as well as developed countries. One illustration of paper-based technologies for diagnostics is the remarkable freeze-dried, paper-based biomolecular platform, capable of being deployed in the field to detect the Zika virus, developed by MIT bioengineering professor and Wyss faculty member Jim Collins and his colleagues (Pardee et al. 2016). The Zika virus is perinatally transmitted and has a devastating impact on the developing fetal nervous system (Driggers et al. 2016); in 2016, it was declared a public health emergency by the World Health Organization (WHO). The freeze-dried, paper-based technology makes it possible to bring programmable RNA sensors, called toehold switches, into low-resource field settings. These devices sense a particular RNA sequence, as in the Zika virus, by a color change in a paper disc.

Why Is It So Difficult to Emulate Muscle-Tendon Units?

Muscle-tendon units are a marvel of nature's engineering, making it possible to generate force by dissipating energy across scales. The emergent properties of these dissipative structures across scales are one reason it has been so difficult for engineers to emulate muscle-tendon units. There are at least five additional reasons for this challenging engineering task, which highlight biological muscle's multifunctional, nonlinear, self-stabilizing, context-sensitive, and state-dependent properties (see also Nishikawa et al. 2007) (see Tables 4.3 and 4.4). First, muscle is multifunctional. Muscles are fundamentally motors that generate force by shortening to perform mechanical work. However, muscles may also act as springs (in fly-flight control muscles), as brakes (during running of cockroaches), as stabilizers or struts at the joints, and as energy storage devices (tendons and aponeuroses of turkeys during running) (Dickinson 2000).

Second, muscles are nonlinear devices for translating control signals into mechanical output, as evident in the force-length behavior of the muscle-tendon unit. The ability of a muscle to control the length

Table 4.3 Multiple functions of muscle

Function	Description
Actuator (motor)	Muscles are considered actuators when they shorten or lengthen to produce or absorb energy, doing mechanical work.
Brake	Muscles may slow the swing of a limb, as in the leg extensor of cockroaches during running.
Spring	Muscles may act as controllable springs to direct the forces of much larger power muscles, as in the control muscles of flies.
Strut	Muscle fibers may be isometric, or shortening while a tendon stretches, permitting the elastic tendons to store and release energy, as in running turkeys and hopping wallabies.
Shifting function relative to what the body is doing	By being activated at particular times relative to an ongoing body motion, muscle may serve different functions. For example, in fish, axial muscle may be a force generator or a force transmitter, depending on when it is activated with respect to the undulatory wave that passes along the body.

Source: M. Dickinson, C. Farley, R. Full, M. A. R. Koehl, R. Kram, & S. Lehman (2000), How animals move: An integrative view, *Science, 288,* 100–106.

of its fibers in relation to the stretch of its tendon depends upon the architecture and physiological properties of its fibers (Biewener 2016; Higham and Biewener 2011). For example, parallel-fibered muscles favor increased shortening and velocity at the expense of force, but highly pinnate muscles enhance force at the expense of velocity and length control (Wilson and Lichtwark 2011). The nonlinear behavior of a muscle is also a reflection of its structural heterogeneity: different muscle segments within a single fascicle may exhibit different mechanical output, and different muscles innervated by the same nerve may exhibit different functions (Biewener and Roberts 2000).

Third, the degree to which sensory information is used to control muscles is dependent upon the task. Rapid, repetitive muscle activations, such as those seen during running in cockroaches, are dominated by the mechanical feedback from the limbs, called "preflexes" (Proctor and Holmes 2010). By contrast, slow, precise, and relatively novel movements involve reflexive neural feedback (Full and Koditschek 1999).

Table 4.4 Muscle as a nonlinear device

Characteristic	Description
Motor unit	A motoneuron and all of the muscle fibers that it innervates. The motor unit is the basic functional unit of the neuromuscular system. Motor units may vary widely in size, having innervation ratios (i.e., number of muscle fibers per motoneuron) ranging from a few to thousands of fibers per motoneuron.
Size principle	When a task demands an increased level of muscle force, motor units are typically recruited in an orderly fashion, from slowest to fastest. However, motor unit recruitment may also be dependent on task factors—e.g., in tasks with shortening-lengthening cycles, such as a cat's paw-shake, recruitment of faster motor units may be advantageous. This suggests that changes in motor unit recruitment may have an underlying mechanical basis, at least for certain locomotor tasks.
Force-length behavior of muscle fibers	The force-length relation of muscle function is described by curvilinear behavior. When fibers are shorter than optimal length, the force a muscle generates when maximally activated increases with fiber length. Beyond the optimal length, the maximal active force that a muscle generates decreases with fiber length, and the muscle generates passive force. Near optimal fiber length, the muscle generates relatively constant force when maximally activated.
Relation of muscle force to muscle fiber velocity	When fibers shorten during activation (concentric), force generation decreases with increasing shortening velocity. When fibers lengthen during activation (eccentric), force generation increases with lengthening velocity.

Sources: L. Mendell (2005), The size principle: A rule describing the recruitment of motoneurons, *Journal of Neurophysiology, 93,* 3024–3026; and A. Biewener (2016), Locomotion as an emergent property of muscle contractile dynamics, *Journal of Experimental Biology, 219,* 285–294.

Fourth, the degree to which muscles are tunable by environmental input depends upon their location along a proximo-distal gradient of motor control (Daley, Felix, and Biewener 2007), as well as their control potential (Sponberg et al. 2011). For example, Daley et al. (2006) found that in guinea fowl running over uneven and unpredictable surfaces, muscles at the distal joints were inherently more sensitive to altered loading, and exhibited more rapid proprioceptive feedback regulation, because the distal-most joints interact directly with the terrain and are the first structures in the kinematic chain to obtain environmental information about surface changes. By contrast, proximal muscles at the hip and knee joint exhibit mechanical performance that is insensitive to load and so are controlled largely with little feedback regulation (Daley et al. 2006). One implication of the proximo-distal gradient of motor control is a need to develop neuromechanical control architectures for synthetic muscles with a capability for regulating the balance of feedforward and feedback control (see, for example, Revzen, Koditschek, and Full 2009; Revzen et al. 2013).

Fifth, an animal's coordinated body motion is a function of the relation between forces generated by muscles and those exerted by the environment (Miller et al. 2012; Tytell, Holmes, and Cohen 2011; Tytell et al. 2010). To illustrate, a common mode of propulsion by animals in fluids, on land, and in sand is undulation, a gait in which thrust is produced in the opposite direction of a traveling wave of body bending (Ding et al. 2013). Neural activity passes from head to tail, activating muscles in a wave along the body. However, the speed of the mechanical wave depends on the characteristics of the body as it interacts with the medium. Muscle size in relation to environmental forces also influences behavior. The muscles of small animals, such as insects, are subject to different gravitational and inertial influences than the muscles of large animals, such as horses or humans. In large animals, the control of muscles can take advantage of momentum—that is, a brief action potential at swing initiation is sufficient to drive leg motion through a complete swing cycle (Hooper 2012). By contrast, stick insects generate negligible momentum during walking such that if the agonist muscle stops generating force, the antagonistic muscle force abruptly stops leg motion (Hooper 2012).

Bioinspired Processes for Building Soft Sensors, Actuators, and Soft Robots

Biological processes have also inspired new technologies for building soft sensors, actuators, and logic gates. For example, in the emerging field of stretchable electronics (Rogers, Someya, and Huang 2010) there have been several solutions discovered for fabricating circuitry that is able to bend and stretch, as the human skin does. For the critical area of mechanotransduction, there are many challenges in building strain sensors enveloped in soft matrices that may assume arbitrary shapes and are both conformal and extensible. An example of one solution—a hyperelastic pressure sensor from the laboratory of Harvard engineer and Wyss faculty member Rob Wood—is to embed a conductive liquid metal within molded soft elastomer microchannels (Park et al. 2010). More recently, Jennifer Lewis and Rob Wood have combined their talents for a new method, called embedded 3D printing, or e-3DP. This method, an extension of the work described earlier, uses a special nozzle to directly deposit a viscoelastic ink into an elastomeric reservoir (Muth et al. 2014).

Soft actuators are biologically inspired: during the earliest periods of animal evolution, nature had already harnessed soft materials, plus the internal pressure of fluids for bending, in structures called hydrostatic skeletons (Kier 2012). Bending actuation by means of a hydrostatic skeleton was well adapted for the fundamental action systems of basic orienting (maintaining stability with respect to some perturbing force), locomotion (changing the location of the body in a structured environment), reaching and grasping (bringing something closer to the body or moving it away from the body), and feeding (ingesting something and converting its form into components for usable energy) (Goldfield 1995). Critical to the success of a hydrostatic skeleton is its architecture. For example, the bending motion of polyps, such as the sea anemone, is based on the arrangement of muscles in the walls of a hollow column: bending occurs due to simultaneous contraction of longitudinal muscles on one side and contraction by circular muscle fibers (Kier 2012).

Elastomers are silicone rubber materials, and one in particular, polydimethylsiloxane (PDMS) has been remarkably well suited to soft lithography. Elastomers have been used to fabricate soft pneumatic actuators, called "Pneu-Nets" (networks of small channels embedded in elastomeric structures and inflated with air at low pressures) (Mosadegh, Polygerinos, et al. 2014). Elastomers also exhibit the property of hyperelasticity, capable of stretching by as much as 1000 percent, making them a choice material for stretchable electronics, as in the fabrication of hyperelastic pressure sensors worn on the skin surface (Park et al. 2010).

Pneumatic networks (Pneu-Nets) are actuators fabricated from soft materials (such as elastomers or paper) that expand when filled with pressurized air (Ilievski et al. 2011). The first Pneu-Nets consisted of repeated chambers with walls of varying thickness. When filled with compressed air, the chambers expanded in regions that were most compliant. By selecting material compliance and fabricating chambers with specific wall thickness, Pneu-Nets could be fabricated so that pressurization would result in forms with predictable curvatures (Ilievski et al. 2011). The design was extended by using composite structures formed by adhering one layer of a stiffer material onto a more compliant one and shaping the multi-layered material into particular geometries. For example, a geometry in which finger-like shapes of layered elastomer curled around a perpendicular axis formed a Pneu-Net "gripper" (Ilievski et al. 2011). Other work by the Whitesides group has embedded folded paper structures in elastomer polymers so that the pattern of folds in the paper formed pleated actuators, including accordion-like structures, bellows structures, bending bellows, and twisting actuators (Martinez et al. 2012). To generate a simple three-dimensional bending motion in soft cylindrical forms, or "tentacles," Martinez et al. (2013) created a geometry consisting of three individual pneumatic channels parallel to a central core. Then, separating each of the three channels into sections, each independently controlled by an external source of pressurized air, opened the possibility for multiple bending modes, and remarkable flexibility for softly grasping delicate complex forms, such as flowers (Martinez et al. 2013).

Designs of soft fiber-reinforced bending actuators begin with models (for example, analytical or finite element) of the relation among pressure input, geometrical properties, and output (Polygerinos et al. 2015). The geometric properties of these actuators include wall thickness of the air chamber, length of the actuator, diameter of the hemicircular chamber shape, and fiber winding pitch and orientation (Polygerinos et al. 2015). The approach of an analytic model is to consider material properties of, for example, the walls (made of incompressible rubber) and of the actuator (the top wall extends while the bottom layer does not because it is constrained by an inextensible layer), and then relate these to the relationship among input pressure, bending angle and output force. A finite element model (FEM) is able to go beyond this analysis with a set of element equations to characterize the nonlinear response of this system and provide visualization of system behavior, such as local stress / strain concentrations as variables are manipulated. Polygerinos et al. (2015) then compared performance of the actuators with predictions from each of the models, identifying, for example, the minimum and maximum pressure locations of actuators with particular geometries from the FEM model. And finally, the modeling results guided the development of a feedback control loop to track angle signals.

The relation between air pressure input and force output of these bending actuators was quasi-monotonic, as in contraction of muscles during voluntary actions during grasping an object with the fingers. However, this is not the only type of force generated in nature. The work of Mahadevan (Forterre et al. 2005) and others has demonstrated ways in which nature harnesses instabilities, such as snap-through transitions, to rapidly promote large changes in force output. Can soft actuators similarly generate sudden, large changes in force output? Overvelde et al. (2015) have developed a new class of soft actuators consisting of interconnected fluid segments that are able to instantaneously trigger large changes in internal pressure, extension, shape, and exerted force.

Origami, Programmable Matter, and Printable Manufacturing

The challenge of manufacturing three-dimensional structures from planar materials is exemplified by the traditional process of folding sheets into intricate forms, called origami (Demaine and O'Rourke 2007). Classic origami involves basic techniques, from the simplest valley-fold that leaves a crease in the sheet, to ever more complex inside and outside reverse folds, crimp-folds, pleat-folds, and sink-folds (Engel 2009). Over many sequential folds, it is possible to produce a remarkable range of objects, especially birds. A goal for manufacturing complex structures using principles from origami is to devise various means for sheets to be "self-folding." Collaborations between Rob Wood and MIT professors Eric Demaine and Daniela Rus have adapted the techniques from human-folded constructs into self-folding programmable matter, including a wonderful self-folding crane, an inchworm, and other self-folding robots (Felton et al. 2013; Hawkes et al. 2010).

Programmable matter is a material with universal crease patterns (interconnected triangles) whose creases are capable of actuation and may be programmed to achieve particular shapes upon command (Hawkes et al. 2010). Programmable matter is of particular interest for building devices because it transforms a single sheet of material into multiple forms. An "end-to-end" fabrication process consists of two steps. First, a set of planning algorithms determines a plan for folding certain creases so that all of the goal shapes can be generated by just one sheet. Then, stretchable electronic circuits and shape memory polymer actuators are fabricated at the creases layer by layer to implement the folding commands. Hawkes et al. (2010) demonstrate how the same sheet of programmable matter is able to transform itself into an origami folded boat, return to its sheet configuration, and then transform itself into an airplane.

Is it possible to use folding processes to create micron to centimeter scale micro-electro-mechanical system (MEMS) devices such as motors, transmissions, linkages, and rotary servo elements, all components of microscale robots? Whitney et al. (2011) have developed a new method, called "pop-up book MEMS," for fabricating MEMS devices and microstructures by means of folding multilayer, rigid-flex laminates. Multiple layers of bonded materials are aligned, and a laser is used to machine pairs of layers—one rigid and one flexible—so that they create a flexure crease, making it possible for the bonded materials to fold like a page in a "pop-up" book. Selective application of adhesives and layers with a range of elastic properties are used to create more complex multilayer folds. The potential for using self-folding with more complex materials, such as paper-elastomer composites (Martinez et al. 2012) has resulted in an additional set of techniques called *printable manufacturing* (Felton et al. 2013). Printable manufacturing rapidly turns digital plans into physical objects, including self-folding planar sheets that form more complex three-dimensional surface shapes (An et al. 2014), an origami worm robot (Onal, Wood, and Rus 2013), and walking and gripping robots (Sung et al. 2013).

A benchmark in the evolution of machines is the ability to morph into a form with one or more emergent functions. To do so requires components designed so that they connect with each other in some new way as the machine changes from one state into another—for example, from a planar geometry into a complex three-dimensional form. The new functional connectivity of parts emerging from the change in their spatial relationship results in the emergence of a functional capability not obvious when the machine is its planar form. Felton et al. (2014) have developed a robot with just this capability: a self-folding machine that transforms itself from a flat sheet with creases in an origami pattern, and the incorporation of embedded electronics for actuation and power, into a quadruped that walks. This remarkable achievement was made possible by a planar linkage design that anticipates the connectivity of parts required for locomotion when the machine transforms into its three-dimensional form. The design includes special four-edge, single-vertex folds to compress a large area into a small volume and linkage assemblies driven by a motor that actuates both a front and rear leg when linkage and motor are engaged by folding (Felton et al. 2014). An obvious feature of the self-folding machines discussed here, thus far, is that their structural support and functionality are designed around rigid materials. Other kinds of machines capable of self-assembly and emergent functionality become possible with the use of soft materials.

Soft Robots

Soft robots have the potential to introduce a profound impact on next-generation systems used in a wide range of applications for remodeling injured nervous systems. Soft robots are characterized by "computational materials" (Correll et al. 2014; Hauser et al. 2013), in which the properties and architectural structure of the materials perform some of the computation currently done by central processing. In biological systems, the "outsourcing" of some intelligence to the body, called "morphological computation" or "intelligence by mechanics" (Blickhan et al. 2007), is possible because the mechanical properties of the body have evolved with inherent multistability and multifunctionality. For example, during the transition from walking to running, insects and vertebrates alike undergo gait transitions with increased velocity because the dynamics of the walking coordination pattern becomes unstable and switches to a more stable one (Holmes et al. 2006). The process of switching between coordination patterns does not require a decision by a centralized controller: the intrinsic dynamics of the body interacting with environmental forces provides all of the intelligence required (Dickinson et al. 2000).

Compelling demonstrations of robots with soft architectures include Trimmer's caterpillar robot, GoQBot, which, like a biological caterpillar, redistributes mechanical energy stored in elastic tissue by leveraging the environment as a skeleton (Lin, Leisk, and Trimmer 2011). The soft architecture and shape memory alloy (SMA) coil actuators of GoQBot allow the robot to transform its elongated narrow body into a circular form. By rapidly (within 100 ms) transforming from a linear to a circular body form, GoQBot is able to produce a ballistic rolling motion, generating adequate acceleration to roll along at a linear velocity of 200 cm / s (Kim, Laschi, and Trimmer 2013). Correll et al. (2012) take another approach to building a soft rolling robot: a traveling wave over the length of a chain structure. When the two ends of the structure are connected to form a tread, controlled filling of its component cells at particular points of tread curvature causes the structure to roll along a surface. Flexible electronic circuits wrapped around the soft tread include computation, communications, and power components, which allow each cell to communicate with its left and right neighbor. The result is a pneumatic belt that rolls autonomously on a flat surface using closed-loop control.

At the Wyss Institute, a soft robot capable of undulation and crawling gaits was designed around a Pneu-Net architecture of chambers embedded in a layer of extensible elastomer bonded to an inextensible layer (Shepherd et al. 2011). The Pneu-Net chamber architecture has made it possible to emulate a key feature of quadrupedal posture required for producing distinctive gaits: lifting off the ground any of its four legs, while leaving the other three legs in contact with the ground surface for stability. Moreover, with four legs on the ground, a fifth Pneu-Net is able to independently raise the main body from the surface (Shepherd et al. 2011). A key advantage of robot fabrication from thin layers of elastomers is that depressurization creates a flat surface that makes ground contact over a large part of the body. When starting from a depressurized state, sequences of pressurization-depressurization make it possible for the robot to undulate underneath an obstacle. Moreover, Majidi et al. (2013) have demonstrated that the undulation mode opens new possibilities for achieving controlled use of friction for locomotion on surfaces that vary in composition, rigidity, and texture. In addition to adapting to surface traction, the soft robot quadruped is able to change its appearance in response to surface coloration and patterning, such as a rock bed or leaf-covered concrete slab, through the use of microfluidics networks within silicone sheets placed on top of the soft robot (Morin et al. 2012). Further integration of microfluidics with soft robots is the development of a controller with multiple pneumatic actuators and valves (Mosadegh, Mazzeo, et al. 2014).

A step toward autonomy for soft robots is to be self-powered. Emancipation from a connection with off-board sources of power is a nontrivial matter for soft robots because they must continue to be able to perform their tasks (for example, locomoting, exploring environments) while carrying the additional load of miniature air compressors, battery, valves, and controller. Moreover, design solutions achieved are constrained by physical law, such as the consequences of increasing size for weight (scalability) and a change in morphology to accommodate newly designed internal structures. With these caveats, the Whitesides lab emulated evolutionary strategy: rather than start from scratch, they modified their existing robot (Tolley et al. 2014). Along with increasing the robot's size, they used an elastomeric material that could better withstand higher pneumatic pressures but that, at the same time, would be lighter. The elastomer was made stronger by embedding polyaramid fiber in the silicone, and was made lighter by incorporating hollow glass microspheres into the silicone. And to accommodate new internal components, they modified the configuration of the Pneu-Net chambers. The result of untethering the soft quadrupedal robot and, in the process, making it better able to withstand increased forces and energies imposed on its body, was that it became better adapted to a wider range of environments, including locomoting in snow, surviving exposure to an open flame, and being run over by a car (Tolley et al. 2014). Despite their impressive early achievements, soft robots have, thus far, largely been solitary machines. New possibilities for functionality emerge when robots, hard and soft, are allowed to interact with each other.

At-Scale Locomoting Microrobots: Running for Cover, Making a Splash, and Causing a Flap

Nowhere to hide. One of the earliest land animals, the myriapod, is distinguished by a multisegmental, flexible body, with most segments bearing two sets of legs on each (Grimaldi and Engel 2005). A multisegmental body, compared to a rigid one, confers several advantages for locomoting, including high stability, impressive climbing and agility, and robustness (adaptability of behavior) if there is loss of one or more legs (Hoffman and Wood 2011). One of the many challenges of building a microrobot that captures these advantages of a multisegmental body for locomoting on land is to design and fabricate a means of mechanical coupling that coordinates rotating and stepping motions between the segments. To address this challenge, Hoffman and Wood (2011) built a centipede micro-robot with a flexible backbone and three segments. The flexibility of their robot derives from a unique mechanism that transmits the force generated by each of its piezoelectric actuators to a rotational force, or torque, in a horizontal plane. A second challenge was to manufacture many copies of each segment, and this was accomplished with a fabrication technology that uses smart composite microstructures (SCMs), developed by Wood et al. (2008). Each SCM is fabricated by lamination and micromachining, folding, and assembly: a flexible sheet of material is sandwiched between two rigid ones; laser micromachining then unlocks flexure-based mechanical joints between the rigid sheets. The planar rigid sheets are folded along the flexures and assembled into three-dimensional forms.

One rigid-bodied insect, the cockroach, is an impressively fast and agile runner, especially during escape from predators (Kubow and Full 1999). When running, the mechanical feedforward "preflexes" of their hexapedal body architecture provide a means for dynamic self-stabilization (Ahn and Full 2002; Proctor and Holmes 2010). Moreover, their antennae leverage mechanical interactions with the environment to sense obstacles (for example, walls) and execute rapid turns (Mongeau et al. 2013). Creating an at-scale (<2-gram) rigid-body micro-robot with the locomotor capabilities of a cockroach presents particular challenges for fabrication because macroscale assembly techniques are too large, and traditional MEMS processes are too time-consuming and costly.

Baisch et al. (2014) address these fabrication challenges in a quadrupedal robot (HAMR-VP) by adopting a revolutionary method for constructing complex, three-dimensional MEMS devices and microstructures, called pop-up MEMS (see earlier work on pop-up techniques by Whitney et al. 2011) (see Figure 4.3). As its name implies, pop-up MEMS are reminiscent of the pop-up books of childhood: when closed, the pop-up appears to be an ordinary book, but when the pages are opened, complex three-dimensional structures arise from the page. In pop-up MEMs, the mechanical linkages are distributed among its multiple layers, and precise alignment is required to fold and assemble the many interconnected components into a functional whole (Whitney et al. 2011). Retaining all the wonderment of those childhood books, the mechanism for pop-up MEMS is a hidden prestressed layer of laser-cut flattened springs that is held under tension and then released (Whitney et al. 2011). The HAMR-VP drive train uses a pop-up assembly that couples contralateral legs so that they move simultaneously in opposite directions, like the hexapedal cockroach (Baisch et al. 2014). Also like the cockroach, HAMR-VP uses elastic elements in its actuators and flexures to leverage mechanical

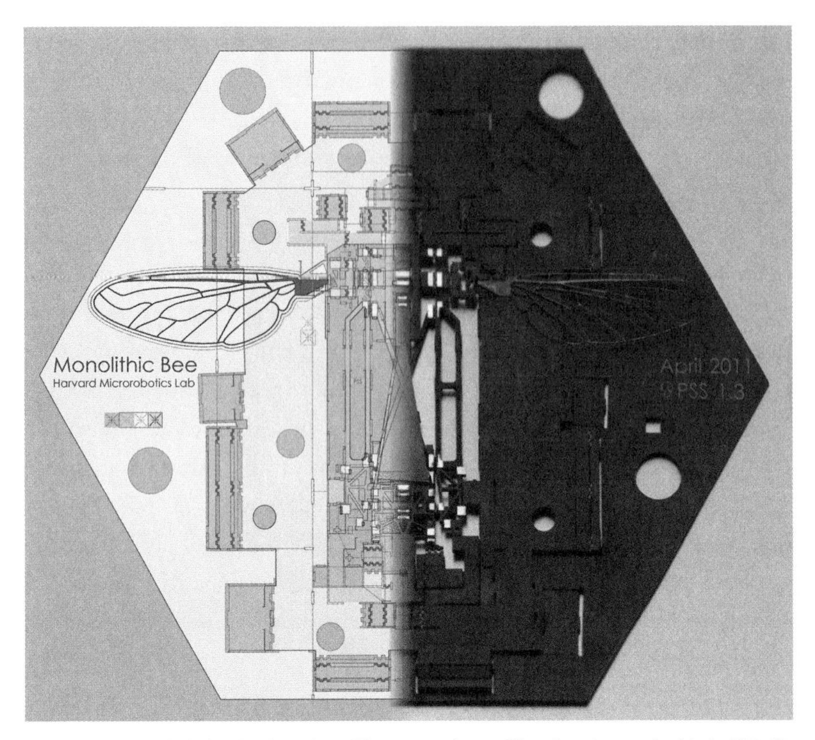

Figure 4.3 Monolithic fabrication of a millimeter-scale machine. A mobee embedded within its folding assembly. Figure by Pratheev Sreetharan, Harvard Microbiotics Lab. Courtesy of Robert J. Wood.

properties for high-speed locomotion near power train resonance (Baisch et al. 2014).

Splash. Observing an insect walking on the surface of a pond or lake may seem impossible from the perspective of our own experience when entering these bodies of water. Humans are able to float if we lie flat to increase the amount of our body surface area in contact with the water surface. However, unlike the gardener, Chance, in the film *Being There,* we are unable to walk on the water's surface. For an insect like the water strider, walking on water is perfectly natural because of its body size relative to the surface tension of water. Propulsive forces for rowing and walking by the water strider are generated by a driving leg striking the free surface with a frequency, *f,* and may be characterized by the dimensionless Weber number, or *We,* the ratio of inertia to curvature forces (Bush and Hu 2006; Hu,

Chan, and Bush 2003). Water striders additionally exhibit a unique mode of escape behavior while on the surface of a pond or lake: they are able to jump on water as high as they can jump when on land (Hu and Bush 2010). What biological mechanism allows the water strider to maximize momentum transfer to water for jumping?

Rob Wood and his colleagues have solved this mystery of nature by painstaking characterization of the way that water striders use their legs to push against water with just the right amount of force to propel themselves above the surface and then verifying their predictions of how this is done by building an at-scale water-jumping robot (Koh et al. 2015). With the aid of high-speed cameras, Koh et al. (2015) found that the insects rotate the curved tips of their middle and hind legs at a relatively low descending velocity, with a force of 144 millinewtons (mN) / sec, a magnitude that creates a "dimpling" depression in the water surface but does not break the surface. These measurements identified specific design characteristics for an at-scale robot capable of jumping on the water surface: an ultralight body; legs capable of rotation so that they can keep pressing on, but not break, the water surface; and leg tips that rotate relative to the formation of the water surface dimple. The design of their water-jumping robot addressed these requirements with a torque-reversal catapult (TRC) fabricated using flexure hinge-based composite structures developed in earlier work (Wood et al. 2008). The TRC mechanism achieved by synthetic means at-scale what the biological structures of the water strider do: it generates an initially small torque that gradually increases through the driving stroke, with high momentum transfer from water surface to jumping body (Koh et al. 2015). This mechanism, paired with shape memory alloy artificial muscles and superhydrophobic material legs with curved tips, makes it possible for the robot to smoothly jump off the water surface, without breaking the free surface or making a large splash.

Jumping and controlled falling for aerial escape from land predators are also of interest because of their possible role in insect flight evolution (Dudley and Yanoviak 2011). There is a consensus that controlled aerial falling by gliding, and use of the sensory and biomechanical properties of the body to orient during freefall, may have preceded the origin of wings in hexapods. Converging findings in arthropod evolution and genetics suggest that the insect wing may have evolved from crustacean thoracic segments, or from modified leg styli (Carroll, Grenier, and Weatherbee 2005; Damen, Saridaki, and Averof 2002; Grimaldi and Engel 2005). If so, the flapping motions of insect wings may have emerged indirectly through actions of the dorsoventral leg muscles that insert on the thorax, as is characteristic of *dipteran* bifunctional muscles (Dudley and Yanoviak 2011). The evolution of insect flight via modifications of thorax structures is a significant finding for bioinspired engineering of at-scale flying insects because it helps to explain how insects are able to derive such power and flexibility with each wing stroke. The thorax is a resonant structure: on each wing stroke, large amounts of kinetic energy may be stored by the thorax as potential energy.

Robobees. The 2008 comedy film *Get Smart* includes a scene in which a tiny flying robotic insect is smashed to the ground by a rather arrogant secret agent. Even as fiction, the idea of an insect-size flying robot seemed as far-fetched in 2008 as the force field creating their "cone of silence." That same year, though, a nonfictional scientific paper by Rob Wood announced the first take-off of a biologically inspired, at-scale robotic flying insect (Wood 2008). There are many challenges beyond take-off for emulating insect flight, including staying aloft for hours and maintaining stability during wind gusts (Wood, Nagpal, and Wei 2013); navigating obstacles and gaps; and switching between the flight modes of gliding, fast forward flapping, slow forward flapping, and hovering (Perez-Arancibia et al. 2015; Reiser and Dickinson 2013). How did these engineer-scientists emulate a biological flying insect taking off?

A first consideration is the forces acting on a half-gram or less robot: at this scale, Newtonian and viscous forces dominate over volume-related ones, such as gravity and inertia (Wood et al. 2013). Nature's solution to generating propulsive forces for locomotion at a Reynolds number of about 100—for example, in the fruit fly *Drosophila*—is a combination of muscle, material, and architectural properties that generate a wing stroke for flapping (Dickinson 1999). Insect wings are diaphanous, veined structures capable of remarkable flexibility of both surface shape and torsion of the joint connecting the wing to the body (Song et al. 2007). The wing membranes and veins undergo significant bending and twisting during flight and may enhance lift by allowing the wings to twist and generate upward force throughout the stroke cycle (Dickinson 1999; Song et al. 2007). Whitney and Wood (2010) and Tanaka, Whitney, and Wood (2011) used a custom micromolding process to create a polymer wing emulating the venation and corrugation of a hoverfly wing. A flapping mechanism was constructed from the micromolded wing, a passive flexure hinge, and a piezoelectric actuator for driving. During tests of flapping of the synthetic wings and hinges, Tanaka et al. (2011) found that when the stiffness of the hinge was similar to the torsional stiffness of the hoverfly wings, the hinge passively pronated and supinated. Thus, both biological and robotic flies may leverage passive pronation and supination for hovering flight.

Through a study of thorax morphology, Finio and Wood (2010) next discovered a clue to nature's solution for stability and control of wing flapping: two functionally and morphologically distinct groups of flight muscles—power muscles to drive the wings at the resonant frequency of the thorax and control muscles to generate asymmetrical motions—are used for wing flapping. Finio and Wood (2010) leverage the elastic properties of the thorax as a starting point for designing a robotic dipteran mechanical structure that drives piezoelectric synthetic muscles to generate a flapping motion. Their experiments produced three designs: one that separates power and control actuation; a second that uses a single, hybrid, two degrees-of-freedom bending actuator for power and control; and a third that uses a hybrid actuator with twisting for control. All three of these designs depend solely on muscle power for actively generating forces and torques. However, in nature, the compliant properties of insect bodies act as a transmission that distributes muscle power to two wings.

To emulate nature's mechanical solution, Sreetharan and Wood (2010, 2011) developed a "mechanically intelligent" device, a single piezoelectric actuator, an underactuated transmission, and passively rotating wings that were successful in passively balancing aerodynamic forces created by two flapping wings. Wing motion consisted of both stroking back and forth and passive rotation, as determined by the interaction of the wing with the air—that is, by inertia—and by the elasticity of the wing hinge (Wood et al. 2013). For such a small flying machine, feedback control of stabilization by sensing and computation creates too great a time delay. To solve this problem, Teoh et al. (2012) once again leveraged body mechanics to develop passive air dampers to stabilize a less than 100-mg flapping-wing robot, called a "robobee," during hovering flight (see Figure 4.4). This design insight provided the desired passive control over yaw and roll, making it possible for the robobee to achieve indefinitely long hovering flight or programmed trajectories.

Due to the multifold layers and sublayers of more complex popups, the development time of early versions of robobees, and of robotic inchworms and roaches, remained long, and implementing new devices has required a high level of expertise (Aukes et al. 2014). To promote

Figure 4.4 Robobees. Five individual robotic flies of identical design are shown alongside a U.S. penny for scale, demonstrating that the manufacturing process facilitates repeatability and mass production. Roll torque is generated by flapping one wing with larger stroke amplitude than the other, inducing differential thrust forces. Pitch torque is generated by moving the mean stroke angle of both wings forward or backward to offset the thrust vector away from the center of mass. To generate yaw torques, the robot influences wing drag forces by cyclically modulating stroke velocity in a "split-cycle" scheme. Photo courtesy Robert J. Wood.

dissemination of the technology, Aukes et al. (2014) have developed a PCMEMS manufacturing process that partitions the sequence of iterative material addition and removal steps into two main cycles. In the first, layers of functional materials are individually cut into patterns and laminated together to form a composite structure. The intricate layers and layer interactions within this composite structure are the basis for the differentiated functions of the device, including, for example, the sensors, actuators, transmission, and wings of the robobee. In the second cycle, materials are selectively removed so that the device is erected, locked into its final position, and freed from the surrounding support material (Aukes et al. 2014). New software, called popupCAD, now makes it possible to create and perform these manufacturing operations on two-dimensional geometric primitives (Aukes et al. 2014). An online suite of popupCAD tools is available at popupcad.org. Even more amazing, is that the pop-up robobee becomes a hovering robot! How does it fly?

Earlier versions of the robobee were designed with a control structure consisting of three independent modules for body attitude, lateral position, and altitude. For example, control of body attitude during hovering flight was based upon (1) stabilizing body torques so that the robobee net thrust vector compensated for gravity and (2) tilting its body in order to generate lateral forces in response to a change in the desired lateral position. To attain control of body attitude in this way, Ma et al. (2013) used a Lyapunov function (a function that takes positive values everywhere but at equilibrium) to derive an attitude control law. This law consisted of two terms: a proportional term that accounts for the error from a reference orientation and a derivative term that opposes angular velocity, providing rotational damping. During test flights, this model-based system successfully controlled hovering flight as well as lateral flight alternating between two fixed points. However, a drawback of a model-based controller when there are independent modules for control of body attitude, lateral position, and altitude is that the modules may interfere with each other in unpredictable ways. As an alternative, therefore, Perez-Arancibia et al. (2015) introduced a model-free approach in which the control system was gradually discovered through experiment. But model-free control requires sensor information. A next crucial step toward the goal of autonomous flight, therefore, was to use sensor information to "close the loop" in the control system.

In the experiments conducted by Perez-Arancibia et al. (2015), sensor information about the robot's center of mass and rotations about pitch, roll, and yaw axes was provided by a multi-camera Vicon motion capture system. Such a system is highly accurate but requires the robobee to remain within the confines of the laboratory. A next step toward autonomous flight, therefore, was to incorporate an onboard sensor to replace the input provided by the Vicon system. Toward that end, the engineers once again turned to nature for biological inspiration. A fly-size robot with flapping wings is intrinsically unstable: its center of mass hangs below the wings, giving rise to pendulum-like dynamics requiring constant corrective feedback with a short time delay (Perez-Arancibia et al. 2011). A close look at a flying insect, such as a fly, shows that their visual organs consist not only of compound eyes, but also of three tiny organs on the top of the head, called ocelli. The ocelli are a second visual system that provides information about levels of light at different, slightly overlapping patches of the upper visual hemisphere (Krapp 2009; Taylor and Krapp 2007). Modeling of the ocelli has determined that these organs use varying luminance in optic flow to encode rate of rotation, reducing latency in responding to attitude disturbances (Gremillion, Humbert, and Krapp 2014). Fuller et al. (2014) have developed an ocelli-inspired visual sensor to stabilize a fly-size robot with flapping wings. The tiny sensor is part of a feedback control system for applying torque in proportion to the angular velocity of light source motion. Application of torque in this way makes it possible to harness the pendular motion of flapping wing dynamics to achieve stable upright attitude—the first known use of onboard sensors for flight control at the scale of a fly.

The evolution of the earliest flight of winged animals from earlier forms of behavior, such as intermittent soaring and landing, directs our attention to the significance of perches between flights: human-built airships land at airports, small soaring mammals (such as squirrels) grab onto tree branches and land in all manners of body orientation, and birds land gracefully on a variety of substances and surfaces. For flying animals at all scales, perching provides enormous advantages in reducing power consumption and mechanical fatigue and in providing opportunities to replenish fuel supplies, serving as a vantage point for exploratory behavior, finding food and mates, or escaping predation. Insects, nature's original "ultralight" devices, are even able to perch on the smooth surface of a ceiling, as flies do, or launch themselves while attached to distant surfaces by silk threads, as spiders do (Kovac 2016). Perching—alighting onto a surface or object and remaining attached—is, thus, a highly desirable, but difficult to achieve, attainment for insect-like aerial robots. One potential solution to perching by insect-like aerial robots, electrostatic adhesion, is made possible because a micro-robot's surface area-to-volume ratio increases with decreasing size. Graule et al. (2016) have designed and built an aerial micro-robot that uses switchable electroadhesion for controlled perching and detachment.

Perhaps the most daunting emulation challenges for building robobees is that nature's bees are smart: not only are bumblebees, for example, able to hover and perch, but they can learn to play golf! In a study by Loukola, Perry, Coscos, and Chittka (2017), bumblebees learned to transport a tiny ball to a defined location to get a sucrose reward. Even more remarkable was their observational learning: for those bees that did not initially succeed during early trials, an experimenter physically demonstrated what they were supposed to do by moving the ball with a tiny putter-like tool. After several of the demonstration trials, the bees succeeded. Here, the bees may have been able to adapt their natural foraging behaviors to a new context, and thus exhibit the kind of flexibility not expected in an insect. What kind of "brain" might robobees require to be able to emulate the flexible takeoff and landing, navigational, and foraging behaviors of their natural counterparts?

As an initial step toward the goal of building robobees capable of some of the learning behaviors of bumblebees, Clawson, Ferrari, Fuller, and Wood (2016) have implemented a neuromorphic controller modeled by a leaky integrate-and-fire (LIF) spiking neural network (SNN). The SNN is a network with input, hidden, and output layers. It uses reward-modulated Hebbian plasticity (see Chapter 7) in order to learn from a reference input, a Linear Quadratic Regulator (LQR) controller with known performance guarantees, to hover and land. Clawson et al. (2016) have conducted experiments to demonstrate that the SNN could closely mimic the way that the LQR controlled the robot. In one experiment, for example, waypoints were selected to guide the sequential behaviors of hovering and landing: the first waypoint guided the robot to hover slightly above the ground, and then a second waypoint caused the robot to slowly lower to a surface while remaining upright. This and other experiments have demonstrated the ability of the SNN to learn from a reference input to perform the kinds of flexible changes in flight control seen during the natural behaviors of biological bees. By emulating the circuit motifs discovered in other insects, such as *Drosophila*, sensory-based switching between modes of behavior at particular waypoints may be possible (see Chapter 5).

From Octobots Down to Where There Is Room at the Bottom

To conclude the chapter, I consider one grand challenge for emulating life's devices posed at the outset: constructing a robot that has a soft body, a circulatory system for internal transport of body fluids, vesicles that provide a milieu for an autonomous power-generating plant, and a soft controller. The closest device yet to meet this challenge is a soft, eight-limbed robot with an elastomeric body, called "octobot"—a new class of soft robot fabricated by combining microfluidics, embedded 3D printing, and power supplied by decomposition of monopropellant fuel in catalytic reaction chambers (Wehner et al. 2016) (see Figure 4.5). Octobot pneumatic actuation is controlled through microfluidic logic rather than by means of electrical circuits: a prefabricated microfluidic controller provides autonomous regulation of the flow of fuel and combustion by-products through a fuel matrix, a kind of circulatory system. This embedded 3D printed matrix internally transports fuel, catalyst, and venting of by-products (gases) through the elastomer body. Embedded 3D printing is also used to print each of eight hyperelastic pneumatic actuators, as well as an interconnecting actuation network.

A catalyst decomposes a monopropellant, producing a gas, which flows through the hyperelastic pneumatic actuation network (Wehner

Figure 4.5 The octobot. Wehner et al. (2016) have made an octopus-shaped robot that is constructed completely from soft materials. The body houses a liquid-fuel supply and a fluidic system that controls a cyclic pattern of leg movements. Actuators that cause the legs to lift are visible as purple rectangles in the legs. Body (including legs) is approximately 55mm in both length and width. Photo: Lori K. Sanders, Ryan Truby, Michael Wehner, Robert J. Wood, and Jennifer Lewis/Wyss Institute for Biologically Inspired Engineering at Harvard University.

et al. 2016). Actuation occurs when the interconnected channels of the network are inflated by the incompressible gas. Differences in modulus between the actuator elastomeric material and the surrounding body matrix material result in actuator bending upon inflation. The flow of gases through the microfluidic control system promotes oscillation cycles between two actuation bending states. Each cycle of fuel decomposition-actuation-venting begins when fuel is injected into a system of pinch and check valves that ensures a balance among the gas supply, actuation pressure, and exhaust rate. Flow rate and switching frequency are functions of upstream pressure and downstream impedance (resistance to flow). Vents in the system must be sufficiently small to allow complete actuation and, at the same time, be sufficiently large to allow timely venting (Wehner et al. 2016). With more advanced fluidic circuits, it will be possible to provide this generation of soft robot with a more varied set of movements and adaptive decision-making. Then, when placed in a realistic econiche, this autonomous soft machine may begin a journey that blurs the boundary between biological and synthetic devices.

Octobots are indeed tiny. However, as the brilliant physicist Richard Feynman noted in a 1959 lecture, "There's plenty of room at the bottom" (Feynman 2011). Heading toward the bottom, molecules are orders of magnitude smaller than octobots. Is it possible to build robots at the scale of molecules such as DNA? Decades after Feynman's prescient lecture, in 1995, George Whitesides offered a vision for leveraging molecular self-assembly to build nanomachines (and see his beautiful 2007 book with Felice Frankel, *On the Surface of Things*). Then, in 1999, Nadrian Seeman introduced a molecular assembly toolkit for creating shapes from nucleic acids. Now, in the early twenty-first century, DNA engineering has matured sufficiently to make possible the arrangement of thousands of nucleotides into nanometer-scale devices, using a procedure called scaffolded DNA origami (Castro et al. 2011). In this technique, a multiple-kilobase, single-stranded DNA scaffold is converted to a desired shape by means of hundreds of short complementary oligonucleotide strands, called "staples" (Ben-Ishay, Abu-Horowitz, and Bachelet 2013).

Working in collaboration with Wyss faculty George Church and William Shih, Shawn Douglas (now at the University of California at San Francisco) developed a critical computer-assisted drawing tool for the design of 2D and 3D DNA origami shapes, called caDNAno (Douglas et al. 2009; Douglas, Bachelet, and Church 2012). Douglas et al. (2012) have used caDNAno to create an autonomous DNA nanorobot for transporting and delivering molecular payloads to a cellular site. Each hexagonal barrel-shaped nanorobot transports and delivers a payload by folding on hinges, like a clamshell, so that it may be in closed or open states. While holding its payload, the robot remains in its closed state by means of a clasp system consisting of DNA locks and keys. The robot opens in response to antigen keys in order to release its payload. The possibility that such robots may one day roam throughout an animal's body has been advanced with a DNA origami robot that interacts with the cells inside living cockroaches (Amir et al. 2014), and with an amoeba-like molecular robot that changes its shape in response to signal molecules (Sato et al. 2017).

PART TWO

Structure-Function, Development, and Vulnerabilities of Nervous Systems

---

A speck of mouse brain is magnified by a factor of a billion to see its nerve fibers, dendrites, and synapses, all in colors that distinguish each. *In vitro* mouse brain tissue is made transparent in order to see large portions of its intact structures and how they interconnect. Selective wavelengths of light are projected into the brain of a mouse in order to activate genetically altered neurons and change the animal's behavior. Over a period of hours, noninvasive *in vivo* recordings are made of the slow, large-scale morphogenetic changes of zebrafish neural tissue. MRI is used to generate color flowlines identifying white matter tracts that distinguish the brain of a prematurely born infant from a child born at term. All of these advances in neuroimaging techniques have made it possible to reveal the microscopic, mesoscopic, and macroscopic complexity of the nervous systems of animals and humans, as well as the processes governing their development. For example, a remarkable characteristic of the human neocortex is that its constituent neurons are generated in proliferative regions outside the cortex itself. Neural stem cells must travel along glial fibers to arrive at particular locations in order to build the characteristic six layers of the human neocortex. Prenatal neuronal migration may be disrupted by errors in the genetic regulation of brain structure formation. Brain networks initially self-assemble to form overabundant synaptic connections, and postnatal experience transforms these circuits by a remodeling process. The chapters of this section relate the processes that build human nervous systems to their vulnerabilities to injury.

• 5 •

Nature's Nervous System Networks

How do brains work? To answer this question requires not only advanced twenty-first-century technologies, but also theoretical and methodological approaches that make it possible to reveal the multiscale processes from which our thoughts, feelings, and actions emerge (Swanson and Lichtman 2016). *Emergence,* in this context, refers to forms and functions that arise from the occurrence of complex interactions among component parts at multiple scales but are not present in the individual elements (Alivisatos et al. 2012; Yuste 2015). A dazzling illustration of emergent neural functioning is presented in the 2015 Pixar animated film, *Inside Out.* At the level of emergent behavior at a macroscopic scale, our protagonist, Riley, an 11-year-old girl, struggles emotionally with her family's geographic relocation. The component "states" guiding her behavior, a set of embodied emotions, reside in "headquarters." The dominant emotion is Joy, evident in flashback at birth in the eyes and face of the newborn Riley, and as she grows, in the fluid balletic motions of the embodied state of Joy in headquarters. Soon enough, though, Fear, Anger, Disgust, and Sadness (differentiated by color and affective tone in their interactions with each other) weigh in when Riley experiences some inevitable disappointments during childhood.

The developmental process itself is brilliantly captured by the self-assembly, dissolution, and reorganization of "islands" of personality function, emerging through core experiences with other people, especially her family. Quite compelling for me was the seemingly limitless color palette of tiny spheres collected together in transparent columns, like the structural organization of columns in the developing cerebral cortex (Rakic 2009). The use of multiple colors to distinguish these neuron-like spheres was a neat way to present interacting individual units because in a biological nervous system, neurons are characterized by overlapping axons and dendrites that are difficult to distinguish.

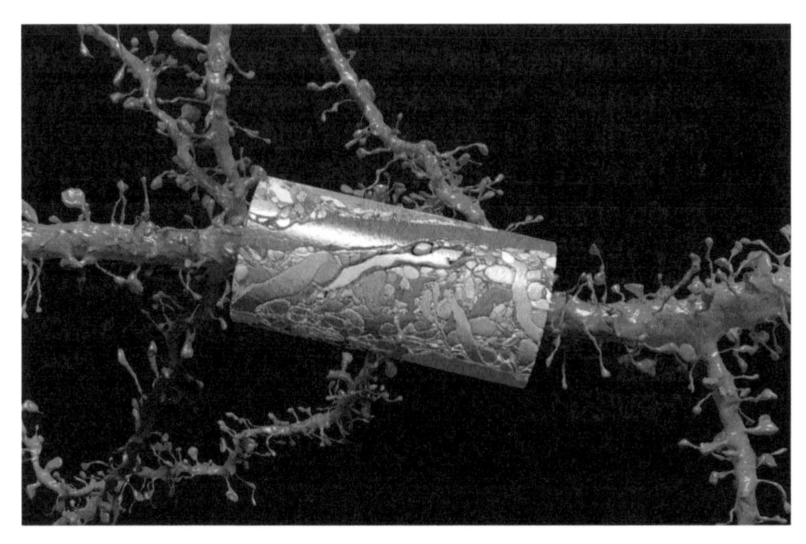

Figure 5.1 Lighting up the intricacies of the brain. A composite of artificially colored EM images reveals details within a cylinder of mouse brain tissue, smaller than a grain of sand, that contains 680 nerve fibers and 774 synapses. Image: Bobby Kasthuri, Daniel Berger, and Richard Schalek / Lichtman Lab at Harvard University.

A similar, color-based solution to distinguishing real neurons in a biological brain, called "brainbow," has made it possible to disambiguate one neural process from the thousands of axons and dendrites it closely approaches over its entire length (Lichtman, Livet, and Sanes 2008). Brainbow is a combinatorial color method that marks individual neurons in one of more than 100 colors through combinations of the expression of fluorescent protein variants, called XFPs (Cai et al. 2013) (see Figure 5.1).

In addition to the significant technical challenges posed by the attempt to understand the emergence of brain function, there are at least two other issues that add layers of complexity to the picture of how the brain works. First, neural circuits are subject to neuromodulation (for example, by chemical neurotransmitters), allowing the same neurons to generate multiple activity patterns (Bargmann 2012). Connectivity diagrams establish only potential configurations, and it is neuromodulation that gives rise to specific output (Bargmann and Marder 2013). Therefore, without identifying all of the neurotransmitters present in each neuron, any account of circuit function will be incomplete. Second, nervous systems evolve and develop as intrinsic parts of bodies behaving in the physical environment: they are embodied. For this reason, a complete account of the function of nervous systems will require the recording and control of neural circuits in freely moving animals exhibiting a range of behaviors in their typical ecological niches.

The urgency of advancing our knowledge about the brain is evident in worldwide, government-funded, as well as private, initiatives (Grillner et al. 2016). The Allen Institute for Brain Science, which opened in 2003, has produced the Allen Mouse Brain Atlas, a comprehensive cellular-level atlas of gene expression, and a companion three-dimensional reference atlas, as well as brain atlases for the developing mouse and for the adult and developing nonhuman primate. Its current efforts include Project MindScope, a focus on the neocortex, characterizing, for example, all of the cortical cell types of the mouse cortex by building distinct experimental-computational platforms, called brain observatories, for studying behaving mice and to construct cortical network models (Hawrylycz et al. 2016). The European Union's Human Brain Project, started in 2013, places its emphasis on modeling and simulation of microcircuits as well as global aspects of brain function. The China Brain Project, starting in 2017, examines the neural basis of cognitive functions. Japan's Brain Mapping by Integrated Neurotechnologies for Disease Studies, begun in 2014, includes an effort to map the brain of a small monkey, the marmoset, as a step toward better understanding of the human brain for diagnosis and understanding of psychiatric and neurological disorders. And the mission of Israel Brain Technologies, started in 2011, is to accelerate brain-related innovation and commercialization.

In 2013, President Obama took action to fund an American Brain Research through Advancing Innovative Neurotechnologies (BRAIN) initiative, a 10-year project (through 2025) for technology development and integration that will allow fundamental new discoveries about the brain. The BRAIN initiative addresses emergence of dynamics in neural circuitry, neuromodulation, different types of cells (for example, glia), and much more (see Table 5.1). The major goal of the BRAIN initiative is as follows:

Table 5.1 Microcircuit temporal characteristics as the basis for distinctive neural functions

Circuit	Temporal characteristics	Function
Hippocampal	Sharp-wave ripple oscillations sustained by tri-synaptic loop are comprised of parallel subpathways.	Ripple oscillations may reactivate prior experience.
Spinal motor	Spinal cord consists of rhythm-generating and pattern-generating circuits. The latter are reciprocally connected via ipsilateral inhibitory interneurons.	Rhythm- and pattern-generating circuits together may reconfigure activation of muscle groups into different functional combinations.
Olivo-cerebellar	Olivary axons (cerebellar climbing fibers) generate bursts that modulate the complex spike response in Purkinje cells.	Inferior olivary bursts may regulate plasticity and timing in the cerebellar cortex.
Thalamocortical	Spikes of the pulvinar, a thalamic nucleus, selectively synchronize activity in visual cortex.	Pulvinar may regulate information transmission across the visual cortex.
Insulo-cortical	Information is multiplexed in the temporal structure of neuron firing.	Activity of a system of parallel circuits may integrate gustatory and multisensory inputs.

Sources: S. Arber (2012), Motor circuits in action: Specification, connectivity, and function, *Neuron, 74,* 975–989; and G. Buzsaki (2006), *Rhythms of the Brain*, New York: Oxford University Press.

> Our charge is to understand the circuits and patterns of neural activity that give rise to mental experience and behavior. To achieve this goal for any circuit requires an integrated view of its component cell types, their local and long-range synaptic connections, their electrical and chemical activity over time, and the functional consequences of that activity at the level of circuits, the brain, and behavior. (Bargmann et al. 2014, 12)

The BRAIN initiative begins to address how "attractors," "oscillation," and other neural motifs interact with obtained sensory information, remembering, and action. It also recognizes that neural dynamics are modulated by slower-acting chemicals that may remodel circuits to produce different patterns of activity and the role of different types of cells, not only neurons but also glia, for brain function. Most significant for human health, the BRAIN initiative is translational: it is expected to accelerate the understanding of brain disorders as a means of repairing damage to the brain, including, for example, stroke, traumatic brain injury, and spinal cord injury, resulting in the loss of perceptual function, remembering, and using the body for walking, talking, and acting on the world with our remarkably dexterous hands.

The research strategy for such an ambitious enterprise calls for an approach that is fundamentally comparative, placing the size and complexity of the human nervous system within the context of other animals. There are many reasons for this, including current limitations on recording technologies and ethical issues in experiments that use "perturbation" techniques, such as optogenetics, to understand how behavior emerges from neural circuitry. Fundamentally, inferences about how the brain works depend upon what is measured. With the emergence, context-dependence, and embodiment themes from the BRAIN initiative as overarching themes, this chapter begins by taking a comparative approach to consider what we already know about circuits in small nervous systems. The neurosciences bring a rich set of technologies from decades of research on invertebrates and nonhuman vertebrate nervous systems, including the nematode worm *C. elegans* (Wen et al. 2013), the virtually transparent zebrafish (Fidelin and Wyart 2014; Portugues et al. 2013), transgenic mice (Fenno, Yizhar, and Deisseroth 2011), and nonhuman primates (Wedeen et al. 2012). By expanding our scope from humans to a widely comparative perspective, it may be possible to achieve new insights into the relation among large populations of neurons, connectivity, and function.

Diversity of Cell Identity

The approximately 10^8 neurons and the even larger number of glia comprising the mouse brain are characterized by diversity: each cell is genetically, anatomically, physiologically, and connectionally distinct (Jorgenson et al. 2015). It is this diversity that has guided the choice of the first research priority in the BRAIN initiative: to identify different brain cell types (Jorgenson et al. 2015). The exceedingly difficult technical challenge involved in addressing this priority, and the implications of study findings, are evident in work from the Lichtman laboratory on the first nanometer-resolution reconstruction of a tiny volume of mouse neocortex (Kasthuri et al. 2015). This study has been able to identify the geometries of all excitatory axons and their synaptic and nonsynaptic juxtapositions with *every* dendritic spine. Another approach is the "barcoding" of the brain, by Wyss Founding Core faculty member George Church and colleagues, by inserting unique nucleotide barcodes in neurons to map cell networks (Underwood 2016).

Emergent Functions

An understanding of the entire brain is obviously beyond the scope of a single chapter (see a perspective on where we stand on this question in Lisman 2015). Instead, the focus of this chapter, and of Part II as a whole, is on the relation among brain, body, and behavior and on the *emergent functions of the nervous system.* These are core questions for understanding how we may build devices for restoring function of injured nervous systems, addressed in Part III of this text. This chapter begins by considering the quest for the brain's connectome and why wiring diagrams alone are inadequate to explain even the behavior of insects, worms, and crustaceans. For example, I summarize the comprehensive work on the stomatogastric nervous systems of lobsters and crabs by Eve Marder, which illustrates that (1) neuromodulatory substances reconfigure circuit dynamics, so that individual neurons can switch among different functional circuits (Marder and Bucher 2007; Marder 2012), and (2) that homeostatic regulation ensures stable network performance, despite the very large number of sets of parameters that can produce similar output patterns (Prinz, Bucher, and Marder 2004; Goaillard et al. 2009; Marder and Taylor 2011; Marder, O'Leary, and Shruti 2014; Marder, Goeritz, and Otopalik 2015). The insights from this work set the stage for discussion of homeostatic plasticity in Chapter 7.

Revealing the structure and function of nervous systems of animals, small and large, is a scientific enterprise being undertaken by thousands of neuroscientists. The MindScope and other initiatives of the Allen Institute as well as other international efforts are part of a new era in nervous system mapping. Through the use of light-sheet microscopy, for example, fluorescence emitted by a stack of illuminated samples can now generate three-dimensional images of the entire zebrafish brain or of large neuronal populations of mouse olfactory cortex (Keller and Ahrens 2015). Optical imaging, genetic tools, brain clearing, and computational techniques are part of the new armamentarium for interrogating intact brains of mice (Renier et al. 2016). And the magnetic resonance imaging (MRI) and measurement of multiple modalities, or types of measurements of the brain—for example, myelin density and cortical thickness—from the Human Connectome Project has made it possible to identify 180 areas in each hemisphere of the human brain (Glasser et al. 2016) (see Figure 5.2).

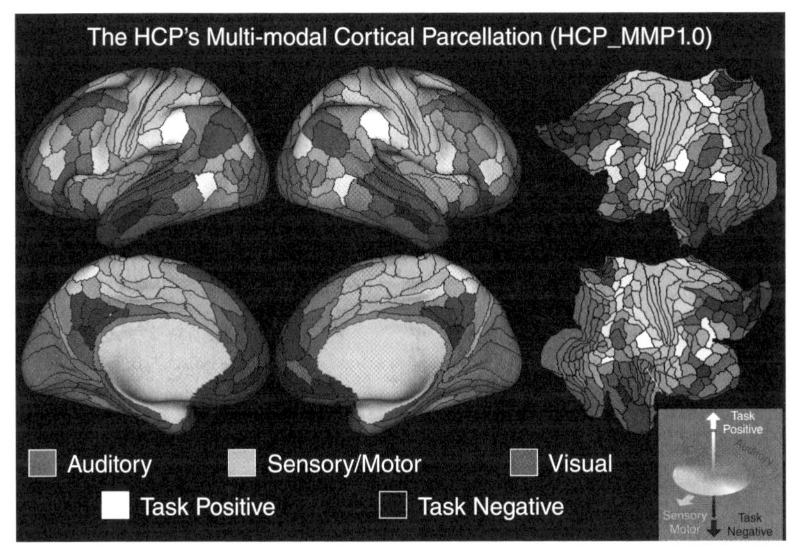

Figure 5.2 Multimodal parcellation. The 180 areas delineated and identified in both left and right hemispheres are displayed on inflated and flattened cortical surfaces. Black outlines indicate areal borders. Colors indicate the extent to which the areas are associated in the resting state with auditory (red), somatosensory (green), visual (blue), task positive (toward white), or task negative (toward black) groups of areas. The legend on the bottom right illustrates the three-dimensional color space used in the figure. Data at http://balsa.wustl.edu/WN56. Matthew F. Glasser, Timothy S. Coalson, Emma C. Robinson, Carl D. Hacker, John Harwell, Essa Yacoub, Kamil Ugurbil, Jesper Andersson, Christian F. Beckmann, Mark Jenkinson, Stephen M. Smith, and David C. Van Essen (2016). Images reprinted with permission from Nature Publishing Group / Macmillan Publishers Ltd.

How may we understand the extent to which both small nervous systems and larger ones are based upon the same principles?

Honeybees (with about 1 million neurons) and humans are able to navigate through complex econiches to find food (Srinivasan 2010) as well as discriminate between Monet and Picasso paintings along the way (Wu et al. 2013). Crows and humans will use a short tool to get a longer one and then use the longer tool to retrieve food (Taylor et al. 2007). Humans and soft-bodied cephalopods learn by observing (Hochner 2012). This chapter continues the themes from the first section of the book and considers *each animal's nervous system as a species-specific soft machine capable of transforming its body into a variety of special-purpose devices adapted to its ecological niche* (Goldfield 1995; see also Shenoy, Sahani, and Churchland 2013). What is common in all such species-specific nervous systems?

In both human and smaller nervous systems, (1) functional networks form and dissolve in the population dynamics of fluctuating energy flows at multiple time scales (Perdikis, Huys, and Jirsa 2011; Bullmore and Sporns 2012; Sporns 2014), (2) each neuronal population dynamical state determines the future evolution of that system and its response to inputs (Shenoy et al. 2013), (3) networks are modulated to form different combinatorial patterns (Bargmann and Marder 2013), (4) nervous system oscillations harness the body's resonances to perform different kinds of energy exchange in the environment (Tytell, Hsu, and Fauci 2014), and (5) behavioral dynamics emerge via the interactions of the animal's dynamical systems with the structured information fields of the environment (Kato et al. 2015; Warren 2006).

The Connectome

The connectome is a comprehensive map of neural circuit synaptic connectivity in the form of networks (Sporns 2011). The mapping of the brain as a complex system with emergent properties has become a multiscale enterprise, ranging from the microscale (individual neurons and their synaptic connections), to the mesoscale (short- or long-range connections among distinct cell populations), to the macroscale (anatomically distinct brain areas, the structural pathways connecting them, and their functional interactions) (Craddock et al. 2013; Kim, Chung, and Deisseroth 2013; Oh et al. 2014). There is an important distinction between the connectome as a "wiring diagram," or map, and the dynamic patterns of neural activity and interactions of the connectome, called "functional connectivity" (Sporns 2013a,b). Functional connectivity is not based on measurements of synaptic connections, but rather on statistical measures derived from neuroimaging-based brain activity (Biswal et al. 2010; Van Dijk et al. 2010; Vertes et al. 2012). Brain activity during a particular state, such as attention or rest, identifies a functional connectome (Biswal et al. 2010). Indeed, there is a new appreciation of the intrinsic states of brain functioning (Fox and Raichle 2007; Raichle 2010), with functional networks continually pulled by intrinsic states toward multiple configurations (Deco, Jirsa, and McIntosh 2013) (discussed later).

One of the most important reasons for studying the way that neurons become interconnected is to better understand the dramatic changes in wiring that occur during development (see Chapters 6 and 7). The processes underlying transition from exuberant initial wiring to remodeling with experience are inaccessible using techniques that sample from only a few cells at a time: answering questions about remodeling requires a network level of analysis (Morgan and Lichtman 2013). The murine neuromuscular circuit provides one starting point for a network-level wiring diagram because it is easy to visualize the neuromuscular junction and because of the possibility of tracking individual axons (Srinivasan, Li et al. 2010). For example, by using serial electron microscopy to image fluorescently labeled motor units (motor neuron axons projecting to neuromuscular junctions), Tapia et al. (2012) have found that before birth, there are many axons converging at a neuromuscular junction, but they occupy only a small percentage of the available receptor sites. This indicates that the axons make only weak connections. However, two weeks later, only one axon occupies all of the receptor sites at each neuromuscular junction. Many synaptic branches are lost, and the remaining ones become strongly innervated (Tapia et al. 2012). The transition from limited intrauterine opportunities for using the muscles before birth to the rich opportunities for behavior during postnatal life promotes a remodeling of the neuromuscular circuits to make possible a connectome based upon individual experience within a scaffold of species-specific constraints.

A second illustration of the value of understanding the process of wiring neurons in forming a connectome is work on the mouse retina by Kim et al. (2014). They address the question of how wiring between star amacrine cells (SACs) and bipolar cells (BCs) support the emergence of direction sensitivity of retinal networks. The task begins with scanning electron microscope (SEM) images of individual neurons. A major challenge in reconstructing the complex wiring of the retinal connectome is the process of tracing the unpredictable paths and unexpected branching of neurites through the maze of other neuronal processes. A notable feature of this work is the use of an online community of "citizen neuroscientists"—a "crowd" working together with a small number of laboratory experts using game-like software, called EyeWire, to build a three-dimensional reconstruction of this retinal connectome for direction sensitivity (Kim et al. 2014). A major finding of this groundbreaking research relates to a particular functional characteristic of each of the many types of BC: each type has a *different time lag* of visual response. Connections between neurons create a network capable of detecting spatiotemporal change due to motion as a moving object appears at a different location after a time delay (Kim et al. 2014). The EyeWire results support the predictions made for the wiring of this network: different locations on the SAC dendrite are wired to BC neuron types characterized by different time lags.

Along with these and other advances in reconstructing connectomes, laboratories are building computational models to understand the relation between connectivity and its underlying biophysical processes (see, for example, Deco et al. 2013; Kopell et al. 2014). There is a growing consensus that brain network oscillations may control the flow of information through anatomical pathways (Buschman et al. 2012; Siegel, Buschman, and Miller 2015). For example, oscillatory synchronization of local field potentials (LFPs) within each frequency band may create temporary circuit connectivity for information transmission—that is, for temporary neural ensembles (Buschman et al. 2014). And because local circuits perform highly specialized computations, different behavioral tasks may be achieved by different combinations of regions working together (Akam and Kullman 2010, 2014). The fast time-scale dynamics of brain oscillations may support a variety of functions, such as the means by which gamma-band activity binds together different regions through coherence (Buzsaki and Wang 2012; Fries 2005). But a remaining challenge is an understanding of how particular regions are selected for different tasks and how functions emerge from combinations of parallel and serial computations.

The conservative nature of evolution promotes some optimism for the view that revealing the connectome of the brain of an insect, such as *Drosophila,* may help to guide the search for general principles by which the brains of humans are also built. Are there, for example, principles of information flow that span the connectomes of flies and humans? To address this question, Shih et al. (2015) assembled 12,995 images of neuron projections of the female *Drosophila* brain *Fly-Circuit* database into a wiring diagram. From this database, they divided the brain into functional local processing units (LPUs)—populations of local interneurons whose fibers are limited to a particular region and that deliver or receive information from other units. Shih et al. (2015) adopted the assumption that major information is transmitted in a neuron from dendrite to axon and asked whether loops of activity between LPUs, or recurrent circuits, might be the foundation for the millisec-long reverberations that we call "memory." Their network simulations provided evidence for recurrent circuits with multiple loops and similarities between the network structure of the *Drosophila* and mammalian brain.

The Role of Neuromodulation in the Emergence of Behavior: *C. Elegans* and *Drosophila*

Publication of the *Drosophila* connectome is a remarkable achievement. However, there is considerable evidence that understanding the emergence of function in anatomical connectomes will require the further inclusion of the neuromodulation of structural circuits (Marder 2012). There are at least three functions of neuromodulation that endow even the smallest nervous systems with the emergence of different functions: (1) *context sensitivity*, where anatomical connections represent a set of potential connections that are shaped by both context and internal states to allow different paths of information flow; (2) *multifunctionality*, where neuromodulators appear to switch circuits between alternative functional states (Bargmann 2012) [for example, a neuropeptide receptor, called npr-1, regulates *C. elegans* aggregation versus avoidance behavior (Macosko et al. 2009), and monoamines are involved in selectively orchestrating the activity of specific circuits during escape, as in the way that tyranine activates a receptor that facilitates contraction of ventral muscles to allow *C. elegans* to turn and resume locomotion in the opposite direction (Donnelly et al. 2013)]; and (3) *selection*, where neuromodulators select a set of functional synapses among a greater number of anatomically specified possibilities (Bargmann 2012).

Consider the role of neuromodulation in the escape behavior of *C. elegans*. Its anatomical connectome was deciphered more than twenty-five years ago, but the processes by which neuromodulators recruit neural circuits for behaviors are only now coming to light (see, especially, optogenetics, discussed later). For example, the escape response is a behavioral sequence by which *C. elegans* navigates away from certain threatening events: a gentle touch to the head results in the animal quickly moving backward and making a "deep turn" that allows it to change directions and move away (Donnelly et al. 2013). Context sensitivity, multifunctionality, and selection are all at play in the escape response of *C. elegans*. Context sensitivity is made possible, in part, by proprioceptors that measure the stretch or dynamic forces generated by the musculoskeletal system, such as the torso-bending movements that propel *C. elegans* (Goulding 2012).

State-dependent modulation of neural circuits is also evident in the flight behavior of the fly *Drosophila*. There is growing evidence that the biogenic amine octopamine orchestrates physiological changes throughout the body of *Drosophila* during flight (Suver, Mamiya, and Dickinson 2012). A type of large-field visual interneuron, known as vertical system (VS) cells, encode optic flow information to stabilize component flight systems (neck and wings). VS cells exhibit a "boost" in their response to visual motion during flight, compared to when they are in a quiescent state (Suver et al. 2012). Suver et al. found that pharmacological application of octopamine in quiescent flies causes an increase in VS cell response to motion, suggesting that octopamine plays an important role in modulating physiology during flight. Finally, circadian rhythms play a role in modulating circuits governing behavior, such as scheduling feeding, sleep, sex, etc. In *Drosophila*, a central brain clock consists of an oscillating biochemical circuit that is modulated during the unfolding of the day to adapt the fly's behavior (Griffith 2012).

A Revolution in Genetic and Molecular Tools for Imaging and Recording from Nervous Systems

Two-Photon Microscopy. New methods for probing neural circuit functions have emerged in the last decade Yang & Yuste, 2017). These methods include genetically encoded fluorescent proteins, optical imaging, optogenetics, chemical genetics, and biosensors (Murphey, Herman, and Arenkiel 2014). Genetic encoding makes it possible for protein "reporters" to respond to neural events, including intracellular calcium dynamics (Lin and Schnitzer 2016). Neuronal calcium tracks action potentials as well as presynaptic and postsynaptic calcium signals at synapses, providing useful information about both input and output signals. Optical imaging enables thousands of neurons to be simultaneously observed *in vivo* (Ahrens et al. 2013). Two-photon excitation (2PE) microscopy combines photo stimulation of the calcium indicator with a laser scanning microscope to image the fluorescence (Svoboda and Yasuda 2006; Yang and Yuste 2017). In 2PE, only a single point in space is excited, providing increased depth penetration and making it useful for imaging at scales ranging from synapse to entire brain (Schrodel et al. 2013).

Optogenetics. A decade-old toolset for neuroscientists, called optogenetics, makes it possible to selectively silence neurons *in vivo* (Boyden 2015). Optogenetics probes and controls neural circuits by means of light-induced neuronal excitation (Tye and Deisseroth 2012). How is it possible to use light to excite neurons? Certain algae living in soil and fresh water, and a type of archaebacteria living in highly saline soda lakes in Egypt and Kenya, have evolved specialized proteins, called *opsins*, that respond to visible light by opening and closing the organism's cell membrane channels (Deisseroth 2010). For example, ChR2 channelrhodopsin allows positive sodium ions to pass through a membrane channel in response to blue light, VChR1 channelrhodopsin responds to some wavelengths of green and yellow light, and NpHR halorhodopsin regulates the flow of negative chloride ions in response to yellow light (Deisseroth 2010). A major breakthrough in optogenetics was to use an established technique called transfection to insert opsin genes into the neurons of mice and then other animals. Transfection involves combining the opsin gene with a promoter (which causes the gene to be active only in a specific type of cell), inserting the gene into a virus, and injecting the virus into the brain of the animal (Zhang et al. 2010). Following opsin expression in the cell population of interest, a light-delivering optical fiber is placed over the cell bodies to target projection neurons (Kim, Adhikari, and Deisseroth 2017). The delivery of light at wavelengths specific to the encoded ion-conductance regulators provides millisecond resolution control of specific projections.

Hundreds of laboratories worldwide now use optogenetics to address questions that seemed out of technical reach only a decade ago. One outcome of this intensive effort has been the combining of optogenetics with other existing, as well as newly emerging, stimulation and recording technologies, including two-photon laser scanning fluorescence microscopy and highly sophisticated computer-controlled tracking systems (see, for example, Zagorovsky and Chan 2013). Techniques from the optogenetic toolbox, used in conjunction with electrophysiology, imaging, and anatomical methods, have led to the major discovery that behavioral states characteristic of a range of psychiatric disorders emerge from the dynamics of multiple neural projection circuits (Kim, Adhikari, and Deisseroth, 2017). For example, Kim et al. (2017) found that in optogenetic experiments with mice, the behavioral state of anxiety is assembled from circuits governing the separable anxiety-related behavioral characteristics of risk avoidance and respiratory rate alterations: three distinct circuit projections from the bed nucleus of the stria terminalis (BNST) each initiated an independent characteristic of the anxiety behavioral state and exerted opposite effects in modulating anxiety. Neuroimaging studies of humans who are depressed and exhibit anhedonia (lack of enjoyment) are characterized by particular resting state correlations between prefrontal cortex and several subcortical regions (Keedwell et al. 2005). By combining optogenetics and functional MRI (see Chapter 5), Ferenczi et al. (2016) were able to demonstrate that the medial prefrontal cortex of rats controls interactions among the subcortical regions governing impaired reward perception and experience, characteristic of anhedonia.

Orchestration of Action: Observing Entire Brains during Behavior

During the early 1960s, an era in which there was great urgency for improved science education the United States, my parents presented me with a very generous gift: a blue metal box containing a Gilbert microscope (although I really wanted a Lionel train set). I still remember my surprise when I opened the box, set up the microscope so that its small mirror could capture the light from my desk lamp to illuminate the stage, added water to the vial of dried shrimp eggs, placed a cover slip on the slide, and brought into focus a field of tiny animals (better than a train set). Now, more than fifty years later, two-photon microscopy and genetically encoded calcium allow scientists to record the neural activity of nearly the entire brain of other tiny animals (see, for example, Grienberger and Konnerth 2012).

Downtown from where I grew up, the Yuste lab at Columbia University has been addressing the Brain Initiative call to understand "emergence" by revealing the relation between nerve net activity and distinct behavioral functions of the tiny freshwater cnidarian *Hydra vulgaris*. Dupre and Yuste (2017) have selected this particular animal for study because its nervous system is composed of between a few hundred and a few thousand neurons, its genome of more than 20,000 genes has been sequenced, it is transparent and small enough for the entire animal to be imaged with a fluorescence dissecting microscope at single-neuron resolution, and it does not age. In order to be able to use fluorescence imaging to reveal the neural circuits of *Hydra* during performance of particular behaviors, Dupre and Yuste (2017) created a line of animals expressing the calcium indicator GCaMP6s in neurons. A major discovery in this work is that the nervous system of *Hydra* includes three major non-overlapping networks (in the sense that any neuron belonging to one network did not belong to the other two networks). Moreover, each of these networks is associated with a distinctive *Hydra* behavior: network RP1 is associated with body elongations, RP2 with radial contractions and the subtentacle local network STN (Dupre and Yuste 2017; Ji and Flavell 2017). The discovery of non-overlapping networks indicates that "evolution has carved out a behavioral repertoire by selectively linking subsets of neurons out of a tapestry of apparently similar cells, as each subset of neurons is associated with a specific behavior. This carving could occur by selectively connecting neurons into subcircuits, or by modifying synaptic strengths" (DuPre and Yuste 2017, 1094). Is there similar evidence of relationship between neuron circuits and behavior in other animals with relatively small nervous systems?

The larva of the fly *Drosophila* performs a rich set of locomotor behaviors, including crawling forward or backward and turning. During foraging, these larvae may stop crawling, reorient themselves, turn, and start crawling again (Reidl and Louis 2012). All of this is done with a nervous system of only 10,000 neurons arranged in three main centers: central brain, subesophageal ganglion, and ventral nerve cord (Boyan and Reichert 2011). Thus, even in the miniature *Drosophila* nervous system, an apparent organizational theme is that more widely distributed circuits, such as the subesophageal ganglion and ventral nerve cord, are capable of spontaneous rhythm generation as well as temporal patterning (for example, selecting right or left sides). The function of the more centralized network, such as the central brain, is to orchestrate the functioning of these networks by using sensory information obtained during exploratory activities within an econiche.

The fly *Drosophila* exhibits a diverse array of behaviors, including social courtship. The way that optogenetics is being used to manipulate behavior of freely flying *Drosophila* is illustrated by a study of courtship behavior. M. C. Wu et al. (2014) combine optogenetics with an automated multilaser tracking system that uses high-intensity laser irradiation to "zap" male flies so that they learn to keep away from virgin females. Two separate light wavelengths are used to target transfected neurons as a way of systematically identifying circuits involved in the social learning of avoidance behavior by the male fly. A technological challenge of using optogenetics in studies of mammals, such as mice, is getting light into the deep tissues of their larger brains. A newly emerging field—injectable, cellular-scale optoelectronics—provides one solution. Microscale, inorganic, light-emitting diodes, or μ-ILEDs, as well as electronic sensors and actuators are injected into deep tissue for stimulation, sensing, and actuation (T. I. Kim et al. 2013; McCall et al. 2013).

This organizational characteristic of the fruit fly nervous system has been demonstrated in experiments with "zombie" larvae. Berni et al. (2012) silenced the activity of the central brain and subesophageal ganglia while fruit fly larvae were foraging (that is, crawling, pausing, head sweeping, and turning). The pause and turn events are considered "decision making" points at which larvae explore available information sources—for example, olfactory and / or thermal gradients—and then select and execute a new movement direction (Berni et al. 2012; Lahiri et al. 2011). Even with only active thoracic and abdominal circuitry active, the larvae were able to crawl, pause, and make turns. This suggests that these distributed circuits generate locomotor rhythms, as well as the capability to pause and turn, and that role of the central brain is to use informational gradients obtained by sensory organs to guide the orientation toward or away from regions being explored and to modify the vigor with which the foraging behavior is performed.

"Flyception" is a *tour de force* method for imaging brain activity during courtship behavior in freely walking fruit flies (Grover, Katsuki, and Greenspan 2016). Grover et al. (2016) have made it possible to conduct simultaneous brain activity imaging in multiple unrestrained freely walking flies by creating a chronic imaging window on the dorsal side of the head of each fly. Odors were presented to flies expressing the calcium indicator GCaMP6s in olfactory projection neurons. A mirror system and multiple blue lasers target a fly head to excite the fluorescent proteins. Grover et al. (2016) report that when a male displayed persistent courtship behavior toward a female, neurons in a particular brain area, the dorsal posterior region, exhibited increased fluorescence. Courtship and mating in fruit flies consist of a sequence of intricate rituals, regulated by a particular gene, called fruitless (fru). The flyception method has been able to identify the specific neurons involved in the sequencing of ritual behaviors.

Distributed circuits organized for performing behavioral sequences are also evident in the escape flight behavior of mature (that is, winged) fruit flies (Card and Dickinson 2008a,b; Card 2012). Escape behavior of fruit flies consists of highly organized motor components used to jump in a direction away from a visually looming (approaching) event: leg repositioning, wing elevation, and leg extension. Prior to executing escape behavior, flies adjust their leg position during an early period in the looming event: to prepare them for escape before actually lifting off from the surface. In the context of courtship behavior of the male, fruit flies also perform a series of highly organized behaviors, including following the female and singing a species-specific courtship song (Coen et al. 2014).

Evidence that organized sequences of behavior are governed by distributed low-dimensional dynamics comes from brain-wide imaging of neuronal activity, using genetically encoded calcium indicators during locomotor action sequences by *C. elegans* (Kato et al. 2015; see also Izquierdo and Beer 2016 for a review). Kato et al. (2015) conducted whole-brain, single-cell resolution imaging (107–131 neurons) with a pan-neuronally expressed calcium indicator. *C. elegans* locomotor behavior is characterized by a sequence of "run" and "turn" states. Remarkably, Kato et al. (2015) found that neural state trajectories traced out a manifold—that is, a subvolume—in principal component analysis space, that took the shape of a cycle of the same successive states. Thus, many interneurons and motor neurons of those recorded during run-and-turn sequences generated a cyclical, low-dimensional, population-state, time-varying signal (Kato et al. 2015).

Bruno, Frost, and Humphries (2015) similarly examined how a distributed network may implement the locomotor behavior of the sea slug *Aplysia*. They used an isolated brain preparation to elicit fictive escape locomotion in the pedal ganglion (a rhythmic series of head reaches and muscular contraction cycles that flow down the body from head to tail), while simultaneously recording individual neurons (57–125 in each of twelve preparations) across the entire ganglion. Their results indicate that the locomotor program is assembled from a set of oscillator ensembles whose population activity reveals a rotation in low-dimensional dynamical space. They hypothesize a model for *Aplysia* locomotor control based on a cyclical attractor network in which "activity in each portion of the loop recruits motorneurons projecting to different muscle groups" (313). As we will see in Chapter 8, loops of activity in cortical attractor networks are characteristic in nonhuman primates as well as in small networks of the nervous systems of worm and sea slug.

Degeneracy

In multifunctional systems, the same components participate in different functional groupings through modulation (for example, participation of the same neurons in different functional groupings by means of modulation). Components, such as combinations of muscle groups, may rapidly switch membership to participate in one or another grouping through parameter settings that change the relative coupling strength of the module with one or another grouping. A profound consequence of flexibility in combining components for the design of devices is that it provides *degeneracy*: multiple solutions produce similar outputs (Marder and Taylor 2011). By providing more than one way to achieve different functions, it is possible for some to change while others maintain smooth operation of the rest of the system.

In addition to the work on the locomotor behavior of flies and worms, several decades of work on crustaceans has revealed multifunctional neural circuits for more complex sequences of feeding behavior (Marder and Bucher 2007). In lobsters and crabs, the stomach is a complex mechanical device that grinds and filters food. The stomatogastric ganglion (STG) is one of four ganglia comprising the stomatogastric nervous system (STNS). The STG controls movements of more than forty pairs of striated stomach muscles. Some of the muscles move a stomach structure, called the pylorus, while others move the gastric mill. Each of these structures has a characteristic rhythm, attributed to specific circuitry (Marder and Bucher 2007). Grinding and filtering functions are made possible by cooperative rhythmic movements of gastric mill teeth and pylorus. The faster pyloric rhythm has a single pacemaker neuron within the STG. By contrast, the gastric mill rhythm exhibits a number of patterns, depending upon the phase relationship of its muscles and the neurons innervating them. There are extensive interactions between pyloric and gastric neurons, and the activity of pyloric neurons may entrain and reset the gastric mill rhythm. As a consequence of these interactions, the gastric neurons are fundamentally multifunctional: they may switch between being members of the pyloric or gastric networks (Marder and Bucher 2007).

Spinal and Brainstem Circuits

Another personal perspective. As a postdoctoral fellow, I had the privilege of participating in a medical school neuroanatomy course, which included a human nervous system dissection laboratory. The course was taught by a very eminent MIT neuroanatomist, Walle Nauta, and we had weekly guest lectures by a veritable "Who's Who" of scholars in the field. One of our first dissection assignments was to examine spinal cord sections, relative to a famous set of archival slides from the Harvard library. I was as unprepared for the surprises ahead in dissection as I was for the nearly simultaneous birth of my daughter, Anna. When I held my newborn daughter for the first time, I marveled at her well-organized movements, as she tracked my face with her eyes. Then, in lab, I had the chance to hold in my hands a pencil-thin human spinal cord and couldn't imagine how this humble looking piece of tissue could be part of the system that generated Anna's already well-organized movements. The focus of my research since then has been on infant sensorimotor behavior, but I have never lost my fascination for how spinal circuits work.

Intensive study of spinal circuits. In the more than two decades since those unforgettable events in my personal life, neuroscientists have developed new molecular and genetic techniques (Goulding 2009) that have promoted many astonishing advances in the understanding of spinal cord structure and function, including the identity of the interneurons that are the source of excitation for generating locomotor rhythms (Arber 2012; Eklof-Ljunggren et al. 2012); the connectivity of microcircuitry for generating patterns of activity, even in the absence of descending sensory inputs (Bagnall and McLean 2014); the role of descending signals for selective recruitment of motor neurons (Wang and McLean 2014); and the contribution of ascending circuits in motor control (Azim et al. 2014). Consider the intrinsic rhythmic patterns generated by the spinal cord. The discovery that intrinsic spinal cord networks generate rhythmic patterns of locomotor activity (Buschges 2012), swimming in vertebrates (Ekloff-Ljunggren et al. 2012), and gut movements in crustaceans (Marder and Bucher 2007) has led to the widely adopted concept of a spinal "central pattern generator" (CPG). How do spinal networks generate intrinsic clock-like activity patterns? Goulding (2009) identifies three fundamental characteristics of the putative CPG: (1) motor neurons are grouped into discrete functional units, called "motor pools," each innervating a single muscle; (2) the graded recruitment and activation of motor neurons within a pool is the basis for variable changes in muscle properties that are necessary for postural control and movement; and (3) fast synaptic and slower modulatory interactions between locomotor interneurons "sculpt" the convergent inputs to motor neurons in order to pattern muscle activations for smooth movements.

The "holy grail" for many neuroscientists studying spinal CPG circuits is to be able to record circuit activity as well as the behavior of the animal. One way to do this is by preparing an isolated spinal cord and recording its activity during what is called "fictive" locomotor activity. An example is the work of Iwasaki and Chen (2014) on the CPG and other structures underlying rhythmic swimming behavior of the leech, an invertebrate. The leech swims by undulation of its segmented body, propagating traveling waves from head to tail. The isolated nerve cord of the leech exhibits fictive swimming, which closely resembles swimming in intact animals (Iwasaki and Chen 2014). A different approach to revealing the relation between spinal circuitry and behavior is illustrated by research on fictive swimming in another swimming animal, the larval zebrafish (Ahrens et al. 2012; Portugues et al. 2013; Portugues et al. 2014; Portugues et al. 2015). In this approach, rather than remove the cord, the intact zebrafish larva is paralyzed and placed in a virtual reality environment, while the activity of large populations of neurons throughout the nervous system is recorded. For example, Ahrens et al. (2012) used a virtual reality environment to present the zebrafish with an enveloping virtual water flow that simulated swimming backward. To test sensory adaptation to optic flow at different velocities, a genetically encoded calcium indicator was used to image a portion of the brain in individual fish and a composite of the entire brain from multiple fish.

New technologies for *in vivo* extracellular recording combined with advances in light sources (for example, light-emitting diodes) have made it possible to develop wireless micro-LED devices that are incorporated with electro-physiological sensors (Jennings and Stuber 2014). These dual function systems now permit the control and monitoring of circuit activity during complex behavioral tasks. For example, Jin, Tecuapetla, and Costa (2014) combined *in vivo* electrophysiology with optogenetic identification to examine basal ganglia circuitry as mice learned to perform a rapid sequence of locomotion, grooming, and bar pressing. They were able to confirm a long-standing model of the role of basal ganglia for organizing functionally specific activities into organized sequences (Graybiel 2008) (see later discussion as well). On the technological horizon are combination electrical-optical devices for *in vivo* closed-loop optogenetic control of behavior using fully implantable wireless devices for optogenetic stimulation (Grosenick, Marshel, and Deisseroth 2015).

Rhythm and pattern. Spinal networks support two basic functions: rhythm generation (clocking) and pattern generation (rhythmic activation of motor neurons, left-right alternation, and for limbed animals with multiple joints flexor-extensor coordination) (Kiehn 2011, 2016). On the basis of the expression pattern of genetic transcription factors, it has been possible to identify the major classes of interneurons comprising the pattern generating region of the spinal cord: dorsal (dl1–dl6) and ventral (V0-V3 and HB9) interneurons and spinal motor neurons (Arber 2012; Kiehn 2011). Talapalar et al. (2013) examined the role of V0 interneurons for limb alternation at different locomotor speeds in wild-type compared to genetically altered mice. The genetic alteration involved a procedure of selectively breeding the mice so that (1) it was possible to maintain their viability after ablating V0 interneurons and (2) it was possible to manipulate excitatory or inhibitory V0 interneurons (see Menelaou and McLean 2012 for more on the procedure). Intact, wild-type mice walk with an alternating gait at low (approximately 2-Hz) locomotor frequencies and trot or gallop at frequencies greater than 10 Hz. In an initial experiment in which the entire population of V0 neurons was ablated, the genetically altered mice consistently exhibited a symmetric hopping gait. Then, in additional experiments, Talpalar et al. (2013) selectively ablated either the excitatory or inhibitory V0 interneurons. When the excitatory interneuron was targeted, left–right alternation was present at low speeds, and hopping was apparent only at medium to high locomotor frequencies. By contrast, selective ablation of inhibitory V0 interneurons resulted in absence of left–right alternation patterns at low locomotor frequencies, mixed coordination at medium frequencies, and alternation at high frequencies. Thus, one functional subpopulation active at slow speeds engages cross inhibition, while a second subpopulation active at fast speeds engages crossed excitation (Talpalar et al. 2013). Zhang et al. (2014) have found that in mice, two types of ventral interneuron, V1 and V2b, are also part of the core elements of a distributed inhibitory spinal network involved in the reciprocal pattern of flexor–extensor motor activity that mice use for locomotion. V1 and V2b gate excitatory inputs to motor neurons. Zhang et al. (2014) propose that V1 and V2b inhibition arose as a result of the evolutionary organization of the rhythm generating circuit, necessary for the emergence of movable bilateral appendages (see Table 5.2).

In addition to the essential role of spinal interneurons for rhythm and pattern generation, Arber (2012) highlights two characteristics of the spinal cord circuits involved in the production and regulation of mammalian motor behavior: ascending and descending communication with supraspinal centers in the brainstem and higher areas and sensory feedback systems constantly monitoring the consequences of

Table 5.2 Spinal network functions

Function	Description
Rhythm and pattern generation	Rhythm generation is the clocking function. Pattern generation has three functions: rhythmic activation of motoneurons, left-right alternation, and (in animals with multiple joints) flexor-extensor patterning.
Function of interneurons	There are five major subclasses of interneurons, called V0, V1, V2, V3, and Hb9. V2 interneurons are an intrinsic source of excitation. Almost all V0 and most V3 are commissural, crossing the midline to coordinate activity on the two sides of the body.
Organization of circuits into functional subcircuits	On each side of the spinal cord, individual neural networks exist that can generate rhythmic motor activity for the motoneuron pools independently.
Recruitment	In legged vertebrates, the configuration of the V0 left-right alternating network changes with speed. Inhibitory V0 neurons are recruited first, during slow locomotion, followed by excitatory V0 neurons recruited at higher frequencies. Similarly, in larval zebrafish, there is a topographic recruitment order for dorsoventrally arranged spinal premotor interneurons and motoneurons during changes in swimming speed: ventrally located motoneurons are active at lower swim frequencies; more dorsally positioned motoneurons are engaged at progressively higher swimming frequencies.

Source: O. Kiehn (2016), Decoding the organization of spinal circuits that control locomotion, *Nature Reviews Neuroscience, 17,* 224–238.

motor action. With regard to the former, it has been known for at least fifty years that there are both direct corticomotoneuronal and indirect lateral descending brain stem pathways to the spinal cord, mediating the use of the hand for reaching and grasping (Alstermark and Isa 2012; Zhou, Wolpert, and De Zeeuw 2014). An unanswered question, however, is the role of brain stem–spinal connectivity that makes forelimbs more capable than hindlimbs for fine grasping. To address this question, Esposito, Capelli, and Arber (2014) examined the role of the medullary reticular formation ventral part (MdV) of the brain stem in control of forelimb use in mice. MdV neurons receive input from the motor cortex and cerebellum and, in turn, project directly to forelimb-specific spinal motor neurons and interneurons (Zhou et al. 2014). They adopted a research strategy with mice combining virus tracing, genetic manipulations, and behavioral studies to determine the role of the brain stem in use of the hands for fine motor behaviors. To assess the role of MdV in motor tasks, Esposito et al. (2014) used diphtheria toxin injections to ablate the MdV neurons and compared the performance of these and control mice reaching for small food pellets. Compared to controls, the mice with ablated MdV neurons exhibited specific defects in fine paw placement and closure of the fingers. Esposito et al. (2014) conclude that forelimb-dominated brain stem nuclei are involved in control of complex sequences of muscle contractions, providing access to specific spinal motor neuron pools for achieving coordination.

Motor neurons and segmental interneurons receive input from cervical propriospinal neurons (PNs), which are intermediary relays for descending motor signals during reaching behavior (Azim et al. 2014; Azim and Alstermark 2015; Zhou et al. 2014). PN circuits send out two outputs: one axon branch projects to forelimb-innervating motor neurons, and the other projects to the lateral reticular nucleus (LRN), a pre-cerebellar relay. Azim et al. (2014) explored two genetic strategies of perturbing V2a PN function: acute ablation of V2a neurons, eliminating both axon projections, and selective optogenetic stimulation of just the pre-cerebellar projection. For the latter, Azim et al. (2014) expressed channelrhodopsin (ChR2) in PNs via viral injection to selectively activate the pre-cerebellar axon projection in order to perturb forelimb movement in a group of mice trained to reach for a food pellet and compared their reaching behavior with a control group of wild-type mice. In the ChR2 expressed group, application of light pulses resulted in the disruption of forelimb reaching movements, indicating that the pre-cerebellar projection of PN circuits were blocked from providing continuous updating of reach trajectories.

Proprioceptive sensory neurons innervate sense organs in the muscle and transmit information about muscle contraction to the spinal cord. Muscle spindle afferents are a subset of proprioceptors contacting muscle spindle sense organs. They establish synaptic contacts with motor neurons and various classes of interneurons and, therefore, are ideally suited to convey direct excitation to spinal circuits relevant to regulating motor behavior (Takeoka et al. 2014). Absence of proprioceptive sensory feedback degrades mouse locomotor patterns (Akay et al. 2014). Similarly, experimentally inactivated spinal interneurons involved in cutaneous sensory feedback during grasping disrupt modulation of grip strength in response to increasing load (Bui et al. 2013).

Cerebellar Function

The well-characterized microcircuitry of the cerebellar cortex consists of a tri-laminar structure composed of molecular, Purkinje cell (PC), and granular layers (Kano and Watanabe 2013). PCs, the sole output neurons of the cerebellar cortex, project GABAergic axons to deep cerebellar nuclei (DCN) and vestibular nuclei; extend well-arborized dendrites in the molecular layer; and are innervated by climbing fibers originating from the inferior olive of the contralateral medulla oblongata and by mossy fibers originating from spinal cord, pontine nuclei, and reticular formation (Kano and Watanabe 2013). PCs are unusual because they generate two distinctive types of action potential: *simple spikes,* which fire spontaneously or by activation of the mossy fiber-granule cell-parallel fiber pathway at high rates (30–100 Hz), and *complex spikes,* approximately 1-Hz spikes consisting of an initial action potential usually followed by a series of smaller spikelets (Cerminara et al. 2015). The cerebellum is also characterized by rich variations in the expression of proteins, including bands of Purkinje cells expressing Zebrin II alternating with Purkinje cells that are negative for this protein (Witter and De Zeeuw 2015). At a more detailed level, these zebrin zones may be further divided into microzones, characterized by a particular functional response of climbing fibers firing in synchrony (De Zeeuw et al. 2011). Thus, from cytoarchitectural and protein data, there is growing evidence that the cerebellum is organized into functional zones connected to other brain areas in a zone-specific manner (Witter and De Zeeuw 2015). But what do these microzones do?

De Zeeuw and ten Brinke (2015) propose that cerebellar microzones have characteristic temporal spiking frequency domains that are specialized for particular tasks, including execution of limb and finger movements, trunk movements for balance, compensatory eye movements about particular spatial axes, reflexes of the facial musculature, homeostasis of particular autonomic processes, and time-sensitive decision making. The signature feature of all functions of the cerebellum is its ability to control timing at a high resolution, a precision of about 5 msec across periods of hundreds of milliseconds, crucial for the plasticity underlying motor learning. An illustration of motor learning in a zebrin-positive zone is control of the amplitude (that is, gain) of the vestibulo-ocular reflex (VOR), a reflex movement of the eyes elicited by vestibular stimulation in which the eyes move in a direction opposite to that of the head to ensure that the retinal image is kept stable (Gao, van Beugen, and De Zeeuw 2012). This is called phase reversal adaptation and is established by moving the mouse's head in phase with (that is, in the same direction as) vestibular stimulation, but with greater amplitude. A period of "mismatch training," during which retinal slip reverses direction, forces the mouse to make compensatory eye movements during vestibular stimulation that are opposite in direction to those before training. In mutant mice lacking a certain receptor, the NR2A subunit of NMDA receptors, and reduced climbing fiber plasticity, there is a reduced capacity for eye movement compensation, confirming the role of the cerebellar cortex in this form of learning.

Basal Ganglia and Motor Programs: Selection

Motor programs have been defined in many ways, but a classic neuroscience perspective is that they are neuronal networks used to generate functionally specific patterns of behavior, including locomotion, posture, eye movements, breathing, chewing, swallowing, and expression of emotions (Grillner et al. 2005). In vertebrates, selection of one motor program (for example, locomotion) from among others (for example, feeding or orienting the eyes and head) involves basal ganglia networks that are believed to inhibit the circuits underlying one program and facilitate others. The basal ganglia (BG) are seven deep brain nuclei that modulate motor output (Chakravarthy, Joseph, and Bapi 2010). BG circuitry is present in the phylogenetically oldest vertebrates, such as lamprey (Grillner et al. 2008; Sarvestani et al. 2013; Stephenson-Jones et al. 2011). The selected program is initiated by brainstem networks, and circuits of the central spinal network then generate intrinsic patterning of muscle activations to produce movements.

Among the many challenges for this conception of motor program is to understand how the coordination between circuits allows *simultaneous behaviors to occur without interfering with each other.* For example, in rodents, *orofacial behaviors*—such as sniffing, chewing, licking, swallowing, and whisking—may occur in parallel, and brainstem circuits may prevent these activities from interfering with breathing (Moore, Kleinfeld, and Wang 2014). What kind of control structure may make this possible? One model posits that *higher-frequency brain stem pattern–generating circuits* for these orofacial behaviors are *modulated at a slower time scale* by multiple intrinsic (for example, cortical) sources converging on brainstem motoneurons, well as extrinsic sources of sensory input (Moore et al. 2014). During exploratory whisking, for example, the cycles of slow time-scale sensory input from the vibrissa informs the modulation of brain stem–mediated fast time-scale cycles of vibrissa contact with objects (Nguyen and Kleinfeld 2005). The intrinsic modulation of brain stem rhythm generation apparently decreases the vibrissa velocity to increase the duration of its contact with the object and enhance the active obtaining of information (Grant et al. 2009).

The approach taken in this book is that perceiving and acting are *embodied* so that the process of selecting and initiating particular behaviors must be considered in the context of potential action opportunities directly specified by the environment—the Gibsonian "affordances" introduced in Chapter 3. As we consider devices that work with the body for shared control of a flow of actions in an ever-changing environment, the question of how to embody motor programs as part of a shared cyberphysical system becomes crucial. Embodied decision making is placed in the context of attractor dynamics, in which goals are attractors and the environment may be defined as a potential field with regions of attraction and repulsion and bifurcations (Cisek and Pastor-Bernier 2014). Here, in attractor dynamics, we begin to have a language for shared decision making for modes of action by both the nervous system and devices. That language is more fully discussed in Part III of the book. Still missing in this chapter, though, is the relation between the nervous system and the forces acting on the body.

Neuromechanics: Embodied Nervous Systems in Insects, Invertebrates, and Small Vertebrates

Chapter 3 discussed how scaling laws reveal a fundamental relationship between body size and brain volume. Nested within that scaling relationship are other lawful relations such as how the body's mechanical properties influence an animal's behavior in a gravitational field, the consequences of increasing nervous system complexity and architecture for time delays in conduction velocity, and the role of sensory information in amplifying fluctuations (Tytell, Holmes, and Cohen 2011). This is apparent in locomotion by animals in water, on land, and in the air. Consider a fish, such as the anguilliform (that is, eel-like) lamprey *Icthyomyzon unicuspis,* hunting for a meal while swimming in a lake. Observed curvature patterns in swimming *behavior* are the combined effects of cephalocaudal waves of *neural activity* that generate muscular forces; the *mechanical properties of the body* (for example, stiffness); and the *fluid dynamic forces acting on the body* to produce a net body curvature, unfolding over time (Tytell et al. 2010). The lamprey uses arrays of mechanoreceptive organs to modulate neural activity under changing fluid flow conditions, such as the direction and magnitude of water currents and turbulence in eddies, or the detection of chemical gradients and vibrations indicating the proximity of prey (Lauder 2015). Several fundamental questions are apparent concerning the combined influences of neural and fluid dynamic forces on behavior. How do neural circuits generate the muscle activation patterns underlying changes in body curvature? What is the underlying information structure of the flow pattern that modulates the timing of muscle activations? What is the nature of the coupling between brain networks and the mechanical forces contributing to body motion?

These types of questions about the emergence of behavioral patterns from interactions between muscle forces, inertial, elastic and damping properties of the body and the reactive forces of the environment (for example, flowing water) on the body all fall within the domain of a branch of integrative biology called *neuromechanics* (Holmes et al. 2006; Lauder 2015; Tytell et al. 2011). Neuromechanics takes as a starting point a model, a relatively abstract "template" of the neural and mechanical components, and gradually builds toward more realistic components that are "anchored" to biological systems (Full and Koditschek 1999). For example, in insect locomotion (Holmes et al. 2006; Kukillaya and Holmes 2009; Proctor, Kukillaya, and Holmes 2010), the model of the neuromechanical system includes (1) a body plan, a bilaterally symmetric arrangement of six jointed limbs, a head at one end, and a tail at the other; (2) a feedforward mechanical system in which the joints of each leg are connected to passive linear springs; (3) muscles (Hill-type) arranged into agonist-antagonist pairs; (4) a proprioceptive circuit for reflex feedback of joint torques, arranged in various orientations to render them sensitive to force direction; and (5) a central feedforward pattern-generating network consisting of "bursting" interneuron-like units. When compared with actual insects, such as the cockroach—that is, anchoring the model to the biology of the insect—the model generates behavior similar to that of its natural counterpart, including the ability to steer itself and remain stable when subjected to perturbations (see Table 5.3).

In a more general version of neuromechanical models, the nervous system, body, and environment are part of a system of nested feedforward (mechanical, or preflexive) loops and feedforward (proprioceptive or reflexive) loops (Holmes et al. 2006; Miller et al. 2012). Unlike models that treat central pattern generators as

Table 5.3 Characteristics of neuromechanical systems

Characteristic	Description
Multifunctionality and self-stabilization of muscle	Muscles are multifunctional: they shorten to perform work (i.e., actuate), stabilize motion at joints, and store elastic energy in connective tissue (e.g., tendons). A single neural signal may produce variable mechanical outputs. Muscles are self-stabilizing due to their force-velocity, force-length, and visco-elastic properties.
Sensors are nonlinear and have intrinsic time delays	Sensors within muscles that monitor fundamental mechanical variables, e.g., Golgi tendon organs and muscle spindles, have nonlinear properties. These affect the timing and dynamics of receptor input. These sensors also have lags in their response, e.g., due to damping and inertia, and add to delays in neural processing times.
Control consequences of time delays	In forced oscillatory systems, feedback delay increases system gain, giving it resonant-like behavior. These resonances may be tied to exact delays in the system and reduce the number of solutions to explore in the solution space for control.
Exploration of solution space by trial and error	Neuromechanical systems may effectively accomplish a task by multiple methods.
Emergent behavior	Behavioral dynamics emerges from the complex interplay of loops of brain activity, sensory anticipatory tuning and modulation, muscle properties, body biomechanics, and environmental forces acting on the body.

Source: K. Nishikawa, A. Biewener, P. Aerts, A. Ahn, H. Chiel, M. Daley, T. Daniel, R. Full, M. Hale, T. Hedrick, A. K. Lappin, T. R. Nichols, R. Quinn, R. Satterlie, & B. Szymik (2007), Neuromechanics: An integrative approach for understanding motor control, *Integrative and Comparative Biology, 47,* 16–54.

"prescriptive" (that is, the sole basis for oscillation frequency of rhythmic behaviors, such as locomotion) (see, for example, Buschges 2012; Sarvestani et al. 2013), *neuromechanical models predict that central pattern generators contribute to behavior by matching to (entraining with) the resonance frequency of the body's oscillatory motions*. For example, in lamprey swimming, the speed of the body's mechanical waves through the water does not match waves of neural activation: there is a phase lag between muscle activity and body curvature (Tytell et al. 2010). This means that by simply measuring muscle activations without measuring the relation of these activations to behavior misses an essential aspect of the relation among nervous system, body, and behavior.

A distinguishing characteristic of neuromechanical models is that they are self-organizing systems that spontaneously generate emergent behaviors by means of distributed parallel loops of energy and information flow, occurring at multiple scales of time (Holmes et al. 2006). The self-organizing character of neuromechanical systems generates several testable predictions about the emergence of behavior at multiple time scales. During interactions of the body with the environment at the briefest time scales, such as during the wing beat of flying insects, patterns of locomotion self-organize via the mechanical interactions of the body and environmental media (for example, air during flying). For example, insect wings must flip over at the end of each wingbeat as the wing reverses direction because the lift force depends sensitively on the timing (Tytell et al. 2010). Passive interactions of the wing architecture with the air, rather than a neural signal, drives this rotation (Sane and Dickinson 2002). Mechanical properties of the wing, such as its pitch, may additionally be actively tuned by small muscle activations that harness these passive interactions for turning the body—for example, in fruit flies (Bergou et al. 2010). Together, both active and passive mechanical properties contribute to the emergence of turning behavior during flight.

Crawling in the caterpillar *Manduca sexta* is a curious gait pattern: it combines posterior segments that pivot around attached claspers with anterior segments coupled in phase to store elastic energy (Trimmer and Issberner 2007). Thus, caterpillar crawling is not simply forward progression due to a peristaltic wave, but rather involves the exchange of kinetic energy between segments, similar to locomotion in other terrestrial animals that make contact with rigid surfaces. This is quite surprising for a soft-bodied animal such as a caterpillar because it lacks a stiff skeleton. When *Manduca* uses its proleg claspers to lock onto a substrate, such as a twig, those body segments are able to counteract the forces of muscle contraction by other segments so that the substrate becomes an "environmental skeleton" (Lin and Trimmer 2010). But what drives the body forward? Simon et al. (2010) used phase-contrast synchrotron X-ray imaging and transmission light microscopy to directly visualize internal soft tissue movements in freely crawling *Manduca*. Incredibly, they found that the gut, which is able to slide through the surrounding body wall, moves out of phase with the body wall but moves simultaneously with forward movement of the terminal prolegs. In other words, *Manduca* crawls by means of the motion of an internal (visceral) "piston" that drives the prolegs forward (Simon et al. 2010).

The octopus is a soft-bodied cephalopod distinguished by the remarkable range of behaviors possible with its eight arms, including locomotion, catching prey, reaching, grasping fetching, probing the environment, digging to collect stones, and more (Hochner 2008). The anatomy of each arm consists of incompressible fluids and tissue, making each one a "hydrostatic skeleton" in which force is transmitted by internal pressure (Kier 2012). How is it possible for arms with no internal skeletal elements to be transformed into a multifunctional structure to perform these behaviors? We saw in *C. elegans* that wave propagation creates body curvatures that are used for propulsion. Nature has found a similar wave propagation solution for transforming the octopus limb into a quasi-articulated structure. Recording of electromyography (EMG) at various locations along the arm of an octopus during grasping reveals that there are two waves of muscle activation that propagate toward each other. One propagates from the target toward the base of the arm, while the other propagates from the base of the arm toward the target. A virtual joint is formed where the two waves collide (Sumbre et al. 2005, 2006)!

Evolution of Brain Structures

Basal ganglia. How is the action selection behavior of a mammal, such as a rodent, different from that of a lamprey? The main basal ganglia (BG) input structure is the striatum, a massive system connecting cortical neurons to the BG and thalamus (Fee 2014). The excitatory cortical signals entering the BG spread through direct and indirect pathways (Hwang 2013). BG output nuclei send inhibitory projections to thalamic nuclei, which then send excitatory projections back to primarily the same cortical areas from which the cortico-striatal inputs originated. The flow of excitatory and inhibitory influences between BG and motor cortex may be the basis for prediction of future rewards: comparison of potential actions and their rewards in a given context (Chakravarthy et al. 2010; see also new anatomical findings by Saunders et al. 2015). How is the behavior of a rodent different from that of a human? Compared to most human behavior, behavior of rodents is predominantly habitual, characterized by complex repetitive acts that emerge in ordered structured sequences, are prone to being elicited by a particular context, and can continue to completion without constant conscious oversight (Graybiel 2008).

Humans under stress, or with certain neurological conditions, may act habitually in a manner similar to rodents (Graybiel 2008). However, healthy, rested humans are able to switch from a habitual mode to one that is evaluative. The resources and structure of the foraging environment, as well as the needs of the animal or human (for example, for nutrition) relative to the costs of action, constitute a sensorimotor space of "affordances" (Gibson 1986)—that is, opportunities for action in relation to both the layout of the environment and the relative value of potential pursuit activities (Goldfield 1995; Warren 2006). Humans not only explore for affordances, but are able to go beyond the available information to invent new ways to act, create new environments for action, and design and build new tools and devices to extend our capabilities and the physical limitations of our bodies.

Cerebral cortex. A significant evolutionary advance enabling nonhuman primates and humans to go beyond the capabilities of rodents is the growth of particular regions of the cerebral cortex. Given the conservative nature of evolution, how has that evolution occurred? As I discuss in greater detail in Chapter 6 on development of the human nervous system, the cerebral cortex is assembled via a process of proliferation of certain neural progenitor cell populations, which then migrate to form the characteristic cortical layers. The cells that are both progenitors and migratory guides for the developing cortex are called *radial glia* (Rakic 2009; Bystron et al. 2008). It is changes to the division of radial glia, the length of their cell cycle or rate of proliferation, that influence the size, composition, and functional repertoire of the neocortex (Lui, Hansen, and Kriegstein 2011b).

Emergence of Structure and Function in Larger Nervous Systems

There have been major advances in the past decade in our understanding of the contribution of structures at the microscopic scale, dendritic spines, to the emergence of behavioral changes with learning (Muller and Nikonenko 2013; Yuste 2011). Spine morphology, growth, and retraction is studied with devices that include electron microscopy, specialized high-resolution light microscopy, and two-photon calcium imaging (Araya, Vogels, and Yuste 2014; Bosch and Hayashi 2012). Spines make contact with as many different axons as possible, and regulation of spinal electrical compartmentalization is hypothesized as the basis for integrated and precise control of synaptic strength (Yuste 2011) (see Chapter 7). Inputs are integrated independently, and linearly, in a "great synaptic democracy" (Yuste 2011, 2013).

A powerful method for revealing the emergent properties of neural circuits is through modeling by means of artificial neural networks (ANN), such as recurrent networks (Mante et al. 2013; Sussillo 2014; Sussillo and Barak 2013; Sussillo et al. 2015). A recurrent neural network is a type of ANN with feedback connections (Sussillo and Barak 2014). The recurrent network consists of pyramidal neurons connected to themselves through recurrent axons with changing synaptic weights (Yuste 2015; see Chapter 7). When the network receives a set of external inputs and generates an output, the activity becomes attracted to particular stable states (Yuste 2015). How might an animal's behavior be related to the emergent states of brain networks? An intensely studied behavior, movement planning during spatial navigation, appears to emerge from the dynamics of recurrent networks in the posterior parietal cortex. Harvey, Coen, and Tank (2012) used a virtual reality system to project a T-maze within which mice were trained to turn left or right at the T-intersection to receive a water reward. The mice ran on a spherical treadmill linked to the projection of the virtual T-maze, and there was repeated recording of spatiotemporal dynamics of cells in the posterior parietal cortex, measured by means of two-photon calcium imaging. When the multineural activity is condensed into three-dimensional plots of the principal component axes, the temporal sequences of firing were predictive of the behavioral choice at the T-intersection.

Efforts to understand the long-range and local connectivity of circuits shifts the scale of investigation to mesoscopic cell assemblies (Buzsaki, Anastassiou, and Koch 2012). Cell assemblies are "coalitions of neurons" that bring together sufficient numbers of peer neurons so that their collective spiking discharges a postsynaptic neuron (Buzsaki and Wang 2012). Cell assemblies may be examined under light microscope using neural tracers—for example, genetically encoded fluorescent proteins introduced via injected viruses (Kim, Chung, and Deisseroth 2013). At the macroscale, long range, region-to-region connectivity is based upon neuroimaging visualization of distinct brain regions, white matter fiber tracts, and correlated activity patterns in the living brain (Buckner, Krienen, and Yeo 2013). Extracellular measures of electrical activity, including electroencephalography (EEG), magnetoencephalography (MEG), electrocorticography (ECoG), local field potential (LFP), and voltage-sensitive dye imaging (VSDI) are (additionally) used to identify scalp, subdural, or deep brain electrical activity oscillation patterns from which dynamical systems measures may be derived (see, for example, Deco, Jirsa, and Macintosh 2011).

Each recording method has its advantages and disadvantages. Starting with the most invasive, VSDI, as I have already discussed in small animals, membrane-bound voltage-sensitive dyes or genetically expressed voltage-sensitive proteins are used along with high-resolution, fast digital cameras to optically detect neuronal voltage changes. The advantage of VSDI is that it directly measures localized transmembrane voltage changes, rather than extracellular potential (Buzsaki et al. 2012). LFP uses electrodes or silicon probes into deep brain locations to record a wide-band signal, containing both action potentials and other membrane potential-derived fluctuations in a small neuronal volume. A disadvantage is that many observation points, with short distances between recording sites, are required to achieve high spatial resolution (Buzsaki et al. 2012). ECoG uses subdural electrodes to record electrical activity directly from the surface of the cerebral cortex, bypassing the signal-distorting skull and intermediate tissue. The least invasive methods are EEG and MEG. EEG uses arrays of electrodes on the scalp surface to provide a spatiotemporally smoothed version of the LFP, integrated over an area of 10 cm^2 or more (Buzsaki et al. 2012). MEG uses superconducting quantum interference devices (SQUIDs) to measure minute magnetic fields outside the skull from neuron-generated currents. MEG has a higher spatiotemporal resolution (1 ms and 2–3 mm) than EEG, and its magnetic signals are much less dependent than EEG on the conductivity of the extracellular space, and thus, contain less distortion (Buzsaki et al. 2012) (see Table 5.4).

A major methodological challenge for large-scale projects that seek an understanding of brain structural connectivity across scales is to create models of the intact brain that also capture its cellular structure (Devor et al. 2013; Kandel et al. 2013). A sense of the magnitude of this task becomes clear by comparing the complexity of the mouse brain and the human brain: the volume of the human cerebral cortex is about 7500 times larger, and the amount of human white matter is 53,000 times greater (Amunts et al. 2013). One approach to bridging the gap of brain structural complexity across scales, called BigBrain, first used a large-scale microtome to obtain 7400 sections, each 20-μm thick, then stained them for cell bodies and digitized the

Table 5.4 Methods used to record extracellular events

Method	Description
Electroencephalography (EEG)	Scalp electroencephalogram, recorded by a single electrode, is a spatiotemporally smoothed version of the local field potential (LFP, see below), integrated over an area of 10 cm^2 or more.
Magnetoencephalography (MEG)	Uses superconducting quantum interference devices (SQUIDs) to measure minute magnetic fields outside the skull (typically in the 10–1000 fT range) from neuron-generated currents. MEG is noninvasive and has a relatively high spatiotemporal resolution (1 ms and 2–3 mm). An advantage of MEG over EEG is that magnetic signals are much less dependent on the conductivity of the extracellular space and show less distortion.
Electrocorticography (ECoG)	Uses subdural electrodes to record electrical activity directly from the surface of the cerebral cortex, bypassing the signal-distorting skull and intermediate tissue. The spatial resolution of the recorded electric field may be substantially improved by using flexible, closely spaced electrodes.
Local field potential (LFP)	Uses electrodes or silicon probes into deeper brain locations to record a wide-band signal, containing both action potentials and other membrane potential-derived fluctuations in a small neuronal volume. Many observation points, with short distances between the recording sites, are needed to achieve high spatial resolution.
Voltage-sensitive dye imaging (VSDI)	Membrane-bound voltage-sensitive dyes or genetically expressed voltage-sensitive proteins may be used to optically detect neuronal voltage changes. High-resolution, fast-speed digital cameras are used. A major advantage of VSDI is that it directly measures localized transmembrane voltage changes rather than extracellular potential.

Source: G. Buzsaki, C. A. Anastassiou, and C. Koch (2012), The origin of extracellular fields and currents—EEG, ECoG, LFP, and spikes, *Nature Reviews Neuroscience, 13*, 407–420.

histological images to create an ultra-high-resolution, three-dimensional human brain model (Amunts et al. 2013). Through this combination of staining and reconstruction of an intact brain, the Big-Brain model takes a first step at integrating human brain structural complexity across scales. But the process of aligning thousands of images of slices may take thousands of hours of work and may be prone to errors.

Modeling Connectome Topology: Graphs

One approach toward making vast imaging datasets more understandable is to characterize the data topologically by means of graphs. A brain graph is a topological model that simplifies the structure and function of nervous system into nodes (that is, separable units) connected by a set of edges (such as synapses) (Bullmore and Bassett 2011). As a topological object, a brain graph is able to capture the connectivity of edges between the nodes, regardless of their physical or anatomical location, even when the surfaces are transformed by growth. Brain graphs have revealed certain global topological properties of the nervous system, including integrated and segregated communities (with a high density of connectivity among members of the same community and a low connectivity density among members of different communities), hubs (highly connected and highly central brain regions), and rich clubs (sets of highly connected and highly central nodes that integrate information across segregated communities and networks for enabling global communications) (Sporns 2012).

Brain graphs have also revealed that brain networks exhibit "small-worldness," in which "all nodes of a large system are linked through relatively few intermediate steps, despite the fact that most nodes maintain only a few direct connections—mostly within a clique of neighbours" (Bullmore and Sporns 2009, 189). The small world topology, thus, negotiates a trade-off between connection distance and topological efficiency (Bullmore and Sporns 2012). These topological features of brain networks—communities and hubs—support the hypothesis that the brain regulates information flow by balancing the integration and segregation of information flow. What, though, are the dynamics that underlie these topological properties?

Oscillation Frequency Bands and the Formation of Temporary Functional Ensembles

The physical architecture of neural circuits—their synaptic trees and axons—generate loops of electrical activity, which provide a structure for coordinating spiking patterns within and across circuits at multiple time scales (Buzsaki and Draghuns 2004). Despite considerable variation in brain size, many vertebrate species exhibit the same bands of frequency oscillation. Indeed, nature appears to regulate axon caliber size in order to achieve particular conduction velocities and a size-invariant parsing of time (Buzsaki, Logothetis, and Singer 2013). Neuronal temporal organization may be conserved in this way for several reasons: (1) oscillation is the most efficient mechanism for achieving synchrony and is the basis for information transfer and binding of functional networks; (2) plasticity is dependent on spike timing and operates within limited time windows, so it is critical that the timing of pre- and postsynaptic neurons are activated in a similar time window, regardless of the spatial distances of their cell bodies; (3) there is a need to maintain membrane properties because changes may alter time constants and resonance properties of neurons and microcircuits; and (4) temporal coordination may be jeopardized by abnormalities that alter path lengths or conduction velocity (Buzsáki 2006; Buzsaki, Logothetis, and Singer 2013).

Networks of neuronal circuits exhibit bands of oscillation covering the frequency range from approximately 0.05 Hz to 500 Hz (Buzsaki 2006). Each of the bands—slow, alpha, beta, theta, gamma, ripple, and ultraslow—has a characteristic range and well-defined functions. Thalamocortical slow oscillations (0.7–2.0 Hz) are most prominent during sleep. They reflect alternating phases of synchronous cortical neuron depolarization and spiking (called the "up state") and hyperpolarization (or "down state" of silence). Transition from down to up states may trigger a reverberating 12- to 18-Hz sleep spindle. Thalamocortical alpha occurs in a waking, relaxed brain state. For example, one type of alpha is an 8- to 12-Hz rhythm that emerges in occipital neocortex after eye closure. A second type, Rolandic μ (8–20 Hz) occurs in the somatosensory system during immobility but disappears upon actual or imagined movement (Buzsaki, Logothetis, and Singer 2013). Beta oscillations are in the 12- to 30-Hz range. They are most pronounced in motor cortex, basal ganglia, and cerebellum; reflect a disengagement of the motor system from action; and appear to coordinate timing of neurons in widespread areas (Buzsaki, Logothetis, and Singer 2013). Theta rhythms (4–12 Hz) are strongly involved in cross-frequency interaction in both hippocampus and cortex, by modulating the firing rate and spike timing of individual hippocampal or cortical neurons. Gamma oscillations occur in the 30- to 90-Hz range. They are ubiquitous in all structures and brain states and emerge from networks of reciprocally coupled pyramidal cells and parvalbumin-containing inhibitory basket cells (Roux et al. 2014). Transiently emerging gamma cycles may serve as multiplexing and "binding" mechanisms (Akam and Kullman 2014; Nikolic, Fries, and Singer 2013). Fast-ripple oscillations of the hippocampus (130–160 Hz) and neocortex (300–500 Hz) are the most precisely synchronized cortical rhythms and may reactivate prior experience (Buzsaki 2006). Ultraslow oscillations (0.1–0.02 Hz) involve coherent fluctuations of resting-state activity across large areas of the neocortex and subcortical areas (Drew et al. 2008).

Binding by Synchronization and Communication through Coherence

What is the relation between bands of brain oscillation frequencies and the physiology of local regions? The most thoroughly studied brain rhythm is gamma, with its characteristic fast-spiking basket cells and interneurons (see, for example, Buzsaki and Wang 2012; Fries 2009; Womelsdorf et al. 2014). Gamma rhythms may emerge through feedback inhibition from fast-spiking parvalbumin positive (PV+) cells to pyramidal cells (in pyramidal interneuron network gamma, or PING) or to the inhibitory cells themselves (interneuron network gamma, or ING) (Kopell et al. 2014). Because gamma oscillations typically arise locally, how are they synchronized, given the long conduction delays of pyramidal cells? One possibility is that global synchrony is made possible by the high conduction velocity possible by the very thick axons and large diameter myelin sheaths of long-range interneurons (Buzsaki and Wang 2012). One example of long-range synchrony of gamma oscillations is neurons sharing receptive fields in the primary visual cortex of the left and right hemispheres (Engel et al. 1991). The mechanisms for brain-wide synchronization include modulation of gamma power by the phase of slower rhythms and cross-frequency (for example, theta-gamma) coupling (Lisman and Jensen 2013), during which theta entrains and phase aligns gamma (Buzsaki and Wang 2012).

What neural and behavioral functions may be served by these bands of oscillation? One influential idea, called binding-by-synchronization, is that phase coupling between gamma band oscillations in a pair of brain areas may enhance the functional connectivity between the two (Deco and Kringelbach 2016; Fries, Nicolić, and Singer 2007; Fries 2015) (see Table 5.5). The second, communication-through-coherence, hypothesizes that peaks of rhythmic excitability serve as rhythmically recurring temporal windows for communications: phase-locked neuronal groups communicate most effectively because their "windows" for input and output are open at the same time (Fries 2005, 2009; Bastos, Vezoli, and Fries 2015). Other work suggests that beta band synchrony may play a role in maintaining a current sensorimotor or cognitive state (Engel and Fries 2010; Jenkinson and Brown 2011) and convey behavioral context to sensory neurons (Bressler and Richter 2014), while alpha-frequency synchrony may deselect a stronger, but currently irrelevant ensemble (Buschman et al. 2012).

The dynamics of an animal's behavior may follow the same principles of binding and alignment of sensory input, respectively: a continuous rhythm may serve both as a means of motor coordination (binding) via phase locking, and as a "reference oscillation" for aligning parallel loops of perceptual input from multiple sensory modalities (Kleinfeld et al. 2014). For example, in rodent exploratory behaviors, there is evidence of phase locking between sniffing and exploratory whisking (Deschenes,

Table 5.5 Synchronization between networks

Process	Description
Periodic oscillatory synchronization	Periodic synchronization of networks occurs when the spike probability of individual neurons exhibits periodic auto- and cross-correlations.
Synchronization via spike-to-spike synchronous states	In these states, individual neurons spike regularly and are brought into synchrony via chemical or electrical synapses. Whether coupled neurons synchronize in this way depends upon how the activity of each neuron affects the phase of other neurons to which it is coupled (as measured by their phase response curves).
Synchronization in sparsely synchronized states	In these states, individual neurons spike irregularly, but the oscillation emerges in the aggregate firing rate of the neuronal population. Oscillatory dynamics at the network level occur as a result of delayed negative feedback, which may arise either through self-inhibition or reciprocal innervation.
Role of negative feedback time delay	In a network with strong delayed negative feedback, a random fluctuation in excitatory drive will lead to undershoot in firing rates below equilibrium. Because of disinhibition, this will, in turn, lead to an overshoot of firing, leading to population oscillation. Depending on the strength and delay of negative feedback, these dynamics may give rise either to a self-sustaining oscillation or resonance at a particular frequency.

Source: T. J. Buschman, E. L. DeNovellis, C. Diogo, D. Bullock, & E. K. Miller (2012), Synchronous oscillatory neural ensembles for rules in the prefrontal cortex, *Neuron, 76,* 838–846.

Moore, and Kleinfeld 2012), suggesting that the breathing rhythm (and air flow) may function as a reference oscillation for the alignment of olfaction and head oscillations. Similarly, for vibrissa-based touch, there is spiking as a function of physical motion, similar to air flow in olfaction (Diamond et al. 2008). So, both smell and vibrissa-based touch in rodents are phase locked to self-generated oscillatory movements: breathing for olfaction and whisking for touch. There is also evidence that sniffing, whisking, head bobbing, and tasting may transiently synchronize the theta rhythm with respiration, forming memories that involve a confluence of orofacial perceptual information (Kleinfeld et al. 2014); and see Buzsáki 2006 for hippocampus and theta).

Resting-State Dynamics

The brain on music. Music, classical or jazz, is often an exploration of temporal patterns. This is apparent, for example, in the fugues of J. S. Bach. In a fugue, "the theme is stated successively in all voices of the polyphonic texture, tonally established, continually expanded, opposed, and re-established" (Randel 2003, 336). Fugue, in other words, is a *spontaneous, improvisational, exploration of sound produced from a particular point of departure* (Randel 2003). Spontaneous, improvisational exploration from a point of departure is also apparent in jazz music. In the history of jazz, the album *Kind of Blue* is a hallmark, on the order of *Beethoven's Ninth* in classical works (Kahn 2000). One of the masterpiece tracks of *Kind of Blue*, "Blue in Green," evolved from a point of departure provided by legendary jazz trumpeter and sextet leader, Miles Davis, to his brilliant pianist on the album, Bill Evans (Pettinger 1998). The liner notes of *Kind of Blue*, written by Bill Evans, and entitled "Improvisation in Jazz," reveal the structure of "Blue in Green" as a ten-measure circular form. So, from the G-minor and A-augmented notes, provided by Miles, emerged an opportunity for Bill's composition and for all the members of the sextet to mutually explore worlds of timing and sequencing possibilities in the music. Each of the remaining tracks provided the sextet with a different point of departure for exploring other time measures. Presented together, as an album, these explorations ushered in a new era in jazz music in the 1960s.

Exploratory behavior. Like a jazz group, the resting brain is never quiescent: it spontaneously explores a number of functional configurations (Fox and Raichle 2007; Biswal 2012). Even when there is no specific task, recordings from different voxels in a number of regions reveal *spontaneous correlated fluctuations* of functional MRI (fMRI) blood oxygenation level-dependent (BOLD) activity (Biswal et al. 2010; Fox and Raichle 2007; Poldrack and Farah 2015). At rest, the brain spontaneously generates slow fluctuations in the power of alpha and beta-frequency oscillations, which correlate across different brain regions (Deco, Jirsa, and MacIntosh 2011, 2013). These findings suggest not only that brain functional activity is intrinsic to the ever-active brain, but also that spontaneous activity may be shaped by underlying anatomical connectivity (Sporns and Honey 2013). What is the evidence that functional connectivity as measured by slow and indirect fMRI recording is related to direct patterns of neuron firing? A study by Wang et al. (2013) addressed this question in squirrel monkey somatosensory cortex (S1) by (1) obtaining fMRI signals to evaluate resting-state functional connectivity, (2) using fiber tract tracing to reconstruct axonal connectivity, and (3) measuring millisecond resolution firing rates of isolated neurons. The study reports two main "axes of information flow": interdigit interactions and inter-area interactions, evident at both the level of functional connectivity and at the single neuron level, implying a local-to-global multiscale hierarchy (Wang et al. 2013).

Computational modeling studies have raised another fundamental question about the brain's resting states—namely, the roles of a network's intrinsic local and global dynamics for the appearance of the signature, slow-fluctuating patterns in BOLD signals that form anticorrelational states. The computational modeling approach attempts to incorporate into neural networks what is known about the anatomical and functional characteristics of neurons and their interconnectivity and then determine if these models emulate the function of the biological network in question. To illustrate, Deco, Jirsa, and McIntosh (2013) propose that noise (probabilistic spiking times of neurons), coupling (strength of connections between nodes in a network), and time delays (the timing of arriving signals, due to signal transmission differences along axonal fibers between the network nodes) all play a fundamental role in resting-state dynamics. The power of this modeling approach is revealed in Deco et al. (2009), who investigated whether the signature ultraslow (0.1-Hz) oscillations of resting states might emerge from fluctuations between multistable states. Deco et al. (2009) were particularly interested in the interaction between fast local dynamics in the gamma range (40 Hz) and these ultraslow oscillations. They conducted simulations based upon a neuroinformatics tool, called CoCoMac, which is a means for visualizing connectivity data in a vast database of primate brain imaging

(Kotter 2004). They began the simulation by creating a simplified network of thirty-eight noise-driven (Wilson-Cowan) oscillators, which have the characteristic of remaining just below their oscillatory threshold when isolated. They then added time-delay coupling based on lengths and strengths of actual primate cortico-cortical pathways. And finally, they systematically manipulated random fluctuations with uncorrelated Gaussian noise to emulate spiking noise. By selecting certain conduction velocities (1–2 m / s) and very weak coupling between oscillators, Deco et al. (2009) were able to recreate two signature characteristics of resting state networks: (1) there were two "communities" of states that were anti-correlated at 0.1 Hz, and (2) there was a stochastic resonance effect, in which there was an optimum effect of attenuation of the global 0.1-Hz oscillations at a specific level of noise.

Brain, Body, and Behavior: Not Computations Alone

How, then, are the neural circuits of small and large animals related to their behavioral capabilities? One possibility, proposed by Schroter, Paulsen, and Bullmore (2017) is that across spatial scales of neuronal networks there are common *motifs*, or patterns of connectivity, between some of the network nodes. A fundamental principle guiding motif formation is an economic tradeoff between biological (that is, wiring) cost and functional (adaptive) value: integrative components crucial for adaptation are expensively, not minimally, wired. One illustration of the importance of circuit flexibility for balancing wiring economy in small neuronal networks is that hub and rich club organization allow flexible switching between behaviors, as in the forward and backward locomotion of *C. elegans* (Schroter et al. 2017). Another illustration of economic tradeoffs in neuronal network architectures is that the parallel pathway architecture of the optic lobe of *Drosophila* is retinotopically arranged for high-speed information flow (Schroter et al. 2017). And in animals with a large cerebral cortex, such as nonhuman primates and humans, the network motif of reciprocal connectivity between principal cells may contribute to the prolongation of activity for cortical computations as well as to synchronizing functional cell assemblies (Womelsdorf et al. 2014).

However, neural circuit wiring alone (for example, Schroter et al. 2017), and models of computation without consideration of co-evolving bodies (Krakauer et al. 2017), may not provide a complete account of the relation between circuit motifs and behaviors such as selecting forward or backward locomotion of *C. elegans*, or visual control of *Drosophila* flight. Compared to *C. elegans* and *Drosophila*, the diversely interconnected musculo-tendinous elastic body components and capability for information pickup by multiple mobile perceptual systems allow mammals to select from a more flexible set of actions in uncertain environments (Pezzulo and Cisek 2016). By including the mechanical properties of evolving bodies as part of the process by which behavior emerges, we may turn to physical principles of self-organization in neuromechanical systems to account for the assembly of an assortment of bodily devices for adaptive behavior. So, for example, (1) positive feedback loops may amplify individual neuron and circuit signaling beyond their local range, and mechanical prestress may provide pathways for rapid conduction of signals to other neural as well as non-neural cells throughout the body to assemble functionally specific muscle groups (Pezzulo and Cisek 2016; Turvey and Fonseca 2014); (2) patterns of structured energy impinging on receptor arrays may directly specify a layout of environmental attractors as choice points for decision-making neuronal networks; and (3) affective and dispositional states at a particular moment in time, with a developmental history, and with future-oriented goals, may modulate the attractor layout of neural fields (Breakspear 2017; Deco, Jirsa, and McIntosh 2011).

As I discuss further in Chapters 6 and 7, the central place of circuit motifs in both small and large nervous systems appears to emerge from the developmental processes of cell migration and plasticity that assemble them. In *C. elegans*, for example, most neurons that share early birth dates are connected by long-range connections: these become network hub nodes, and are organized as a rich club (Schroter et al. 2017; van den Heuvel and Sporns 2013). In the mammalian cortex, lineage-dependent circuit formations appear to be precursors of the mature columnar function of the neocortex (Gao et al. 2013). However, the developing brain has a body: at the same time that these amazing processes of neuronal development occur, the body is growing, changing form, and epigenetically contributing receptor feedback through spontaneous movements (Gottlieb 2007). Mammalian fetal life occurs within a supportive milieu, and postnatal life introduces dramatic changes in opportunities for actively exploring available information and for all-important social interactions. Chapters 6 and 7 consider how ontogeny and plasticity, respectively, may provide crucial additional insights into the relation between brain and body in the emergence of adaptive behavior.

• 6 •

Human Nervous System: Development and Vulnerability

I had been on the faculty of the Boston Children's Hospital for several years, but that day in early 2013 was the first time that I had worn a hard hat at work. Other colleagues and I were invited on a tour of the first-of-its kind Fetal-Neonatal Neuroimaging and Developmental Science Center at Boston Children's Hospital, opened in 2014. The construction elevator opened to a large space that already revealed its difference from other neuroimaging facilities. The floor plan had the feel of the integrative medicine of the twenty-first century: it was organized so that multiple neuroimaging systems—near-infrared spectroscopy (NIRS), cranial ultrasound (CUS), magnetic resonance imaging (MRI), and magnetoencephalography (MEG)—were within walking distance of each other, and easily accessible for the young patients being transported from the nearby neonatal intensive care unit (NICU). These technologies have been optimized for the pediatric population, including a combination of frequency-domain NIRS and diffuse correlation spectroscopy (P. Y. Lin et al. 2016) and a "BabyMEG" whole head magnetoencephalography system (Okada et al. 2016).

At the Center, Dr. Ellen Grant, along with a cadre of clinician-scientists and engineers, are using these imaging systems to advance our understanding of fundamental neurodevelopmental processes, as well as improve clinical assessment and treatment of developmental nervous system disorders. For example, one line of studies is using diffusion tensor imaging (including high-angular-resolution diffusion imaging, or HARDI) as well as tractography to delineate structural changes that occur in developing fetal (preterm) brains (Song et al. 2014; Takahashi et al. 2010; Takahashi et al. 2011), the spatiotemporal characteristics of radial and tangential migration streams in the subpallial ganglionic eminence (GE) and in the dorsopallial-ventricular-subventricular zone (Kolasinski et al. 2013), the radial coherence of telencephalic white matter (G. Xu et al. 2014), the emergence of cerebral connectivity in the human fetal brain (Takahashi et al. 2012), and the development of white and gray matter pathways and cerebellar connectivity in human brains (Takahashi et al. 2014; Takahashi et al. 2013). The Center is, thus, a mecca for developmental research—for example, for constructing remarkable three-dimensional reconstructions of developing fiber tracts and superimposing multiple imaging modalities.

The development and applications of imaging are an example of developmental translational medicine, the confluence of research and work with clinical populations for advancing clinical practice. Thus, the Center, like Boston Children's as a whole, is both developmentally oriented and translational in approach. This chapter is similarly inspired by a developmental and translational approach: it addresses how the vulnerabilities of the human nervous system are related to its development throughout the life span. One emphasis is on the earliest period of development when the brain develops most rapidly and profoundly and is most vulnerable to injury.

I walk through a connecting corridor and into the hospital neonatal intensive care unit (NICU). My eyes are immediately drawn to the tiny premature infant bathed in the blue glow of bilirubin lights. An infant born at twenty-six weeks gestational age, as is the case for this child, is characterized by poorly developed lungs and gastrointestinal tract and an immature nervous system (including the retina), vulnerable to injury by the fluctuations of the brain's circulatory system and certain molecules circulating through the bloodstream. She is not yet taking oral feedings and is, instead, fed by means of a nasogastric tube. Because the oral-pharyngeal route is bypassed, and there is no need for swallowing, respiration remains stable. Another newborn infant, born at term but waiting for repair of a congenital heart defect, is being fed breast milk from a bottle while being held by a nurse in a rocking chair. A pulse oximeter alarm emits a soft "beep," a warning that the infant's oxygen saturation levels have moved below a set threshold value. As the infant's movements become agitated, she begins to cry. The nurse evaluates the infant's behavior, pauses the feeding, and waits

Table 6.1 Developmental windows of central nervous system vulnerability leading to later disability

Developmental period	Vulnerability	Later disability
Prenatal	Genetic	Fragile X syndrome, autism spectrum disorder, Rett syndrome
Perinatal	Perinatal white matter injury, hypoxia-ischemia Perinatal sensitization of arousal regulation	Cerebral palsy Schizophrenia
Infancy	Dysfunctional cerebral cortex development	Autism spectrum disorder
Childhood	Pediatric stroke	Neuromotor and intellectual disabilities
Adolescence	Concussion and other traumatic brain injury Spinal cord injury	Neuromotor and intellectual disabilities Partial or complete paralysis
Adulthood	Stroke	Neuromotor and intellectual disabilities
Late adulthood	Late-life neurodegenerative diseases (Alzheimer's, Parkinson's)	Neuromotor and intellectual disabilities

until she judges that it is safe to continue feeding the infant. In both of these infants, there is increased risk of injury to the brain because of medical conditions that may precipitate some instability in the availability of oxygen and / or the circulation of inflammatory agents activated by the child's own body. The nervous system vulnerabilities of these children are a consequence of the interdependencies of all of the body organs and of fluctuations in the ability of these systems to maintain functioning within a range of values (see Table 6.1).

Circuit Self Assembly

Neural development is a dynamical process that is both exploratory and selective (Goldfield 1995). As succinctly summarized by Lichtman and Smith:

> More neurons are made than ultimately survive in development, growth cones navigate to targets by exploration of potential directions rather than making a bee line, dendritic branches and spines form and are lost as a dendritic tree matures, and synapse formation and synaptic elimination often occur simultaneously by the same neuron as circuitry is built. (2008, 443)

The exploratory and selective nature of neural development relates to the model of attractor dynamics in an epigenetic landscape, presented in Chapter 2. The attractor landscape consists of a dense population of local minima, or potentials, that provide chemical and mechanical signals to guide and nurture exploratory behavior. Neurons, as single-celled motile organisms that live inside us,

> are engineered to adhere to certain substrates rather than others. They fasciculate with neurons of the same type but sometimes avoid growing along with axons of different types. All of these constraints mean that these single-celled organisms will exhibit exuberant cell dynamics, exploratory behavior, trial and error based refinements, and competitive interactions leading to death of some cells and removal of some processes. Out of all this activity comes a harmonious system that reaches some equilibrium. (Lichtman and Smith 2008, 445)

Imaging the Developing Brain of the Fetus and Neonate: Structure and Function

Advances in *in vivo* MRI of the fetal and neonatal nervous system are now making it possible to reveal how anatomical changes at a microscopic scale are related to the emergence of macroscopic structural and functional networks (Clouchoux and Limperopoulos 2012; Studholme 2015) (see Table 6.2 for a comparison of imaging modalities). Diffusion tensor imaging (DTI), or "tractography," is an *in vivo* MRI

Table 6.2 Neuroimaging modalities for measuring neuroplasticity during development

Modality resolution	Measures	Characteristics / challenges
Transcranial magnetic stimulation (TMS) (≈1cm / ≈1ms)	Nerve conductivity; functional organization	Direct objective measure; sensitive to change in neuroplasticity, but may cause headaches in some children
Diffusion MRI (2–3mm)	Tissue microstructure; structural connectivity	Simple biological interpretation that correlates well with clinical scores, but large lesions may preclude automated analyses
EEG (≥1 cm; <1 ms) MEG (<1 cm; <1 ms)	Neuronal signaling	Able to distinguish between different stages of sensorimotor processing; complex to interpret, sensitive to motion artifacts; MEG scanning facilities are expensive
Structural MRI (1 mm)	Cortical thickness; gray matter volume	Simple biological interpretation; moderately sensitive to changes in neuroplasticity
BOLD fMRI	Functional organization of the brain	Simple to analyze data and visualize results with respect to brain structure; sensitive to motion artifacts

Source: L. Reid, S. Rose, & R. Boyd (2015), Rehabilitation and neuroplasticity in children with unilateral cerebral palsy, *Nature Reviews Neurology, 11,* 390–400.

technology that uses water diffusion in brain tissue to visualize in stunning detail the brain's three-dimensional white matter anatomy (Johansen-Berg and Rushworth 2009; Mori and Zhang 2006; Qiu, Mori, and Miller 2015). DTI is made possible by characterizing water diffusion in tissues by means of a mathematical tool called a tensor, based on matrix algebra (Mori and Zhang 2006; Zhang, Aggarwal, and Mori 2012): (1) a 3×3 matrix, called a diffusion tensor, is used to characterize the three-dimensional properties of water molecule diffusion; (2) from each diffusion tensor, three pairs of eigenvalues and eigenvectors are calculated using matrix diagonalization; and (3) the eigenvector that corresponds to the largest eigenvalue is selected as the primary eigenvector. A "streamline" algorithm then creates "tracts" by connecting adjacent voxels if their directional bias is above some threshold level (see, for example, Mori 2002, 2013). Does the orientation of the primary eigenvector coincide with that of the actual axon fibers in most white matter tracts? Affirmative evidence comes from studies that directly examine the relation between tractography and actual histology. Takahashi et al. (2011), for example, have demonstrated that radial organization of the subplate revealed via tractography directly correlates with its radial cellular organization, and G. Xu et al. (2014) were able to determine that transient radial coherence of white matter in the developing fetus reflected a composite of radial glial fibers, penetrating blood vessels, and radial axons.

It is now also possible to use three-dimensional volumetric MRI to measure fetal brain volumes, to segment white and gray matter, and to plot brain growth trajectories (Habas et al. 2010). Fetal MRI imaging has also been used to quantify cortical folding in healthy fetuses between twenty-five and thirty-five weeks (Clouchoux et al. 2010). The rapid increase in synaptic density, as well as the elongation of axons at around twenty-six weeks, may initiate connectome formation. Increasingly definitive connectome organization is evident around the time of birth during the period corresponding to peak synaptogenesis, completion of cortico-cortical axonal connections, and appearance of the major white matter tracts (Collin and van den Heuvel 2013; Dubois et al. 2006; Tau and Peterson 2010).

As discussed in Chapter 5, resting-state functional MRI (fMRI) is used to measure temporal correlations of low-frequency (<.01 Hz) fluctuations in blood oxygenation level-dependent (BOLD) signal, indicating baseline neural activity in the absence of goal directed activity and stimulation (Fox and Raichle 2007). There is now evidence that by the time of term birth, there are resting-state networks incorporating the primary sensory and motor areas, but these may be quite different from that of adults (Aslin, Shukla, and Emberson 2015; Fransson 2005; Smyser and Neil 2015). The resting state of infants is dominated by correlated activity in sensorimotor regions, with restricted and local connectivity within and across the hemispheres (Fransson et al. 2011). Functional near-infrared spectroscopy (fNIRS) is being used in addition to fMRI to study infant multimodal perceptual processing, social cognition, learning, and memory (Aslin et al. 2015).

An Overview of Human Central Nervous System Development

Induction. Local cell populations, called "organizers," play a central role in inducing central nervous system (CNS) anatomy as part of the initial major event in neurogenesis, called *neuronal proliferation* (Scholpp and Lumsden 2010). Each organizer establishes concentration gradients of morphogenetic signal molecules in particular tissue regions, which induce cells to self-assemble into functionally differentiated collectives (de Robertis 2009; Kiecker and Lumsden 2012; Wolpert et al. 2007). Organizers induce ectoderm lying along the dorsal midline to become the neural plate. Mediolateral gradients of BMP and Shh activity, together with AP Wnt gradient activity, establish a quasi-Cartesian coordinate system across the neural plate (Kiecker and Lumsden 2012). Mesoderm signals also give regional identity to the neural tube: the notochord, a thin rod-shaped structure, induces the lateral folds of the neural plate to roll up and form the neural tube (Kiecker and Lumsden 2012). Signaling centers become stabilized as compartments that subdivide large regions into smaller compartments, and one local signaling center—the mid-diencephalic organizer—forms the "bridal chamber," or thalamus, in the posterior part of the embryonic forebrain (Scholpp and Lumsden 2010).

Neuronal migration. During the next major event of neurogenesis (see Table 6.3), neuronal migration, neurons migrate radially away from the proliferation zones at the inside surface of the tube toward the outer surface and also migrate tangentially—that is, parallel to the surface of the tube (Noctor, Cunningham, and Kriegstein 2013). The directionality of neuronal migration may be guided by a landscape of both mechanical tension and by diffusible and membrane-bound chemical signals (Franze 2013). Until recently, however, it has not been possible to measure mechanical forces guiding axons *in vivo*

Table 6.3 Major events of neurogenesis: From proliferation to network formation

Event	Description
Neuronal proliferation	The neural tube becomes specified into a number of distinct domains, each of which is the precursor of a nervous system area. The neural tube thickens as new neurons are generated.
Neuronal migration	Neurons migrate radially away from the proliferation zones at the inside surface of the tube toward the outer surface. Neurons also migrate tangentially, i.e., parallel to the surface of the tube.
Differentiation	During migration, in response to both intrinsic and extrinsic factors, neurons gradually become specified into different cellular types.
Axon specification	Upon arrival at their destination, neurons begin to produce several undifferentiated neurites. In a competitive process, one neurite becomes specified as the axon, and those that remain differentiate into dendrites.
Neurite elongation and branching	The dynamic behavior of growth cones causes dendrites to branch extensively and gradually form their characteristic morphologies.
Axon guidance	Guided by diffusible and membrane-bound chemical signals, as well as mechanical regulation in their local surroundings, axons continue growing to their targets.
Network formation	Once they have arrived in their target region, axons may branch considerably before terminating to form initial synaptic connections with target structures.

Sources: J. Stiles & T. Jernigan (2010), The basis of brain development, *Neuropsychology Review, 20,* 327–348; Arjen van Ooyen (2011), Using theoretical models to analyze neural development, *Nature Reviews Neuroscience, 12,* 311–326; and K. Franze (2013), The mechanical control of nervous system development, *Development,* 140, 3069–3077.

during nervous system development. In the Franze lab at the University of Cambridge, neuroscientists and physicists using atomic-force microscopy have found that axonal growth patterns of retinal ganglion cells in the *Xenopus* are influenced by the changing mechanical properties of the surrounding tissue. Substrate stiffness has specific influences on growth: on stiffer substrates, axons grow faster, straighter, and more parallel, while on softer substrates, axons splay apart, branch, and form synapses (Koser et al. 2016).

During migration, in response to both intrinsic and extrinsic factors, neurons begin to differentiate—that is, become specified into different cellular types (van Ooyen 2011). Upon arrival at their destination, neurons begin to produce several undifferentiated neurites. In a competitive process, one neurite becomes specified as the axon, a long process extending out from the soma, and those that remain differentiate into dendrites (Ma and Gibson 2013). Each neuron has only one axon, which may reach 1 meter in length in humans as it makes contact with its synaptic targets (Ma and Gibson 2013). Axons may branch considerably before terminating to form initial synaptic connections with target structures, the basis for nervous system networks. During development through the immediate postnatal period and beyond, oligodendrocytes wrap themselves around axons to begin the process of myelination (Emery 2010). A critical characteristic of myelin, from the standpoint of axon injury and regeneration (see Chapter 7), is that myelin is inhibitory to axon growth, due to production of several myelin-associated inhibitors, such as Nogo (Yiu and He 2006).

The Fetal Brain, Prematurity, and Periventricular Leukomalcia

A 2014 editorial in *Science Translational Medicine* notes that for the first time in history, preterm birth rather than infectious disease is the world's number-one killer of young children (Lawn and Kinney 2014). The picture becomes more complex: medical and technological advances have increased the survival rate of infants born increasingly early and small, particularly those born weighing less than 1500 grams and younger than thirty-two weeks. However, the lower birth weight and earlier gestational age of these very preterm survivors has come at the cost of increased morbidity, especially in gray and white matter brain disorders (Salmaso et al. 2014). During typical brain development, there is a doubling in total brain volume, due mostly to growth of gray matter (especially thalamus and basal ganglia; see Chapter 5). However, preterm birth is associated with loss of gray matter volume, and the earlier the gestational age, the greater the reduction in volume (G. Ball et al. 2012). The development of white matter is a process that reflects the generation of functionally mature myelin-forming oligodendrocytes from oligodendrocyte precursor cells (Sherman and Brophy 2005). Severe white matter injury, such as periventricular leukomalcia, or PVL (Khwaja and Volpe 2008), usually results in spastic diplegic cerebral palsy and visual dysfunction, often with deficits in cognition and learning (Back and Miller 2014; Buser et al. 2012).

The preterm neonate born less than thirty-two weeks gestational age is developmentally still at a fetal stage of nervous system development. Technologies for imaging the fetal nervous system *postmortem* and *in utero* are providing important insights into the processes underlying brain development between seventeen and forty weeks of gestation (for example, Takahashi et al. 2012). MRI, an essential tool for imaging the fetal brain, noninvasively measures the signals from ^{1}H (proton) of water molecules because more than 90 percent of protons in the body are located in water molecules (Huang and Vasung 2014). Diffusion-based imaging processes are based on the restrictive influence of water diffusion within a tube. Diffusion-weighted imaging is an *in vivo* technique that allows visualization of the water diffusion constant in different directions. Diffusion tensor imaging characterizes the ellipsoidal shape of the water diffusion using a symmetric positive tensor field derived from diffusion-weighted imaging (DWI) (Qiu, Mori, and Miller 2015). DTI-derived images are obtained from the eigenvalues and eigenvectors of the diffusion tensor, based on the assumption that the direction of the primary eigenvector of the tensor aligns with the orientation of its underlying organized structures (Huang and Vasung 2014). DTI-based tractography, then, refers to the techniques used to connect these primary eigenvectors to reconstruct the pathways of the white matter tracts.

There is a clear developmental pattern of limbic, brain stem, commissural, projection, association, and thalamocortical white matter tracts of fetal brains: limbic fibers develop first, with commissural, thalamocortical, and projection tracts growing from the core to the periphery of the brain (Huang and Vasung 2014). It has been more difficult to identify the radial organization of radial glial fibers (Rakic et al. 2009; Marín-Padilla 2011). High-angular-resolution diffusion imaging (HARDI) is a type of diffusion MRI that uses the directionality of tissue water diffusion to reveal the organization of structural pathways, including radial coherence (G. Xu et al. 2014; see later discussion).

Cerebral Cortex

In a side-by-side comparison with other mammals, the human neocortex is visually distinctive because of how large it is compared to our body size and because of the complexity of its foldings, or convolutions. There is now considerable evidence that the disproportionate growth of the human cortex compared to our body size is due to the proliferative output during neurogenesis of neural progenitor cells lining several compartments, or zones, within the brain ventricles (Geschwind and Rakic 2013; Lui, Hansen, and Kriegstein 2011a,b; Rakic 2009; Sun and Hevner 2014). An initial evolutionary step toward an outsized neocortex, common to all primates, may have been that one region, called the ventricular zone, expanded to increase the number of proliferative cells during early development. Then, the emergence of a subventricular zone above the ventricular zone may have resulted in an increase in the rate of neuron proliferation. But the second distinctive feature of the human neocortex (also evident in the ferret and rat)—its gyrencephalic, or folded, appearance—may not be due to proliferation, per se, but rather to an evolutionary change in the characteristic geometry of the proliferative cells lining a ventricle compartment called the outer subventricular zone (OSVZ) (Lui, Hansen, and Kriegstein 2011a,b).

The mammalian cerebral cortex consists of glutamatergic projection (or pyramidal) neurons and GABAergic neurons (or interneurons) with a distinctive cellular architecture: horizontal layers (laminar structure) intersected by vertical (or radial) columns (Bystron, Blakemore, and Rakic 2008). Projection neurons may be further categorized according to their radial position within the six neocortical cell layers (named layers I–VI) (Franco and Muller 2013). For example, most layer VI neurons form corticothalamic connections, while neurons connecting to basal ganglia, midbrain, hindbrain, and spinal cord are typically found in layer V (Molyneaux et al. 2007). Different subpopulations of projection neurons are born in overlapping temporal waves, with newly born projection neurons migrating past the earlier-born neurons to form the six-layered structure of the mature neocortex in an "inside-out" manner (Rakic 2009).

Cortical neurons are generated within proliferative transient embryonic zones, or "factories," situated near the surface of the cerebral lateral ventricles (Rakic 2009). The progenitor cells undergo symmetric divisions that expand the surface area of the cortex by *increasing the number of progenitors and asymmetric divisions* that produce intermediate progenitors (postmitotic neurons). These daughter cells then produce the outermost zone, the cortical plate, where neurons align in an inside-out pattern to form the six-layered cerebral cortex (Ayoub et al. 2011) (discussed later). Pyramidal cells are generated in the ventricular zone (VZ) of the embryonic pallium (the roof of the telencephalon) and reach their final position by radial migration (Rakic 2007). Radial migration is a process in which radially migrating neurons use radial glial fibers to guide their journey. By contrast, cortical interneurons, born in the subpallium (the base of the telencephalon), reach the cerebral cortex through tangential migration (Hatten 2002). Tangentially migrating fibers, like radially migrating neurons, extend "leading processes" that detect the local environment but do not seem to require glial fibers (Marin et al. 2010). The leading process selects the direction of migration in response to chemotactic gradients and by a process of nucleokinesis, where the nucleus is pulled by motor proteins toward the leading process (Marin et al. 2010).

What accounts for the formation of layers and columns? According to the radial unit hypothesis (Rakic 1988; Jones and Rakic 2010), the tangential (horizontal) coordinates of cortical neurons are determined by the *relative position* of their precursors within the ventricular zone, while their radial (vertical) position is related to their *time of origin (birthdate) and arrival in the cortex* (Rakic 1988; Jones and Rakic 2010). After completion of their last division cycles, postmitotic cells migrate along a common radial pathway to the developing cortical plate, where they form ontogenetic columns (Rakic 2009). How do immature cells arriving in the cortical plate differentiate into the highly specialized cells that will serve area-specific functions? According to the protomap hypothesis, the positional information of the postmitotic neurons is maintained during their migration to the cerebral cortex by the radial constraints in the transient radial glial scaffolding (Rakic 1988). Even as the cortical surface expands, individual cells preserve their laminar and areal positions and act as a template that promotes selective connectivity by attracting specific afferents (for example, from the thalamus) (Rakic et al. 2009). Geschwind and Rakic (2013) propose that migration was probably introduced during evolution to preserve neuronal position from the VZ protomap to the overlaying cerebral cortex. There is growing evidence that area-specific functions of the cerebral cortex may reside in neuronal genetic (transcriptional) programs specific to each of these transient embryonic proliferative zones (Ayoub et al. 2011) but that postmitotic molecular mechanisms may also contribute to laminar position, neuronal identity, and connectivity (Kwan et al. 2008; MacDonald et al. 2013).

Changes in cortical thickness reflect the migration of neurons. MRI measures of fractional anisotropy (FA), or direction-dependent movement of water molecules restricted by fibers, and T1-weighted and T2-weighted MRI signal intensity reveal that (1) the cortical plate is first composed of densely packed postmigratory neurons and (2) after twenty-four weeks postconception, the subplate "waiting compartment" is identified on T1-weighted FA (Vasung, Fischi-Gomez, and Huppi 2013). HARDI tractography reveals complex crossing tissue coherence, even in fetal brains (Song et al. 2014). Radial coherence refers to pathways running across the cerebral mantle, perpendicular to the cortical surface, while tangential coherence refers to pathways running parallel to the cortical surface (Kolasinski et al. 2013). A developmental trend in fetal brain organization is that there is regional regression of radial and tangential coherence and regional emergence of connectivity from posterodorsal to anteroventral with local variations (Takahashi et al. 2011). To illustrate, between nineteen and twenty-two weeks, there are two distinct three-dimensional patterns of diffusion coherence: a radial pattern originating in the ventricular / subventricular zone and a tangential radial pattern originating in the ganglionic eminence. But by week 24, there is a gradual regression of radial pathways in dorsal frontal, parietal, and inferior frontal lobes and an emergence of short-range corticicortical and long-range association pathways. And by week 31, radial coherence in the temporal and occipital lobes is less apparent, and long-range association pathways emerge. By weeks 38 and 40, there is no radial or tangential coherence in evidence. Thus, connectivity emerges as radial and tangential coherence regress.

The Transient Subplate Zone

The exploratory behavior of migrating thalamocortical axons (TCAs) takes place within a transient subplate zone, which begins to form in humans by around fifteen postconceptional weeks (PCW) and disappears at around thirty-five PCW (Duque et al. 2016) (see Figure 6.1). The subplate zone is a dynamically changing neurosecretory, extracellular milieu characterized by transient guideposts and corridors that collectively guide TCAs to their cortical targets (Lopez-Bendito and Molnar 2003; Molnar et al. 2012). During their trajectory out of the

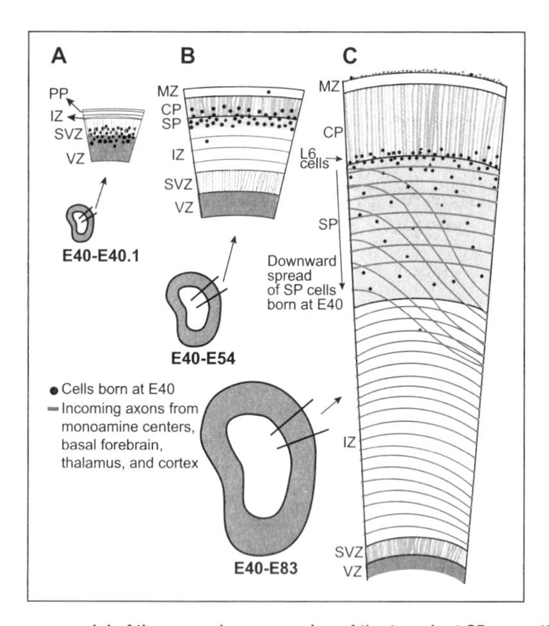

Figure 6.1 Summary model of the secondary expansion of the transient SP zone. (A) E40 born SP cells are visible in the VZ one hour after injection. By E40, there is no SP yet. (B) By E54, these early born cells accumulate in the SP, and a few Cajal-Retzius cells in the MZ become evident. (C) By E83, and later in mid-gestation, these cells become secondarily displaced and widespread into the expanding SP by ingrowth of the neuropil. In particular, monoamine, basal forebrain, thalamocortical, and later on a large contingent of corticocortical axons. Alvaro Duque, Zeljka Krsnik, Ivica Kostović, and Pasko Rakic (2016). Image reprinted with permission of the National Academy of Sciences.

thalamus, through the prethalamus and subpallium, and into the neocortex, TCAs rely on diffusible guidance cues and a series of migrating guidepost cells that act to repel them away from some structures (such as the hypothalamus) to open a permissive corridor along the way (Garel and Lopez-Bendito 2014).

One of the most remarkable findings is that in order for the thalamus and cerebral cortex to establish reciprocal connections, there are "temporal checkpoints," including a waiting period in axon progression to insure that TCAs have reached the proper position (Deck et al. (2013). At around the time of subplate dissolution, another cellular layer called the cortical plate forms and splits the preplate into the subplate below and marginal zone above (Hoerder-Suabedissen and Molnar 2013) (see Figure 6.2). The cortical plate eventually produces laminae II–VI of the mature cerebral cortex, and the marginal zone

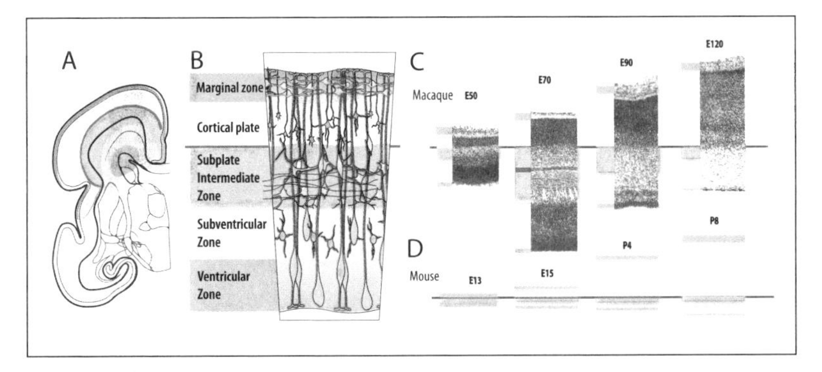

Figure 6.2 Compartments and zones of the developing human cerebral cortex. Schematic coronal section showing the relative location and size of the major compartments within the developing human dorsal cortex at twenty-six postconception weeks (PCWs), at the peak of neurogenesis and cell migration. The germinal zone consists of layers [ventricular zone (VZ) and subventricular zone (SVZ)] in which cell divisions take place. The subplate (SP) and the intermediate zone (IZ) lie between the SVZ and the cortical plate (CP). The outermost layer is the marginal zone (MZ). The inset box provides a higher-powered view of the cellular make-up of the transient developmental zones within this developing cortical region. Radial glial progenitors divide asymmetrically within the VZ while in contact with the ventricular surface, whereas intermediate progenitors divide symmetrically in the SVZ. The SP and IZ boundary cannot be always distinguished on the basis of simple cytoarchitecture. The IZ contains more fiber tracts, whereas the SP is more abundant in postmitotic SP neurons with well-developed cellular processes. Synapses form in this layer. The CP contains migrating and immature CP neurons that are densely packed with only rudimentary cell processes, whereas the MZ contains fiber bundles and Cajal–Retzius cells as well as distal dendrites of cells below the MZ. Figure courtesy of Anna Hoerder-Suabedissen.

ultimately forms cortical layer I. Each layer contains a distinct array of cell types, whose morphology and location dictate the pattern of local and distant projections that each cell may send or receive. Cells in adjacent vertical lamina are organized into functional radial ontogenetic columns, each of which consists of many smaller minicolumns with their longest axes arranged perpendicular to the cortical surface (Kostovic and Vasung 2009).

Cortical Folding

The process of cortical folding in the human brain begins at around gestational week 16, increases rapidly during the first trimester, continues well after birth, and reaches a maximum between sixty-six and eighty weeks postconception (Zilles, Palomero-Gallagher, and Amunts 2013). The distinctive fissures, sulci, and convolutions of the human gyrencephalic brain develop due to differential radial growth of the cerebral wall. The growth is driven not only by gene expression (Bae et al. 2014; Rash and Rakic 2014), but also by the influences that are characteristic of successive periods of afferent innervation, dendritic growth, synaptogenesis, and gliogenesis. To illustrate, there is an initial period during which gyrification is primarily driven by differential amplification of neural progenitor cell proliferation. During this period, folding is due to the proliferation and differentiation of progenitor cells in particular locations relative to the ventricular surface of the developing cortex, including the ventricular, intermediate, inner, and outer zones. Later in development, gyrification is influenced by growth of afferent fibers from thalamic and other sources and by axonal interactions with neurons and progenitors. These secondary and tertiary gyri and sulci appear parallel with the disappearance of the subplate zone and the ingrowth of the long and short cortico-cortical association fibers (Sun and Hevner 2014)

The identities of the genetic regulators of gyral folding have been revealed in studies of surface folding pattern differences associated with genetic abnormalities, such as the relation between a faulty GPR 56 gene and the formation of small polymicrogyria, a pattern in which the cortex folds in on itself to give a thickened appearance (Bae et al. 2014). GPR 56 may be critical for cortical folding because it is required for normal attachment of basal processes to the pial membrane (Bae et al. 2014). But how does gene expression induce cortical tissue to undergo folding? Gene expression may control relative rate of tangential growth of the brain's outer surface gray matter relative to the underlying white matter, such that gyrification is due to a mechanical instability. To test this possibility, Wyss faculty member L. Mahadevan and colleagues (Tallinen et al. 2014) have developed a set of mathematical and physical models of layered gray and white matter. One mathematical model applies a tangential growth profile to a rectangular domain consisting of a deep layer of white matter with a gray matter layer above, each with the same uniform shear modulus. Experiments tested the model by manipulating the tangential expansion ratio; measuring the geometric characteristics (depth and width) of a derived sulcus; and making comparisons with sections of the brains of porcupine, cat, and human.

Tallinen et al. (2014) further modeled the brain in three dimensions and examined the relation between brain radius (R), cortical thickness (T), and tangential expansion (g^2). When they plotted g^2 against relative brain size (R / T), Tallinen et al. (2014) found a clear relation between size and gyrification: when g is not sufficiently large to cause buckling, the physical surface of the brain is smooth, as in the rat. As brain size increases to the intermediate size of lemurs and wolves, sulci are isolated and localized within the gray matter. In a large-size human brain, there is increased folding with sulci penetrating the white matter, and the brain surface displaying complicated patterns of sulcus branching. As a final demonstration of the role of a mechanical instability resulting from the growth of gray matter, Tallinen et al. (2014) fabricated a hemispherical elastomer coated with a top elastomer layer made to swell by absorbing solvent over time. Experiments manipulated exposure time of the elastomer to the solvent and thickness of the top layer of elastomer, with resulting patterns of gyrification similar to rat, wolf and lemur, and human-size brains, respectively. Most recently, Tallinen et al. (2016) compared convolutions of model and actual human brain during development and found a striking correspondence in convolutions of synthetic and biological tissue.

Synaptogenesis and Remodeling of the Healthy Developing Nervous System

The rates of dendritic arborization and synaptogenesis accelerate in the third trimester (40,000 synapses form each second at a peak of thirty-four weeks!) to produce a thickening of the developing human cerebral cortex, coinciding with the appearance of cortical gyri and sulci (Kostovic and Jovanov-Milosevic 2006). As many as half of the neurons produced during neurogenesis are eliminated during two discrete periods of apoptosis, or programmed cell death, the first beginning around GA week 7 and the second peaking at GA weeks 19–23 (eliminating postmitotic neurons within the cortical plate) (Dekkers and Barde 2013; Southwell et al. 2012). It now appears that apoptosis of cortical interneurons is determined intrinsically through a competition for survival signals from other interneurons (Southwell et al. 2012).

In the cerebellum and at the neuromuscular junction, initially exuberant synaptic connections are refined by elimination of weak connections and strengthening of functionally important ones (for a review, see Hashimoto and Kano 2013; Tapia et al. 2012). As in humans, a characteristic of cerebellar anatomy in mice is development of a one-to-one innervation between an excitatory climbing fiber (CF) and a Purkinje cell (PC) (Hashimoto et al. 2009; Kano and Hashimoto 2009). Around postnatal day 3, PCs are innervated by multiple CFs, and during P3–P7, only a single CF is strengthened. How does this elimination process occur? Hashimoto et al. (2009) demonstrate that a single "winning" CF is selectively strengthened, while others are weakened, through a process of activity-dependent competition. The competition involves movement of initial synaptic contacts made on the PC soma through a process of translocation to PC dendrites. At P7–P8, one CF becomes the most prominent because it exhibits the strongest synaptic efficiency. After P9, only this stronger CF translocates to PC dendrites, and the weaker CFs remaining on the PC soma are eliminated.

Further insights into the dynamic processes involved in synapse elimination have been provided by systematic studies of connections between motor neuron axons and their target muscle cells at the neuromuscular junction (NMJ) (see, for example, Walsh and Lichtman 2003). These studies support the hypothesis that synapse loss is driven by competitive events occurring at the NMJ (that is, through local regulation): after axonal inputs are eliminated from the NMJ, the axon that remains increases its synaptic contact area by occupying many of the previously occupied synaptic sites. To elucidate the dynamic process of axon response to vacated synaptic sites, Turney and Lichtman (2012) developed a laser microsurgery technique that removed one of two closely spaced axons innervating the same NMJ. With this technique, Turney and Lichtman (2012) were able to demonstrate that within one day of removal of other axons, a remaining axon (whose elimination was imminent) occupied the sites of those that were experimentally removed. Even axons that had already withdrawn from a site reoccupy it if vacated. As Turney and Lichtman (2012) conclude: "these results strongly support the idea that the process leading to single innervation is competitive: an axon destined for elimination always survives if the other innervating axon is removed" (2012, 9).

Glia: Dark Matter of the Nervous System

Philip Morrison's landmark book and movie, called *Powers of Ten* (1982), opens our eyes to the embeddedness of all things across many orders of magnitude. At the same time, supercomputers and ever-larger telescopes measuring various forms of electromagnetic energy are revealing that the large-scale structure of the universe looks like a diaphanous web (Weinberg 2005). And most surprising of all is the matter that interacts with light—including stars, planets, and us—constitutes just a tiny fraction of all matter in the universe. The remainder is a mysterious "dark matter" (Spergel 2015). Perhaps, then, it should not be so surprising that, until recently, at the

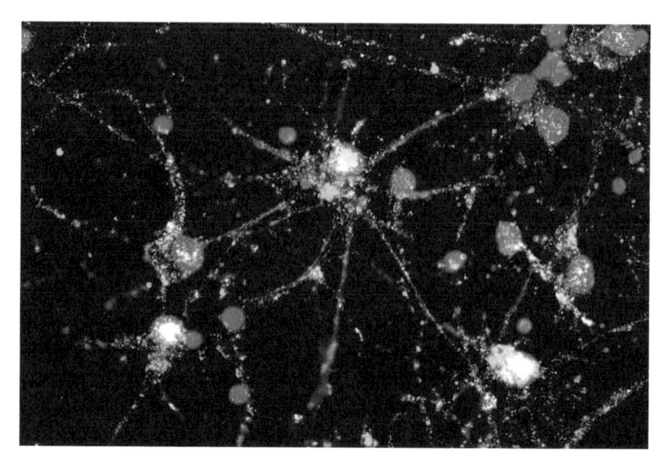

Figure 6.3 Glia interacting with synapses. The complement protein C4 (green) often overlaps with synaptic markers (red and white dots) in this culture of neurons (blue marks main cell bodies), a sign of how it may flag synapses for pruning in brain development and disease. Image by Heather de Rivera / McCarroll Lab, Department of Genetics, Harvard Medical School.

microscale, we have missed the significance of non-neuronal glial cells, which are intimately interconnected with neurons. Glial cells (microglia, oligodendrocytes, and astrocytes) detect and control neural activity, and a growing consensus is that these non-neuronal cells constitute "the other" nervous system (Fields et al. 2013) (see Figure 6.3).

Glial functions. Glia play several important functions during neurogenesis of healthy brains, including guidance of neuronal migration, regulation of the composition of the extracellular environment, modulation of synaptic connections, and clearance of neurotransmitters (see Tables 6.4 and 6.5). Oligodendrocytes form myelin sheaths for electrical insulation, increasing conduction velocity by 50-fold (Nave 2010). They provide vital support for axons and are involved in the inhibition of repair following spinal cord injury (see Chapter 8). Oligodendritic precursor cells (OPCs) are of particular interest because their differentiation into myelin-producing oligodendrocytes occurs during a period of vulnerability to injury as a result of premature birth (discussed later) (Back and Rivkees 2004). OPCs use the vasculature (blood vessels) as a physical scaffold for migration (Tsai et al. 2016). Astrocytes ensheath synapses and regulate neuronal excitability

Table 6.4 Roles of glia in health and disease: Microglia

Role	Description
Survey and sculpt developing synapses	Microglia are the resident macrophages and phagocytes of the CNS. Their processes constantly extend and retract as they survey their local environment as a means of affecting synaptic function.
Contribute to synapse elimination and formation	Phagocytic microglia prune synaptic connections during development by engulfing pre- and postsynaptic elements.
Promote functional synaptic maturation	Microglia regulate functional synaptic maturation.
Regulate synaptic plasticity	Microglia create and remove synapses, as well as change the strength of existing ones.

Source: W.-S. Chung, C. Welsh, B. Barres, & B. Stevens (2015), Do glia drive synaptic and cognitive impairment in disease? *Nature Neuroscience, 18,* 1539–1545.

Table 6.5 Roles of glia in health and disease: Astrocytes

Role	Description
Active participants in synaptic function	Astrocytes have a close physical association with synapses, which affords sensing and modulating synaptic activity, as well as forming and eliminating synapses. Astrocytes constantly modulate their physical contacts with synapses over the course of minutes.
Required for synapse maturation	The formation of mature functional synapses is not an autonomous property of neurons, but rather requires multiple astrocyte-secreted signals *in vitro* and *in vivo*.
Mediate synapse elimination through phagocytosis	Astrocytes express genes implicated in phagocytosis and physically eliminate synapses.
Modulate synaptic transmission and plasticity	Astrocytes participate in synaptic transmission and actively control synaptic plasticity.

Source: W.-S. Chung, C. Welsh, B. Barres, & B. Stevens (2015), Do glia drive synaptic and cognitive impairment in disease? *Nature Neuroscience, 18,* 1539–1545.

and synaptic transmission (Sloan and Barres 2014). They respond to injury by secreting extracellular matrix proteins and are implicated in many neurological and psychiatric disorders (Fields et al. 2013). Microglia are highly motile brain cells that detect pathological tissue alterations (Graeber 2010), but they also play an important role in healthy brain function, including the regulation of cell death, synapse elimination, neurogenesis, and neuronal surveillance (Schafer et al. 2012; Wake et al. 2013). Among the crucial roles played by glia, one of the most crucial is remodeling of the overabundant and relatively crude wiring of the initial synaptic connections made during development of the nervous system.

Important work by Beth Stevens at the Boston Children's Hospital reveals that microglia serve surveillance and scavenging functions in the nervous system: they engulf and clear damaged cellular debris after brain insult (Ransohoff and Stevens 2011; Salter and Beggs 2014). Considerable experimental evidence now supports the hypothesis that microglia also play a crucial role in the developmental wiring of healthy brains (Chung et al. 2015). Research with mice by Paolicelli et al. (2011) initially determined that disruption of the interactions between microglia and synapses delays the developmental process of spine formation in the hippocampus. To do so, Paolicelli et al. (2011) genetically modified a group of mice to prevent microglia activation by eliminating the functioning of a key signaling receptor. The result was that, compared with age-matched controls, the genetically modified mice showed increased density of hippocampal dendritic spines, establishing that microglia were the key mediators of synapse development (Ransohoff and Stevens 2011). But what is the actual function of the microglia at synaptic sites? To address this question, Schafer et al. (2012) conducted experiments on the retinogeniculate system of mice, specifically examining the activity-dependent pruning of initially exuberant connections between retinal ganglion cells (RGC) and relay neurons throughout the dorsal lateral geniculate nucleus (dLGN) of the thalamus. With the aid of high-resolution confocal imaging, Schafer et al. (2012) observed microglia engulfing RGC inputs undergoing active synaptic remodeling. Does this engulfment contribute to the normal process of synaptic pruning? As in cerebellum and the NMJ, it is believed that in dLGN, RGC inputs compete for postsynaptic territory and that less active or weaker inputs lose territory to stronger inputs (Schafer et al. 2012). By chemically injecting mice to manipulate activity in the right or left eye, Schafer et al. (2012) were able to demonstrate that microglia-mediated engulfment of RGC inputs was greater in the weaker eye, supporting their engulfment role in synaptic pruning. Microglia provide the engulfment function in other areas of the brain, as well, including somatosensory cortex (barrel centers, in mice) and motor cortex (Salter and Beggs 2014). In the latter, for example, a microglia-derived nerve factor called BDNF influences formation of synaptic structures in motor cortex during learning of a motor task (Parkhurst et al. 2013). Further, deletion of BDNF in genetically modified mice results in reduction of both dendritic spine dynamics and motor task performance (Parkhurst et al. 2013).

Anticipating Function

Following neuronal migration, but before the onset of sensory experience, there is a developmental period during which neurons exhibit spontaneous rhythmic bursts of high activity levels, followed by periods of quiescence. There is now substantial evidence that this correlated activity—recorded in the retina, cochlea, spinal cord, cerebellum, hippocampus, and neocortex—is the means by which developing neural circuits initially establish maps (see Blankenship and Feller 2009; Khazipov, Colonnese, and Minlebaev 2013; Kirkby et al. 2013 for reviews). To illustrate, early, correlated, spontaneous activity is apparent in the somatosensory system. In the rat, there is somatotopic mapping of the vibrissae (whiskers): the follicle of each whisker contains nerve endings that convert mechanical energy into action potentials, which are conveyed to the brain stem, thalamus, and somatosensory cortex. This latter structure is characterized by specialized barrel cells corresponding to the topological grid of whiskers on each side of the rat's snout (Diamond et al. 2008). During the first postnatal week of rat development, thalamocortical axons grow and project into the neocortex and form synapses with these whisker-specific barrel cortical neurons. Extracellular and patch-clamp recordings reveal two types of transient oscillations in this "critical period" prior to the onset of active perceptual exploration by the animal: spindle bursts (5–25 Hz activity lasting about 1 second) and shorter early gamma oscillations (EGOs) (40–50 Hz activity lasting about 200 msec) (Khazipov et al. 2013). It is notable that, in rats, EGOs are evident in barrel cell recordings only during a brief developmental time window, occurring from birth to postnatal day 7, when thalamic axons are entering the cortical plate and the barrel map is emerging. At day 7, EGOs and immature bursting disappear and are replaced by adult-like whisking behavior and mature gamma oscillations (Khazipov et al. 2013).

In the early developing spinal cord of species ranging from *Drosophila* to mouse, spontaneous electrical activity, along with genetic specification, is responsible for regulating the number of neurons produced and their differentiation before mature synapses are established (Arber 2012; Borodinsky, Belgacem, and Swapna 2012). Furthermore, calcium-mediated rhythmic bursting of electrical activity is important for axonal pathfinding, the process by which spinal neurons extend their axons in search of appropriate targets (Kastanenka and Landmesser 2010). Does spontaneous electrical activity also play a role in the construction of spinal circuits? Classic work on *Xenopus* using immobilizing anesthetics suggested that electrical spiking activity in developing spinal circuits was not necessary for the construction of functional spinal networks: these circuits developed even without activity (Buss, Sun, and Oppenheim 2006; see also Wenner 2012). However, a study by Warp et al. (2012) suggests a role for activity in developing function in nascent spinal networks. Using optogenetics and a genetically encoded calcium indicator, Warp et al. (2012) were able to show that activity over a period of just two hours promoted synchronization from just a few local neurons to a large network.

In the visual system, the immature retina generates spontaneous, periodic action potential bursts that sweep across retinal ganglion (RG) cells in the form of waves, at a frequency of approximately one per minute (Kirkby et al. 2013). RG cells projecting to the thalamus and superior colliculus are relayed to primary visual cortex and downstream so that these retinal waves drive correlated patterns of activity throughout the visual system (Ackman, Burbridge, and Crair 2012; Ackman and Crair 2014). Similarly, prior to the onset of hearing in the auditory system, cochlear inner hair cells fire action potential trains that sweep across spiral ganglion neurons at about three per minute (Clause et al. 2014).

Just prior to the time of birth, the mammalian visual system undergoes a dramatic transition (see Khazipov et al. 2013 for a review). In a landmark study, Colonnese et al. (2010) made evoked potential recordings from primary visual cortex of neonatal rats and premature newborn humans of visual responses to whole-field, 100-ms light flashes. In the rats, they observed three distinct periods, which they call physiological blindness of immaturity, bursting, and acuity. Most notable is the distinction between the bursting mode of visual cortical function and the adult-like acuity mode. During the bursting mode, light flashes reliably produced oscillatory patterns in the range of the gamma band and a second component, called delta brushes. Significantly, the visual response did not depend on the duration of light exposures. By contrast, two days before eye opening, the light-evoked oscillations and slow waves "disappeared" and were replaced by an adult-like visual-evoked potential spike. From these data, Colonnese et al. propose that "the early development of visual processing is governed by a conserved, intrinsic program that switches thalamocortical response properties in anticipation of patterned vision" (2010, 480). This apparent switching on of the adult-like function of vision is attributed to an intrinsic program. However, the nature of this program is not yet known. If visual function is governed by gene regulatory programs with switches, then it may be possible to leverage these to promote remodeling of the visual system in cases of degenerative disease, a possibility I discuss further in Chapters 7, 9, and 10.

Endogenous Brain Vulnerabilities Due to Mishaps in Developmental Processes

Each period in nervous system development is characterized by particular endogenous vulnerabilities because of the "bottom-up" nature of ontogenetic self-assembly: the proliferation, migration, and differentiation of early forming cells are the precursors of eventual fully elaborated brain networks (Sun and Hevner 2014) (see Table 6.6). Earlier sections illustrated how neurons not only multiply, but also follow the guidance of signaling pathways (Chao, Ma, and Shen 2009; Hilgetag and Barbas 2006), become functional under the influence of their more mature peers (Hoerder-Suabedissen and Molnar 2015), and selectively form synapses by through competition (Turney and Lichtman 2012). Glial cells, such as oligodendrocytes that wrap around neuronal axons to form a myelin sheath, form new partnerships with neurons to achieve their useful function (Bau-

Table 6.6 Cost–benefit trade-offs of structure and function in the human nervous system

Structure / Function	Cost	Benefit
Anatomically larger brain	Metabolically expensive (high aerobic glycolysis)	Most advanced for emergent functions
Densely interconnected networks with hubs forming a "rich club"	Increased susceptibility of critical pathways to injury Role of amyloid plaques in disease	High-capacity backbone for global brain communications
Cerebral circulatory system	Susceptibility to hypoxia and hyperoxia (e.g., middle cerebral artery)	Oxygen transport for high metabolic activity
Neuronal migration as the basis for building differentiated cortical layers	Mishaps of migration	Specialization of function
Axon myelination	Oligodendrocytes wrap around axons during a period of vulnerability to periventricular leukomalacia	Efficiency of axon conduction velocity

Sources: E. Bullmore & O. Sporns (2012), The economy of brain network organization, *Nature Reviews Neuroscience 13,* 336–349; and C. Metin, R. Vallee, P. Rakic, & P. Bhide (2008), Modes and mishaps of neuronal migration in the mammalian brain, *Journal of Neuroscience, 28,* 11746–11752.

mann and Pham-Dinh 2001). This section addresses the specific vulnerabilities and consequences of injury characteristic of each of these distinctive processes in nervous system development.

Mishaps in migration. It takes several months of arduous travel for the vast numbers of cortical neurons of the developing brain, each with particular birth dates, to migrate from their birthplaces in the proliferative zones of the cortical wall to their cortical plate destinations (Rakic 2009). Multiple cortical abnormalities result from errors of proliferation and migration, including polymicrogyria, lissencephaly, epilepsy, dyslexia, mental retardation, fragile X, and autism (Noctor et al. 2013). The products of genes associated with these disorders, as well as environmental factors, may disrupt proliferation and neuronal migration (Metin et al. 2008). However, the underlying disruptions that cause these "neural migration disorders" may not be due to a direct deficit in the process of neuronal motility. Changes in cell death, cellular differentiation, or cellular adhesion may alter positioning of neurons by blocking or changing physical pathways, without a direct change in the immature neuron's intrinsic ability to move (LoTurco and Booker 2013).

Why are there so many different disorders in which cell placement is disrupted during neuronal migration? Polymicrogyria (PMG), for instance, is a malformation in which a part of, or the entire, cerebral cortex contains many more sulci and gyri than seen in typical development; these are smaller than usual, and cortical lamination is reduced from six to two to three layers (LoTurco and Booker 2013). One possible explanation is that the highly heterogeneous populations of progenitor cells that form the cerebral cortex "protomap" (Rakic 1988, 2009) become separate targets for mutations of specific genes (Im et al. 2014). Sulcal topographies in PMG have been of special interest because earlier studies suggested that their spatial distributions appear to be related to mutations of particular genes and could help identify how genetic mutations are related to the processes by which the sulci and gyri form (Piao et al. 2004). To examine this possibility, Im et al. compared the sulci of children born with PMG and typically developing children by means of a graph structure: they first identified the deepest points of sulci in MRI images and then used these as nodes to construct the graph. They found that similarities in sulcal graphs between typical and PMG groups were significantly lower than the similarities measured within the typical group.

Im et al. (2014) conducted further network organization analyses, including measures of network segregation (clustering coefficient and transitivity), network integration (characteristic path length), and modularity (where a module is a group of nodes that have strong connections to other nodes within the module but weak connections to nodes outside the module). The global network analyses indicated reduced network segregation in the PMG group (lower clustering) and higher modularity. Im et al. (2014) additionally reconstructed whole-brain white matter fiber tracts and measured fractional anisotropy (FA) in the white matter near the PMG cortex, as well as the length of U-shaped fibers connecting a given gyrus to other adjacent gyri. The white matter network organization of PMG patients was characterized by decreased connectivity of short association fibers linking adjacent primary gyri, decreased FA, reduced connectivity of long association and inter-hemispheric fibers, and decreased connectivity within an involved node. Together, the results indicate that polymicrogyria is a reflection of the continuing roles of gene expression, multiple molecular pathways, and complex cell–cell interactions in abnormal neuronal migration (Rakic 2008, 2009).

Once each fetal neuron arrives in its target region, its axons begin to branch; synapse; and together with others, form an initial, crude, synaptic wiring diagram—the beginnings of neuronal networks (van Ooyen 2011). The events that occur during this early preterm phase in humans (twenty-three to thirty-two PCW age) include growth of thalamocortical fibers into the cortical plate, and elaboration of thalamocortical axons (Kostovic and Jovanov-Milosevic 2006). Crucially, it is also a developmental window of vulnerability for perinatal white matter injury, called periventricular leukomalacia, or PVL. This early preterm phase coincides with a particular moment in the lineage of progenitors of the oligodendrocytes (OL) that wrap around axons to form myelin but precedes myelination (Back et al. 2001). Injury due to PVL is, thus, related to the presence of late OL progenitors in the periventricular white matter (Back et al. 2001).

Complementary perspectives on injury. Brain injury in very low birth weight (VLBW) premature infants (< 1.5 kg) is recognized as a major public health problem, affecting several million infants worldwide (Muglia and Katz 2010). There has been a dramatic improvement in survival over the past decade, with a decrease in the incidence of PVL, but at the same time, an increase in the number of survivors with long-term neurodevelopmental morbidity. These survivors are characterized by a diverse spectrum of cognitive and motor outcomes (Back and Miller 2014). Our current understanding of the causes of PVL rests on a foundation of decades of work by many scientists, including neurologist Joseph Volpe and neuroepidemiologist Alan Leviton and their colleagues and students at the Boston Children's Hospital. Volpe's *Neonatal Neurology* textbook has become the classic reference work in the field, and he has brought attention to the dual roles of PVL and neuronal / axonal disease in brain injury of the premature infant (Volpe 2008, 2009). Leviton's work, such as the ELGAN neuroepidemiological study, for example, has revealed that PVL in extremely preterm infants is associated with inflammatory processes, such as microorganisms in placenta parenchyma (O'Shea et al. 2009). These complementary approaches to understanding brain injury in premature infants, generating vast amounts of neuroimaging, animal model, and neuroepidemiological data, are now converging on a "systems" view of the effects of injury on the interactions between different brain regions (Dean et al. 2013; Leviton et al. 2013; Leviton et al. 2015; Volpe 2009; Volpe et al. 2011).

One of the continuing puzzles about brain injury in premature infants, though, is that even with a decline in the severity of white matter injury, up to 25 to 50 percent of preterm survivors continue to exhibit a wide range of cognitive and social deficits, suggesting crucial roles for gray matter pathology (Back and Miller 2014; Dean et al. 2013) and cerebellar as well as cerebral cortical injury (Limperopoulos et al. 2012; Volpe 2009). In animal models of brain injury in prematurity,

Dean et al. (2013) report that cerebral ischemia disrupts cortical maturation by disturbing dendrite and spine formation, rather than by causing a loss of cortical neurons. Even without overt gray matter injury, the cortex of animal subjects showed evidence of widespread "dysmaturity" of the dendritic arbor of cortical projection neurons (Back and Miller 2014). This work suggests that a next step in understanding and treating brain dysmaturity is to identify the neuronal networks that contribute to these neurobehavioral impairments.

Developmental Emergence of Typical Brain Networks: Five Principles

A starting point for understanding the role of dysmature networks in neurobehavioral impairments is to identify the typical patterns of growth of different brain regions that are the basis for the emergence of function. Research points to five principles underlying the relation between developmental patterns of typical brain growth and the emergence of network properties (see Menon 2013).

Principle 1. The first principle is that *brain networks have an emergent organizational structure of "communities" characterized by "small world" architecture* (as discussed in Chapter 5). The communities are defined by *segregation*, a low density of connections between members of different communities, and *integration*, a high density of connections among members of the same community (Sporns 2013a,b). Three characteristics of brain networks that promote the integration of information are (1) "hubs," highly connected and highly central brain regions; (2) a tendency for network hubs to be mutually interconnected (Sporns 2013a,b); and (3) a connective core, or "rich club," of neural processes that play a role in the coupling of neural resources as a function of task demands (Sporns 2013a,b). Network communities develop functional hubs by age 2, establish more well-developed global brain architecture by age 8, and undergo significant restructuring during late childhood and adolescence (Khundra-Kapam et al. 2013; Hagmann, Grant, and Fair 2012; Power et al. 2010).

Principle 2. A second developmental principle is that *there is heterochronic growth in different brain regions, as well as regional differences in brain wiring,* evident in studies of brain volume changes during early development. Choe et al. (2013), for example, conducted volumetric analysis of typically developing infant brains aged over three to thirteen months and found that many structures exhibited measurable differences in growth across time. During this same period, there is an exuberant increase, and subsequent decrease, in connectivity evident both in measures of synaptic (Paolicelli et al. 2011) and connectome wiring (Tymofiyeva et al. 2012). Also during this early period in development, cortical hubs and their associated cortical networks (Sporns and Betzel 2016; van den Heuvel et al. 2012) are evident mostly in primary sensory and motor brain regions (Fransson et al. 2011). It is not until later in development that these hubs shift to the posterior congulate cortex and insula (van den Heuvel and Sporns 2013).

Principle 3. A third principle in the developmental emergence of networks is that *there is a reconfiguration of subcortical and cortical connectivity during childhood, resulting in new patterns of connectivity with development to adulthood* (Menon 2013). Supekar, Musen, and Menon (2009), for example, compared the functional organization of brain networks of 7- to 9-year-old children with that of adults (ages 19–22). They found that in children, subcortical areas, especially in the basal ganglia, were more strongly connected with primary sensory, association, and paralimbic areas and that there were prominent developmental changes in the degree, path length, and efficiency of wiring connections. By contrast, adults showed stronger corticocortical connectivity among paralimbic, limbic, and association areas. Moreover, in children at around age 7 compared to adults, three prominent functional networks exhibit weaker connectivity: the attention-directing salience network (anterior insula and anterior cingulate cortex), the self-referential default mode network (posterior cingulate cortex and ventromedial frontal cortex), and the decision-making central executive network (fronto-parietal cortex) (Menon 2013). Of particular significance is that functional maturation of the anterior insula pathways is a critical process in development of the more flexible cognitive control processes of adults (Uddin et al. 2011).

Principle 4. A fourth developmental principle is that *excitatory and inhibitory circuits establish a balance during development* (Isaacson and Scanziani 2011; Turrigiano 2011). Inferred from this principle is that *any excitatory-inhibitory imbalance may result in neuropsychiatric disorders* (Deco and Kringelbach 2014; Fornito, Zalesky, and Breakspear 2015). Circuits—the fundamental computational primitives of the nervous system—are based upon dendritic and axonal branching. Dendrites receive input through synaptic connections and sensory terminals, while axons transmit integrated signals from each neuron to the next via synaptic connections (Dong, Shen, and Bulow 2015). Excitation and inhibition are synaptic conductances that, together, orchestrate spatiotemporal cortical function (Isaacson and Scanziani 2011). In the cerebral cortex, interactions between GABAergic inhibitory interneurons and glutamatergic excitatory principal cells are *reciprocal:* interneurons inhibit principal cells and are excited by them. A major discovery about excitation and inhibition is that the latter generated in cortical networks is proportional to, or "in balance with" local and / or incoming excitation (Poo and Isaacson 2009). Changes in the weights of excitation or inhibition are accompanied by compensatory processes that preserve the excitability of cortical networks (Turrigiano 2011). This balance, however, does not mean that excitatory and inhibitory conductances cancel each other out: "despite the overall proportionality of excitation and inhibition, their exact ratio is highly dynamic" (Isaacson and Scanziani 2011, 233).

Principal 5. A fifth, and final, principle for the emergence of networks is that *glia play a crucial role in the developmental process of remodeling via synaptic pruning* (Bialas and Stevens 2013; Chung and Barres 2012; Clarke and Barres 2013; Schafer and Stevens 2013, 2015; Schafer, Lehrman, and Stevens 2012). The initial exuberant synaptic connections of the nervous system formed early in development are remodeled by a process of local activity-dependent synaptic pruning involving microglia (Schafer and Stevens 2013). In the visual cortex, for example, those presynaptic inputs that are less active are removed when microglia are attracted to "eat me" signals initiating phagocytosis, and those that are more active are maintained and strengthened to form eye-specific territories (Schafer and Stevens 2013).

Vulnerabilities and Neuropathologies

Together, these five principles of brain network organization—segregation and integration, regulation of regional patterns of growth, reconfigurations of regional connectivity, dynamic balancing of excitation-inhibition, and synaptic pruning—may help us further our understanding of developmental neuropathologies (for example, autism spectrum disorder and schizophrenia) and neurological diseases, including Alzheimer's disease.

Autism spectrum disorder. Autism spectrum disorder (ASD), for example, is characterized by impairments in social interaction and communication, and by restricted, repetitive, and stereotyped behavior (Zoghbi and Bear 2012). Research shows unequivocally that ASD is caused by genetic mutations (Abrahams and Geschwind 2008). Ebert and Greenberg (2013) review evidence that neuronal activity induces genes to regulate synapse formation, maturation, elimination, and plasticity. They hypothesize that disruption of the activity-dependent genetic regulatory programs controlling synaptic function significantly contribute to the molecular basis of ASD.

These findings of genetic mutations in autism may also underlie early brain overgrowth and dysfunction (Stoner et al. 2014). Stoner et al. (2014), for example, used a procedure called RNA *in situ* hybridization to examine cortical microstructure in postmortem tissue from children, 2 to 15 years of age, with or without autism. Most notable was the discovery of regions of patch-like abnormalities in laminar cytoarchitecture of the prefrontal and temporal cortices (but not occipital cortex). Stoner et al. (2014) suggest that their findings of laminar disorganization in prefrontal and temporal cortices are consistent with an etiology that includes migration defects, or changes, in early neurodevelopmental processes. But what are the consequences of such patchiness?

One possible outcome of a cortical cytoarchitecture with irregular synaptic connectivity is a disruption of synchronization between neural networks located in particular brain regions (Geschwind and Levitt 2007). A study by Dinstein et al. (2011), for example, reports disrupted inter-hemispheric synchronization in the spontaneous cortical activity of naturally sleeping toddlers with autism—but not in toddlers with language delay or typical development. Other studies under active task conditions have also reported abnormal synchronization across brain areas (Weng et al. 2010). Why, then, is weak synchronization related to the aberrant behavior of autistic individuals? As discussed in Chapter 5, synchronization (1) facilitates segregation of responses originating from different neuronal assemblies, (2) focuses spikes to a narrow window of the oscillation cycle to summate more effectively in target neurons, and (3) consolidates synaptic modifications (Uhlhaas and Singer 2015). Reduced synchronization may disrupt the process of communication between brain regions and lead to the kinds of disruptions in language and social behavior seen in autistic children (Edgar et al. 2015).

While most ASD studies have emphasized brain-specific mechanisms involved in ASD, there is also evidence of a contribution of the peripheral nervous system (Orefice et al. 2016). Individuals with ASD exhibit aberrant reactivity to sensory stimulation, especially tactile, and many patients exhibit increased sensitivity to vibration and thermal pain as well as altered somatosensation (Tomchek and Dunn 2007). From work with mice, it is also known that systemic virally mediated replacement of ASD-associated genes, such as Mecp2 (rather than intracranial viral delivery), restores behavioral deficits in male mice exhibiting ASD-like behavior. This suggests a potential role for somatosensory neurons in ASD. By using several ASD genetic models combined with behavioral testing of mice, Orefice et al. (2016) found a mechanosensory neuron synaptic dysfunction underlying aberrant tactile perception, as well as tactile processing deficiency contributing to development of anxiety-like behavior and social interaction deficits in adulthood.

Schizophrenia. Schizophrenia is also a brain disease that dramatically disrupts social behavior. At least four of the preceding principles of brain development have been implicated in the etiology of schizophrenia: integration-segregation, excitation-inhibition imbalance, glia-mediated inflammation, and regulation of regional patterns of growth. First, there may be a disruption of the process during fetal life by which gene expression provides molecular guidance for neurons to migrate during formation of the cerebral cortex: neurons in the cortex of schizophrenics may have migrated in such a way that they *do not establish circuits capable of generating typical segregation and integration of oscillatory patterns* (Ayoub and Rakic 2015). Second, once established, certain network circuits with high energy demands—parvalbumin-positive interneurons (PVIs), in particular—may be susceptible to oxidative damage that *disrupts the balance of excitation-inhibition* (Hensch 2005). PVI networks play a role in synchronizing the excitatory state of large numbers of pyramidal neurons, exerting precise temporal control of information flow (Do, Cuenod, and Hensch 2015).

The third developmental contributor to etiology of schizophrenia is oxidative stress and the *activation of glia.* To support such high-frequency neuronal synchronization, PVI are "energy-hungry" and, therefore, are highly vulnerable to oxidative stress. Oxidative stress, in turn, activates inflammatory microglial cells and damages both PVI and their surrounding perineuronal nets and myelin-forming oligodendrocytes. This damage alters local oscillations, distant synchronization, and critical period timing (Do et al. 2015).

The hypothesized link between the effects of oxidative stress on PVI and a hallmark of schizophrenia etiology—its prolonged period of onset—is a failure of critical period onset and closure during PVI circuit maturation (Do et al. 2015; Takesian and Hensch 2013; Toyoizumi et al. 2013; Werker and Hensch 2015). Do et al. (2015)

propose, for example, that PVI large basket cells are a "pivotal plasticity switch" and that they mature at different rates across brain regions. Oxidative stress on basket cells dysregulates expression of "molecular brakes" on plasticity, thus pathologically prolonging the developmental plasticity of PVI circuits. Over time, this removal of brakes on plasticity, altering of excitation-inhibition balance, and oxidation-based damage to myelin destabilize circuit functioning. By early adulthood, the destabilized circuit functioning results in disturbances in perceiving the emotional content of facial expressions (Azuma et al. 2015), becoming engaged in social coordination (Varlet et al. 2012), and social cognition (Green, Horan, and Lee 2015).

The devastating effects of prolonged plasticity on PVI circuits in schizophrenia at the circuit level is also apparent at the mesoscopic and macroscopic levels of analysis of neural oscillations and synchrony (Uhlhaas and Singer 2010, 2015) and of brain network connectivity (see, for example, Alexander-Bloch et al. 2013; Fornito et al. 2015). Studies of schizophrenic individuals using EEG / MEG have identified abnormalities in the amplitude and synchrony of neural oscillations (Uhlhaas and Singer 2010) and are consistent with the proposal that the balance between neuronal excitation and inhibition is fundamental to disordered behavior. Ongoing changes in oscillatory dynamics may be the basis for the onset of clinical symptoms in the prodromal phase, eventually resulting in the full expression of symptoms in the psychotic phase.

Maternal Immune Activation: Nexus of Developmental Brain Vulnerability

Maternal immune activation (MIA) is a response of the immune system during pregnancy that may be initiated by the mother's autoimmune and genetic predisposition and / or infection (Estes and McAllister 2016). The possibility of MIA during pregnancy places the intrinsic vulnerability of the developing fetal nervous system—and disorders such as cerebral palsy, autism, and schizophrenia—in the broader context of the mother's immune response (Estes and

McAllister 2016; Knuesel et al. 2014). However, MIA potentially provides a more complete understanding of the emergence of developmental disorders because it identifies processes by which nervous system vulnerabilities may be affected *well beyond* intrauterine life, through cascades of responses to environmental stressors and persistent inflammatory effects, or "second hits," during infancy, childhood, adolescence, and adulthood (Estes and McAllister 2016). A key process in the continuing postnatal influence of MIA is the effect of the changes in the expression of immune molecules, including cytokines on the mTOR (mechanistic target of rapamycin) signaling pathway. Such mTOR signaling regulates the assembly and maintenance of circuits and experience-dependent plasticity (Lipton and Sahin 2014). Immune molecule alterations in the mTOR pathway may result in dysregulation of protein synthesis in the brain, disrupting formation of parvalbumin (PV) cells, selectively altered in schizophrenia, as discussed earlier. There is evidence that in adult MIA offspring, there is a specific reduction in inhibition from PV cells onto pyramidal neurons sufficient to enhance anxiety-related behavior (Estes and McAllister 2016). An important implication of the discovery of immune system effects on the developing nervous system is that it may be possible to target therapies to particular stages in development (see Chapter 9).

• 7 •

How Nature Remodels and Repairs Neural Circuits

The Arnold Arboretum is an oasis of trees and ponds open to all and a popular spot for running, biking, or strolling among city dwellers and suburban neighbors. The "Arb" was established in 1872 as Harvard-owned land that is part of the park system of the city of Boston. It was designed by the great landscape architect Frederick Law Olmsted, and, thanks to several ambitious nineteenth-century expeditions, includes many plants from eastern Asia as well as other parts of North America. This diversity of trees, shrubs, and flowers has provided rich habitats for many species of small animal. I have been fortunate to visit many times. On a recent visit, as I walked next to a vernal pond, a salamander skittered by and darted under a rock. Within the pond, a frog used its long legs to swiftly propel itself with brief kicking bursts toward shore, just under the water's surface. A bird in the oak tree above my head sang its species-specific song. And a field mouse poked its head out from one of several visible holes on a nearby hillside, doubtless a part of a hidden maze of underground tunnels.

Although humans share a common evolutionary history with all of these animals, changing the time scale of this reverie by powers of ten reveals a wide range of evolutionary and developmental processes that nature uses to remodel their body plans and neural circuits. The salamander will regrow a tail lost to a predator, but humans do not have this spontaneous regenerative capacity. During the transformation from tadpole to frog, swimming switches from tail propulsion to kicking as the tadpole grows limbs and the tail is resorbed into the body. Early in its life, the songbird learned the song of its species by listening to the singing of its tutor—and did so within a critical period in development. And the exploratory behavior of the mouse is its means for learning the locations of food and shelter. These wondrous evolutionary solutions unfolding in developmental time—a sampling of the ways that nature remodels existing body plans, neural circuits, and behavior—provide invaluable lessons for using bioinspired devices to repair injured neural circuits.

The chapter focuses on nature's remodeling processes in animals and how these processes may inform the loss of sensorimotor function that occurs due to spinal cord injury (SCI), stroke, or loss of vision in humans. Stroke is a major cause of severe disability, robbing individuals of hand function, especially in the elderly, with only limited recovery following large strokes (Zeiler and Krakauer 2013). Nevertheless, a stroke model with adult rats has shown that if a nerve growth-promoting antibody therapy precedes rehabilitative training, forelimb function may be nearly completely restored (Wahl et al. 2014). In loss of vision, brain plasticity is possible, but there are structural and functional "brakes" on plasticity (Takesian and Hensch 2013). Plasticity in adult rodents may be induced through genetic, pharmacological, and environmental removal of these brakes to enable recovery of vision (Bavelier et al. 2010). But much remains to be learned about critical periods for neuroplasticity and what causes critical periods to open and close (Toyoizumi et al. 2013).

The first part of the chapter is organized around four comparative case studies of remodeling and their implications for neuroprosthetic devices, neural repair, and neurorehabilitation: metamorphosis from tadpole to frog in *Xenopus*, regeneration of neural circuitry in salamanders, critical periods for acquisition of birdsong, and plasticity during the exploratory whisking of rodents. What might the regeneration of the salamander's tail teach us about ways to restore function to humans following SCI? Is there a developmental lesson for rehabilitation scientists in the way that the same spinal circuits for the tadpole's tail are repurposed for swimming by the mature frog that has lost its tail? How is motor sequence production in songbirds like the dynamical systems of the human motor cortex? What is the significance of the limited period during which a songbird must hear its species-specific song for understanding the constraints on plasticity? Is it possible to reopen critical periods?

Evolution at the Water's Edge

The water's edge, a transition from a liquid medium to a solid substrate, is a natural laboratory for studying transformations of the vertebrate body plan and behavior for a particular adaptation: how to locomote on land when you begin life swimming like a fish. At the boundary of water and land, the effects of forces acting on the body change from buoyancy when under water to the full effects of gravitational pull on body mass. There is also a change in the effect of moving body appendages against a watery surface versus a solid resistive substrate. Evolutionary biologists have used fossil remains, as well as extant experimental animals, to better understand how changes in the effect of a gravitational field and of reactive forces when the body makes contact with solid surfaces influences remodeling of body form and function.

At an evolutionary time scale, the discovery of the fossilized bones of *Tiktaalik rosae*, a Devonian tetrapod-like fish, is evidence of specific morphological transformations for terrestrial locomotion. *Tiktaalik* fossils are characterized by raised and dorsally placed eyes, a mobile neck, and a pectoral girdle and forefins capable of both appendage support of the body against gravity and complex movements (Daeschler, Shubin, and Jenkins 2006; Shubin, Daeschler, and Jenkins 2014). However, the fossils themselves do not address questions about the process by which behavior at the water–land boundary influenced transformation of body morphology nor how, reciprocally, behavior was transformed by morphology (Gottlieb 2007).

One way for scientists to "animate" the fossil remains of *Tiktaalik* is to study the behavior and morphology of an extant relative, such as the African lungfish *(Protopterus annectens)*, as have King et al. (2011). *P. annectens* are remarkable animals that swim like fish but have lungs for breathing out of water. They are also a useful model for understanding locomotion in the earliest tetrapods because, like their sister species, air-filled lungs provide sufficient buoyancy to lift a heavy body clear of the substrate (King et al. 2011). Do lungfish use substrate-based locomotion? Is the body morphology of lungfish consistent with tetrapod locomotion inferred from fossil trackways? King et al. (2011)

address these questions through behavioral recordings and anatomical measurements. Essentially a fish with lungs, *P. annectens* use substrate-dependent and pelvic-driven bipedal locomotion, with gaits ranging from walking to bounding. A fascinating additional finding concerns the role of the soft and flexible morphology of the fin: as the pelvic fins contact the substrate, they bend along their length to form temporary support regions, similar to the function of a foot. Thus, like the soft architecture of an octopus limb bending to form functional joints (see Chapter 4), nature uses available body materials to temporarily assemble a device for a particular function (Hochner 2012; Levy, Flash, and Hochner 2015). One of the remaining puzzles is why aquatic tetrapods subsequently evolved jointed skeletal appendages when soft fins worked adequately for locomotion.

On the heels of the discoveries with *P. annectens* is a set of findings from a study that experimentally manipulated early locomotor experience in a related species of lungfish, *Polypterus bichir*, capable of breathing and locomoting on land. Standen, Du, and Larsson (2014) reared one group of *P. bichir* on land for eight months and then compared their skeletal anatomy and locomotion on land with a second group of *P. bichir*, reared in their typical aquatic environment. The logic behind the experiment was to determine whether the increased influence of gravity, as well as sensory information from contact of the fins with the substrate, would result in group differences in locomotor kinematics and in the shape of anatomical support structures. When both groups were tested, the treatment group animals walked differently from controls in two ways. First, fins were planted closer to the midline, the head was lifted higher, and there were smaller tail oscillations—all indications that the land-reared animals used a more energy-efficient gait. Second, treatment group *P. bichir* had more predictable patterns of gait timing than control fish: they precisely timed tail and body thrust to occur simultaneously, with the body and head lifted from the substrate by the fins (Standen et al. 2014).

Remarkably, compared to controls, the treatment group also exhibited anatomical changes in response to being reared on land: the structures used for supporting the head and body during locomotion had significantly different shapes. These changes may have been due to the increased gravitational forces acting on bone and soft tissue, as well as the effect of the fins moving through a larger range of motion during walking compared to swimming (Standen et al. 2014). Finally, when these observed group changes were compared to anatomical changes in Devonian stem tetrapods, there were notable similarities: strengthening of ventral bracing through the clavicle, increased flexibility between the pectoral girdle and operculum, and emergence of a functioning neck from the dissociation of the pectoral girdle and head (Standen et al. 2014).

The lessons to be learned about remodeling at the time scale of evolution from these studies at the water's edge, and the ones to follow, include:

1. Nature builds new devices by remodeling well-functioning evolved systems, rather than developing entirely new systems.
2. It may be possible for evolving systems to be remodeled by reengaging developmental processes.
3. The process of remodeling is most effective within econiches that provide an initial opportunity for exploring variability of self-produced behavior.

Metamorphosis and Evolution of Locomotion

Not easy being green. Metamorphosis is defined as a postembryonic transformation of a larva into a juvenile (Laudet 2011). Insects, amphibians, and fish exhibit metamorphosis. However, while metamorphosing *Lepidoptera* (butterflies) remain immobile in a nonfeeding stage in their protective chrysalis, frogs remain active, avoiding predators and catching food. There are, nevertheless, some common principles of metamorphosis, including: (1) it is an ecological transformation in which the larvae and adult do not share the same resources; (2) it is an extensive transformation of the animal, sometimes resulting in a radical change in body plan organization; and (3) it is regulated by hormonal and neuroendocrine systems whose specialized cells detect environmental signals and initiate processes to remodel organ systems so they are better adapted to an ecological niche (Laudet 2011).

Endopterygote insects, including *Lepidoptera,* are among the most successful living creatures in both species abundance and diversity (Grimaldi and Engel 2005). In order to transform within their chrysalis into insects that are capable of flight, *Lepidoptera* undergo complete metamorphosis of multiple internal structures. Lowe et al. (2013) have used high-resolution X-ray microtomography (micro-CT) to document the longitudinal changes that occur during the thirteen days of larval metamorphosis in the butterfly Painted Lady (*Vanessa cardui*). The result of these recorded images is a time-lapsed, three-dimensional characterization of the metamorphosis of certain organ systems inside the living chrysalis. One theme from the study findings is that certain organ systems of flying insects undergo an increase in size and functional capacity to support the emergence of flight. For example, the thorax is the main body unit for both legs and wings and is characterized by openings for respiration, called spiracles. The spiracles attach internally to tracheae, the internal respiratory system. The tracheae are characterized by spines that prevent entry of foreign particles and simultaneously attract water vapor from gases exiting the body (Grimaldi and Engel 2005). Metamorphic changes of the tracheae include an increase in volume and number of the tracheal branches associated with the flight muscles (Lowe et al. 2013).

In its watery ecological niche, survival of a freely swimming and feeding tadpole is dependent on being able to swim with changes of speed and direction. Zhang, Issbergner, and Sillar (2011) have found that during a twenty-four-hour period from the time of hatching to free swimming, there are two changes in the myotomal motorneuron (MN) pool that may endow the larval *Xenopus* with more flexible swimming behavior (see Table 7.1). First, MNs that dictate muscle recruitment order become capable of differentiated firing probabilities, giving the animal greater flexibility because intensity and frequency of

Table 7.1 Metamorphosis in *Xenopus:* Axial, combined, and limb-based locomotion

Stage	Description
Pre-metamorphic axial-based swimming (<stage 53)	Undulatory propulsive movements are generated by alternate bilateral contractions of axial myotomes, with characteristic rostro-caudal delay down the body. Limb circuitry is absent.
Early pro-metamorphic combined axial and hindlimb-based locomotion (stages 54–56)	The hindlimbs are not yet sufficiently developed to contribute actively to locomotion. Swimming is still performed by tail undulations, while the emerging hindlimbs are held against the body. Spinal cord motor output to the tail and hindlimbs remains tightly coordinated in a single rhythm. The future limb-kick CPG remains embedded within the precursor tail swimming network.
Metamorphic climax segregated axial and hindlimb-based locomotion (stages 59–63)	Animals are able to use both tail- and limb-based locomotor synergies, conjointly or independently. In contrast to the fast left–right alternation of tail undulations, rhythmic limb movements are slower and bilaterally synchronous, corresponding to kicks in which the flexor and extensor muscles of each limb are alternately active. The spinal cord can generate separate motor patterns appropriate for both locomotor modes, suggesting the existence of virtually independent CPG networks.
Post-metamorphic froglet limb-based swimming (>stage 64)	The tail has been resorbed, and swimming is now produced exclusively by slower, bilaterally synchronous cycles of hindlimb extension and flexion.

Source: D. Combes, S. Merrywest, J. Simmers, & K. Sillar (2004), Developmental segregation of spinal networks driving axial- and hindlimb-based locomotion in metamorphosing *Xenopus laevis, Journal of Physiology, 559,* 17–24.

muscle activation can change on a cycle-by-cycle basis. Second, the peripheral innervation fields become restricted within the dorsoventral plane of each myotomal muscle block, making possible more differentiated control of pitch and roll during swimming. Zhang et al. (2011) propose that these two developmental changes in MNs are the substrate for an early evolutionary achievement, conserved in other vertebrates: a device from which different combinations of locomotor speed and direction may be selected—*a locomotor rheostat.* There is evidence for this claim in similar topographic recruitment order of spinal premotor interneurons (INs) and MNs during changes in swimming speed in larval zebrafish (Ampatzis et al. 2014; Kinkhabwala et al. 2011). This device, like other special-purpose devices and smart perceptual instruments in Nature's toolkit, as discussed in Chapter 3 (for example, muscles coupled with spring-like tendons, pumps, measurement tools, and navigation systems), becomes available for assembly into a variety of organs for adaptive behavior in unpredictably changing ecological niches.

The metamorphic transformation of amphibians is a window into the evolutionary and developmental processes by which a body plan adapts to living in an environment at the boundary between water and land. In amphibian metamorphosis, we see a complex interplay among morphology, nervous system organization, biomechanics, and behavior that promotes locomotion better suited to land and the effects of gravity in a gaseous medium, but also capable of swimming. This metamorphosis, involving resorption of the tail and emergence of coordinated use of the legs for jumping in air and swimming in water, has profound implications for understanding the adaptive nature of nervous system circuits for driving excitable biomechanical systems.

The developmental nexus for frog metamorphosis is the coexistence of two pattern-generating networks: early-developing axial circuits exhibiting relatively faster (3–5 Hz) oscillations of the tail, and circuits for slower (1–2 Hz) appendicular (limbed) kicking, both under neuromodulatory influences (Sillar et al. 2008). Whereas axial circuits make possible alternating contractions of trunk muscles on the left and right sides of the body, the later developing appendicular network for control of the hindlimbs is characterized by bilaterally synchronous activity (Combes et al. 2004). During metamorphic climax (stages 58–63), there is a developmental sequence in which appendicular circuits are (1) present but not functional, (2) functional but not separable from the axial network, (3) functionally separable from the axial network (with the possibility of switching between or combining them), and (4) wholly capable of propulsion even though the tail has been resorbed (Rauscent et al. 2006).

What, then, is the basis for selection of the axial or appendicular networks for locomotion when both are functional? From a physiological perspective, selecting particular circuits is achieved through neuromodulation by various amines, neuropeptides, and other molecules

(Bargmann 2012; Marder 2012). For vertebrate locomotor control networks, in particular, the monoamine 5-HT is a phylogenetically old, conserved neuromodulator of spinal cord circuits (Miles and Sillar 2011; Sillar, Combes, and Simmers 2014). Interestingly, another neuromodulator of spinal locomotor circuitry, noradrenaline (NA), exerts an influence on period and duration of motorneuron bursts opposing the action of 5-HT. A study by Rauscent et al. (2009) examined the role that opposing aminergic modulation might play in controlling the motor output of coexisting axial and appendicular circuitry. In experiments that applied 5-HT to isolated spinal preparations, they found that (1) there was an acceleration of appendicular rhythms and a slowing axial activity and (2) 5-HT induced a cycle-by-cycle coupling of appendicular and axial rhythms into a single, combined rhythm. By contrast, application of NA had the opposite effect: cycle period of axial MN bursts decreased, and cycle period of appendicular bursts increased, with the result that the coupled rhythms separated into distinct patterns of motor burst activity.

Animals and robots. From a dynamical systems perspective, the selection of axial or appendicular rhythms from the same substrate may be indicative of bi-stable attractor dynamics driven into one or the other state by a control parameter—in these experiments, the complementary actions of the neuromodulators 5-HT and NA. What is the consequence for remaining lumbar spinal segments and the muscles formerly controlling axial propulsion when the tail is resorbed? Beyeler et al. (2008) addressed this question through the use of electromyography (EMG) recordings of muscles controlling straight-ahead swimming of adult frogs. In complex systems, when an existing component is no longer required for its original function, it may be transformed for other functions (Goldfield 1995). In the case of metamorphosis, axial tail motion during the larval tadpole stage is driven by lumbar muscles. When the caudal spinal cord segments that control axial movements disappear with tail resorption, the segments above the lumbar enlargement are released to perform other functions. Beyeler et al. (2008) report that when legs sprout and are used for propulsion by means of kicking, the lumbar muscles begin to take on a postural support function, helping to stabilize the body for swimming.

There have been new collaborations between developmental scientists and experts in robotics to test hypotheses about neural control and environmental interaction mechanics during swimming in water and locomoting on land (Ijspeert 2014). An ongoing project at the Wyss Institute, a swimming robot called *Metamorpho* (Goldfield 2016), has been exploring how component oscillation frequencies of the tail and legs may contribute to the development of coordinated swimming behavior. One possibility, for example, is that there are preferred frequency couplings of the tail and legs for swimming and that the transition from swimming to locomoting on land is related to symmetry breaking dynamics (Collins and Stewart 1993; Golubitsky et al. 1999; Holmes et al. 2006; Pinto and Golubitsky 2006; Turvey et al. 2012). Another question is how animals living in water adapted to the transition from water to land. One hypothesis is that once a well-functioning system has evolved, such as the eel-like swimming of lamprey, it is easier to modify it to adapt to a change in ecological niche, such as walking on land, than build a new system from scratch (Bicanski et al. 2013a).

Salamanders have been chosen as a model system for understanding nature's remodeling of existing circuits because these animals face the problem of generating propulsive forces in water and on land—two media with large differences in the density, viscosity, and gravitational load they impose during locomotion (Bicanski et al. 2013a; Ijspeert 2008). The robot body consisted of axial segments and rotational limbs, with sensory feedback provided by stretch sensors on both sides of each joint. The circuit architecture of the salamander spinal cord was emulated by a system of coupled, nonlinear oscillators (see Chapter 2) implemented in the robot's tail and limb controllers (Crespi et al. 2013). Ijspeert et al. (2007) were able to demonstrate that the robot's oscillator network could be set to produce traveling waves, emulating the way that real salamanders propel themselves in water (see Figure 7.1). During walking, by contrast, strong coupling from limb-to-body oscillators forces the body to oscillate at a low frequency with

Figure 7.1 *Salamandra Robotica,* a salamander robot that can swim and walk. This robot was designed to test hypotheses about the organization of salamander spinal circuits and the mechanisms of gait transition. The waterproof robot is equipped with eight motors for spine undulations and four motors, one per leg, for leg rotation. Photo courtesy of Auke Ijspeert.

an S-shaped standing wave. Moreover, these traveling waves are not observed simultaneously with limb movements, supporting the main hypothesis that two modes of locomotion are achieved by switching between traveling waves (for swimming) and standing waves (for walking). The possibility of developing robotic controllers with coupling parameters that allow a collection of local limit cycle oscillators to interact in controlled fashion opens the door to the use of wearable robots for remodeling injured nervous systems, a possibility considered in Chapters 8 and 9.

Regeneration

Because humans are not capable of spontaneously regenerating severed spinal cord connections, spinal cord injury is a devastating event. Spinal cord injury has motivated laboratories around the globe to intensively study regeneration in different animal species for insights into the means by which nature regrows body parts that are partially or completely severed from the nervous system. *Regeneration* may be defined as "reacquiring nervous system function after injury or disruption, with or without the need to replicate the original structure, or its full functionality" (Tanaka and Ferretti 2009, 713). Studies using experimental amputation with animals such as salamanders address questions that include why amputation starts regeneration and how a limb "knows" what parts to regenerate (Tanaka 2016). For example, following experimental amputation, salamanders regenerate the severed tail as well as the lost spinal cord with the correct number of segmental vertebrae and myotomes (Mchedlishvili et al. 2012). Two fundamental questions that may be addressed by comparative research on regeneration are (1) how the process of regeneration is related to the embryogenetic mechanisms that initially built the structure (Nacu and Tanaka 2011) and (2) how factors such as inhibitory molecules, glial scar formation, and axonal changes in myelination with development may impair regrowth of injured axons within the adult central nervous system. By learning more about nature's processes for regenerating severed connections, we may be inspired to emulate those processes in remodeling the injured nervous system.

How is regeneration of the spinal cord in the salamander related to its initial development? There is now evidence that in the salamander, the steps of progenitor cell-patterning and controlled neurogenesis that naturally regenerate a severed tail largely *recapitulate* the steps followed during early embryonic development to initially build the central nervous system (Nacu and Tanaka 2011; Nacu et al. 2016). For example, ependymal cells are descendants of radial glial cells retained from the earliest developmental stages in regenerating vertebrates. The ependymal tube that gives rise to regenerated spinal cord following salamander tail amputation is very similar in appearance to the early structure of the neural tube of developing amniotes (Tanaka and Ferretti 2009). But how does that recapitulation occur? By using a transgenic axolotl that expresses green fluorescent proteins (GFPs), Mchedlishvili et al. (2012) further examined the regenerated spinal cord by replacing a segment of the spinal cord from a typical animal with a piece of the spinal cord from a GFP-expressing animal—that is, one with green fluorescent cells. They found that the implanted cells in the experimental animals regenerated a green spinal cord! Thus, regeneration may be a more neural stem cell–like, or pluripotent, state as a response to injury (Diaz Quiroz and Echeverri 2013).

Glial scars. Until recently, it was widely believed that following central nervous system (CNS) injury in mammals, damaged glia release inhibitors of axon regeneration and that astrocytic scar formation prevents transected axons from regrowing spontaneously (Liddelow and Barres 2016). However, experiments with adult mice by Anderson et al. (2016) have demonstrated that contrary to this prevailing view, astrocyte scar formation actually aids CNS axon regeneration. Anderson et al. (2016) used genetically modified mouse models to prevent formation of reactive astrocytes—either by selectively killing these cells or by deleting a transcription factor necessary for astrocyte formation—and found no spontaneous regrowth of damaged axons. Instead, removal of the astrocytic scar promoted cell dieback. Indeed, when the scar was present and was additionally injected with a gel laden with growth-promoting factors, there was actually axon regrowth. Together, the results of these experiments provide evidence that astrocyte-scar formation *aids,* rather than inhibits, postinjury axon regeneration (Liddelow and Barres 2016).

In a separate set of experiments, Clifford Woolf at Boston Children's Hospital and a large consortium of investigators across the country have demonstrated that following a preconditioning lesion, a particular mouse strain, CAST / Ei, is able to regenerate injured CNS neurons (Omura et al. 2015). This work draws upon an insight from the regenerative capacity of the peripheral nervous system (PNS): preconditioning PNS neurons by means of a peripheral axonal injury prepares them to regenerate more vigorously to a second injury (Tedeschi 2011). Why are CAST / Ei mice given a preconditioning lesion able to similarly regenerate injured CNS? Omura et al. (2015) have found that an Activin signaling cascade in these mice initiates transcription to orchestrate a regenerative response.

Myelin. Another approach to better understanding functional recovery of the CNS following injury is to examine whether developmental mechanisms are reused during oligodendrocyte recovery (Gallo and Deneen 2014). Injury to oligodendrocytes (OLs) due to preterm birth has a dramatic impact on myelination (as discussed in Chapter 6). Animal models of perinatal periventricular white matter injury indicate that changes in the levels of specific growth factors induced by white matter injury are similar to those seen in preterm infants with periventricular leukomalacia (PVL) and that injury may reactivate endogenous developmental programs for OL recovery (Back and Rosenberg 2014). The adult brain contains populations of undifferentiated oligodendrocyte precursor cells (OPCs), able to divide and regenerate myelinating OLs, promoting the hope of recapitulating developmental programs to treat white matter injury (Gallo and Deneen 2014). Furthering this hope is that several intrinsic and extrinsic regulators of OL development have been identified as regeneration enhancers. Like the injured preterm infant brain, specific adult brain growth factors activate and direct endogenous pools of OPCs from proliferative areas to injury sites to promote recovery (Gallo and Deneen 2014).

Hair cells. There have also been breakthroughs in understanding hair cell regeneration. Loss of mechanotransducing cochlear hair cells due to acute noise trauma results in hearing loss, motivating decades of research on how to regenerate hair cells (Hudspeth 2014). Limitations of endogenous mammalian hair cell regeneration are related, in part, by the complexity of the cochlea, which would be disrupted by a regenerative response to insult (Fujioka, Okano, and Edge 2015; Zhao and Muller 2015). The characteristic "checkerboard" architecture of hair cells and supporting cells, though, offers a potential point of entry for attempts in regenerative medicine to differentiate undifferentiated epithelial cells into hair cells and potentially restore hearing (Fujioka et al. 2015). During development, epithelial cells acquire different fates, becoming either hair cell or supporting cell, through a process of lateral inhibition. Notch signaling (which modulates molecular differences in controlling how cells respond as part of developmental programs) prevents supporting cells from differentiating into hair cells (Artavanis-Tsakonas, Rand, and Lake 1999; Kelley 2006). With a mouse cochlear damage model, Mizutari et al. (2013) have demonstrated that by using an inhibitor to manipulate notch signaling, it is possible to transdifferentiate supporting cells into hair cells. Moreover, the new hair cells contributed to a partial reversal of their hearing loss.

Birdsong: Critical Periods for Exploratory Behavior

Birdsong consists of sounds generated during expirations, with silent periods (called minibreaths) between phonations (Trevisan, Mindlin, and Goller 2006). There is every indication in the data on birdsong that a generative dynamical system for sequence production and remodeling may be common to species as diverse as humans and primates. For example, the complex temporal structure of the song is due to oscillatory dynamics in the songbird analog of the motor cortex, the premotor HVC (used as a proper name) (Lynch et al. 2016), as well as to the rapid switching between expiration and inspiration, to phonation and filtering, and to resonation and modulation of articulators with exquisitely controlled biomechanical properties (Amador et al.

2013). HVC has a central role in controlling the temporal structure of birdsong, by producing a sequence of bursts (Long, Jin, and Fee 2010). The anterior forebrain pathway pathway (AFP) includes HVC neurons, the striatal region called Area X, the thalamic nucleus DLM, and the telencephalic nucleus LMAN, which provides input to the song motor pathway (SMP) via its projection to RA (Tschida and Mooney 2012a,b). How do developmental changes in HVC promote changes in motor sequencing in songbird vocal development? To address this question, Okubo et al. (2015) examined singing-related firing patterns of HVC projection neurons as they relate to developmental emergence of three stages of birdsong: subsong (highly variable, akin to human babbling), protosyllable (incorporating a variety of syllable types), and motif (comprised of reliable syllable sequences) (see Figure 7.2). A notable characteristic of the developmental sequence of these stages is the emergence and splitting of distinct syllables from a common undifferentiated precursor, suggesting that new syllables may emerge through interactions of changing networks of neurons.

Zebra finch vocal behavior has also been used to model the way that birds learn to modify their own song by imitating the song of a conspecific tutor during a juvenile-sensitive period for vocal learning (Olveczky and Gardner 2011) (see Table 7.2). Only male zebra finches develop a functional song circuit, beginning in the embryo and continuing after hatching (Olveczky and Gardner 2011). Juvenile songbirds learn to sing during two phases of a sensitive period, both dependent upon auditory experience. In the first phase of perceptual learning by listening to others, the juvenile bird listens to and memorizes one or more tutor songs. Then, in the second phase of sensorimotor learning, by listening to its own sounds, during tens to hundreds of thousands of song repetitions over many weeks, the juvenile uses auditory feedback to compare its song to the memorized model (Brainard and Doupe 2013; Mooney 2009). Song variability over these thousands of repetitions (called plastic song) is believed to be essential for the juvenile to learn the tutor song and produce it in stereotypic fashion (Mooney 2009). The song system consists of two distinct neural pathways: SMP and AFP (Tschida and Mooney 2012a,b). The former includes the premotor HVC

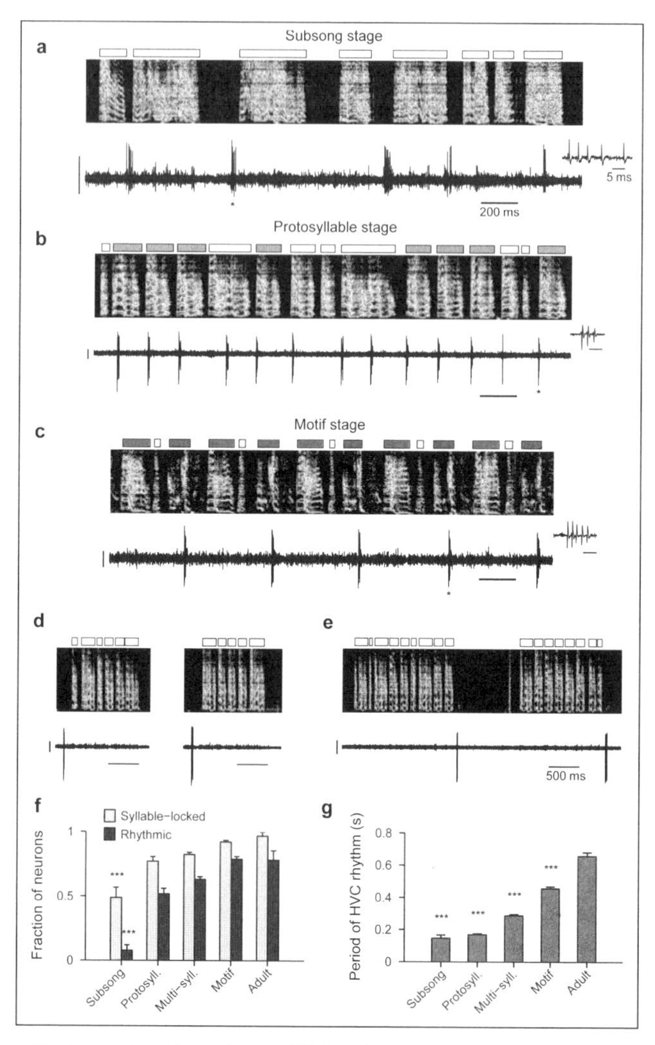

Figure 7.2 Singing-related firing patterns of HVC projection neurons in juvenile birds. (a) Neuron recorded in the subsong stage, before the in juvenile formation of protosyllables (RA-projecting HVC neuron, HVCRA; 51 dph; bird 7). Top, song spectrogram with syllables indicated above. Bottom, extracellular voltage trace. (b) Neuron recorded in the protosyllable stage (HVCRA; 62 dph; bird 2). Protosyllables indicated (gray bars). (c) Neuron recorded after motif formation (HVCRA; 68 dph; bird 8). (d) Neuron bursting exclusively at bout onset (X-projecting HVC neuron, HVCX; 61 dph; bird 2). (e) Neuron bursting exclusively at bout offset (HVCRA; 65 dph; bird 2). (f) Developmental change in the fraction of neurons locked to syllable onsets (gray) and fraction of neurons with rhythmic bursting (black) (mean ± s.e.m.; n = 39, 135, 565, 378, and 32 neurons, respectively). (g) Mean period of the HVC rhythmicity as a function of song stage (n = 3, 70, 356, 298, and 25 neurons, respectively). ***$P < 0.001$, post-hoc comparison with the adult stage. Spectrogram vertical axis 500–8,000 Hz. ms; panels d–e, 1 mV, 500 ms. Inset in show zoom of bursts indicated by an asterisk; scale bar, 5 ms. Tatsuo S. Okubo, Emily L. Mackevicius, Hannah L. Payne, Galen F. Lynch, and Michale S. Fee (2015). Image reprinted with permission from Nature Publishing Group/Macmillan Publishers Ltd.

Table 7.2 How juvenile song birds produce subsong through vocal babbling and behavior of mature birds

Component	Function
Babbling, or subsong	Juvenile birds learn the causal relation between actions and the effects of those actions by producing highly variable, exploratory behavior
Developmental progression in song	Subsong (babbling) occurs from 30 to 40 days posthatch (dph), followed by plastic song and gradual appearance of distinct and identifiable, but variable, vocal elements (syllables). By 80 to 90 dph, plastic song is gradually transformed into highly complex stereotyped motifs (sequences of syllables that constitute adult song), called crystallized song.
Anterior forebrain pathway	The output of AFP through the forebrain nucleus (LMAN) modulates or instructs the motor pathway (premotor circuit) and plays a role in producing song variability. Thus, juvenile singing is driven by a circuit (LMAN), distinct from that which produces adult behavior.
Continuing plasticity of crystallized song in mature birds	Mature birds are able to learn to modify small variations in the pitch of their vocalizations in response to experimental disruptions. This may reflect an adaptive process in adult birds that helps them to maintain stable, learned song, despite changes to the vocal control system arising from aging or injury.

Sources: D. Aronov, A. Andalman, & M. Fee (2008), A specialized forebrain circuit for vocal babbling in the juvenile songbird, *Science, 320,* 630–634; and M. Brainard & A. Doupe (2013), Translating birdsong: Songbirds as a model for basic and applied medical research, *Annual Review of Neuroscience, 36,* 489–517.

neurons, the downstream song premotor nucleus RA, and projections to vocal-respiratory structures in the brainstem that control song production.

Early in song learning, the variable activity in the AFP drives vocal exploration required for trial-and-error learning (Woolley et al. 2014). As vocal learning progresses, HVC axons start to make functional connections in RA, and it is the experience-dependent refinement of synaptic connections within the motor pathway that is believed to be the neural substrate for the gradual convergence to imitated song sequences (Olveczky and Gardner 2011). To determine how learning during the sensitive period is related to properties of the dendritic spines of neurons implicated in correlating the song of tutor and learner, Roberts et al. (2010) used two-photon *in vivo* imaging to measure dendritic spine dynamics of HVC neurons and used intracellular recordings before and after the first day of tutoring to measure target neuron activity. For birds with higher levels of spine turnover before tutoring, there was increased stabilization and enlargement of dendritic spines, as well as enhanced synaptic activity, suggesting that learning occurs with the strengthening of synaptic connections.

The findings from birdsong on the significance of variability are directly relevant to learning in humans and other mammals, as is apparent in parallels with young children's opportunities for producing and listening to their exploratory babbling (Lipkind et al. 2013), in the continuing role of motor variability for adult learning (H. G. Wu et al. 2014), and in the importance of variability for learning by rats following experimental spinal cord transection (Ziegler et al. 2010) (see Chapter 8). The common thread in these studies relating to the findings in birdsong is that the motor system actively engages in sensory-motor exploration, possibly sacrificing accurate performance, in order to facilitate learning. This may be true not only for birds acquiring birdsong, but also for other instances of learning guided by the underlying attractor dynamics of brain and body, including rodent "whisking," human suckling, babbling, and walking (see Table 7.3).

Toward Control of the Opening and Closure of Critical Periods

It has long been known from human clinical cases, as well as classic animal experiments, that occlusion of one eye during early development (monocular development, or MD) leads to loss of visual acuity, or amblyopia, through the deprived eye into adulthood (see, for example, Hensch 2005; Morishita and Hensch 2008). However, it has only been during the past decade that the processes governing regulation of neural circuits underlying the critical period for visual development have been elucidated (LeFort, Gray, and Turrigiano 2013; for a review, see also Takesian and Hensch 2013). The development of retinotopic maps in the visual cortex (V1) has been well studied in mouse, rat, ferret, cat, and monkey and is similar in humans (Espinosa and Stryker 2012). Development of V1 involves the establishment of oc-

Table 7.3 Exploration of the nonlinear attractor dynamics of the body's neuro-mechanical systems

Behavior	Attractor
Rodent whisking	Rodents use their vibrissal sensory system to explore the nearby environment. Exploratory behavior typically consists of bouts of simultaneous whisking and fast breathing, or "sniffing." Each whisk is phase locked with breathing.
Birdsong	The avian vocal organ, the syrinx, is a nonlinear device that is capable of generating complex sounds even when driven by simple instructions. The syrinx exhibits nonlinear oscillatory dynamics. Zebra finches produce rapid song modulations that reflect transitions in the dynamical state of the syrinx.
Human suckling	Sucking and respiration exhibit nonlinear dynamics due to the mechanical properties of the chest wall, tongue, pharynx, and jaw. Infants explore their phase locked sucking and breathing in order to find locations in time when they may safely swallow, without aspirating liquid into the airway.
Human prosody and babbling	The articulators of the human vocal tract are characterized by nonlinear dynamics. Infants regulate the degree of their mouth opening to explore vowel sounds, and during repetitive jaw oscillations, use the oral articulators to selectively obstruct airflow to produce and explore babbling sounds.
Human learning to stand and walk	The oscillations of the body center of mass exhibit nonlinearities due to the mechanical and elastic properties of the body in a gravitational field. Infants and toddlers explore the oscillatory behavior of the body center of mass and use different muscle groups to selectively produce forces that cancel some forces and allow others to accelerate body motion.

Sources: D. Kleinfeld, M. Deschenes, F. Weng, & J. Moore (2014), More than a rhythm of life: Breathing as a binder of orofacial sensation, *Nature Neuroscience, 17,* 647–651; J. Goldberg and M. Fee (2011), Vocal babbling in songbirds requires the basal ganglia-recipient motor thalamus but not the basal ganglia, *Journal of Neurophysiology,* 105, 2729–2739; and W-H. Hsu, D. Miranda, D. Young, K. Cakert, M. Qureshi, and E. Goldfield (2014), Developmental changes in coordination of infant arm and leg movements and the emergence of function, *Journal of Motor Learning and Development,* 2, 69–79.

ular dominance, the relative strength of connections from either eye to individual cortical cells. The cortical column is a fundamental unit of organization of the mammalian neocortex, including V1, and thalamic inputs from the right or left eye are spread within the cortex by local excitatory connections and long-range inhibition. With some early visual experience at eye opening, there begins a critical period in which the selective properties of V1 neurons are refined *to make them similar through the two eyes*. This stage is considered a critical period because "visual deprivation causes rapid and dramatic changes in the strength and organization of inputs from the two eyes to cortical cells" (Espinosa and Stryker 2012, 232). A key finding is that sensitivity to monocular deprivation in mice is restricted to a critical period that begins about one week after the eyes open (postnatal day 13) and peaks one month after birth and that it causes amblyopia only during the critical period (Hensch 2005).

Takao Hensch and colleagues at the Boston Children's Hospital have identified some of the molecular mechanisms that control the initiation and closure of "windows for cortical plasticity" (Toyoizumi et al. 2013; Werker and Hensch 2015). A set of proposals from the Hensch lab concerning critical periods includes the following: (1) critical periods themselves exhibit plasticity, distinguished by states of onset, maintenance, and closure; (2) parvalbumin (PV) cells (see Chapter 5) are a fundamental plasticity "switch"; (3) the excitation-inhibition (E-I) balance of PV cells determines the timing of critical periods; (4) critical periods open when molecular triggers, in response to sensory input, promote PV cell maturation and GABA function; (5) in an immature state, molecular brakes prevent precocious plasticity; (6) during the maintenance period, neural circuits are wired in response to sensory experience; and (7) during the closure period, molecular brakes consolidate neural circuits from a plastic to a stable state.

The discovery that inhibitory interneurons act as a "switch" for controlling critical period timing has motivated the experimental procedure of transplanting inhibitory interneurons to induce new plasticity (Levine, Gu, and Cang 2015). Southwell et al. (2010), for example, transplanted cortical inhibitory interneurons from embryonic visual forebrain into the visual cortex of young, pre-critical-period mice. They found that the transplanted interneurons made inhibitory synapses with cortical neurons and promoted plasticity when they reached a cellular age equivalent to that of endogenous interneurons (Southwell et al. 2010). Davis et al. (2015) transplanted PV interneurons in adult mice and additionally used a genetically encoded calcium indicator to demonstrate that the transplanted cells, like endogenous ones, were broadly tuned for stimulus orientation. To determine whether this procedure could repair amblyopia, Davis et al. (2015) monocularly deprived mice for two weeks (including the critical period). They then transplanted interneurons in the visual cortex and found, remarkably, that the deprived eye's acuity completely recovered to the normal level. Thus, transplantation of embryonic inhibitory interneurons is able to rescue the deficits caused by visual deprivation during the endogeneous critical period.

From Mouse to Human: Cortical Plasticity

A definition. At a 2009 National Institutes of Health workshop on translation of neuroplasticity research into effective clinical therapies, there was a consensus on the following definition of neuroplasticity:

> Neuroplasticity can be broadly defined as the ability of the nervous system to respond to intrinsic and extrinsic stimuli by reorganizing its structure, function, and connections; can be described at many levels, from molecular to cellular to systems to behavior; and can occur during development, in response to the environment, in support of learning, in response to disease, or in relation to therapy. (Cramer, Sur, and Dobkin 2011)

It is important to distinguish plasticity at the organizational levels of the nervous system introduced in Chapter 5: the macroscale level of spatiotemporal patterns of activation of different brain regions, the mesoscale level of long-range and local connections among distinct types of neurons, and the microscale cellular level of synapses (Kim et al. 2014; Oh et al. 2014). Here I highlight the distinction at the microscale level between *Hebbian* forms of activity-dependent plasticity and *homeostatic* plasticity, the regulatory process for balancing synaptic excitation and inhibition (Turrigiano 2012) (see Table 7.4). Hebbian plasticity refers to correlations between pre- and postsynaptic activity

Table 7.4 Synaptic Plasticity: LTP, LTD, STDP, and AMPARs

Plasticity	Description
Long-term potentiation (LTP)	Rapid, long-lasting increase in synaptic strength induced by a specific neural activity pattern. Based on usually brief, strongly correlated pre- and post-synaptic activity.
Long-term depression (LTD)	Rapid, long-lasting decrease in synaptic strength, induced by a specific neural activity pattern. Based on usually sustained, weakly correlated pre- and postsynaptic activity.
STDP	Induction of LTP and LTD by pairing pre- and postsynaptic action potentials is critically dependent on spike timing dependent plasticity, the order and relative timing of single spikes, down to the millisecond scale.
AMPARs	Excitatory synaptic transmission between neurons is typically mediated by the neurotransmitter glutamate. The AMPA family of glutamate receptors, or AMPARs, act postsynaptically to mediate much of the fast excitatory transmission at central synapses. AMPAR postsynaptic accumulation, regulated by neuronal activity, plays a fundamental role in synaptic plasticity.

Sources: D. Shulz & D. Feldman (2013), Spike timing-dependent plasticity. In J. Rubenstein & P. Rakic (Eds.), *Neural Circuit Development and Function in the Brain: Comprehensive Developmental Neuroscience* (Vol. 3)., New York: Elsevier; and D. Feldman (2009), Synaptic mechanisms for plasticity in neocortex, *Annual Review of Neuroscience 32,* 33–55.

by which some synapses grow stronger while ineffective ones grow weaker, as captured by the phrase "synapses that fire together, wire together" (Katz and Shatz 1996). Hebbian plasticity is a positive feedback process, while homeostatic plasticity is a negative feedback mechanism that maintains activity levels within a dynamic range (Toyoizumi et al. 2014; Turrigiano 2011, 2012). In classic long-term potentiation (LTP) and long-term depression (LTD), a change in synaptic weight depends on correlated pre- and postsynaptic activity (Harvey and Svoboda 2007). In the late 1990s, several groups discovered that the induction of LTP and LTD by pairing pre- and postsynaptic action potentials depended critically upon both the order and relative timing of individual spikes, down to the millisecond scale (for example, Markram et al. 1997). This form of temporally precise bidirectional Hebbian plasticity, as when a presynaptic spike in a pyramidal cell leads a postsynaptic spike by up to 10–20 msec to induce an increase in postsynaptic strength, is called spike-timing-dependent plasticity, or

Table 7.5 Relation of dendritic spines to plasticity at multiple time scales

Plasticity (longest to shortest time scale)	Dendritic spine behavior
Spine formation/elimination	There is a balance maintained between spine formation and elimination, related to glucocorticoid oscillation.
Synaptic remodeling	There is selective elimination of previously existing spines, leading to significant remodeling.
Sensory experience and critical periods	The dynamics of spine growth and retraction is regulated by sensory experience during critical periods.
Learning	New spines form rapidly as animals learn new tasks.
Development	Spines change their morphology and turn over rapidly during development but become more stable in adults.

Sources: R. Araya, T. Vogels, & R. Yuste (2014), Activity-dependent dendritic spine neck changes are correlated with synaptic strength, *Proceedings of the National Academy of Sciences, 111,* E2895–E2904; and X. Yu & Y. Zuo (2011), Spinal plasticity in the motor cortex, *Current Opinion in Neurobiology, 21,* 169–174.

STDP (see Shulz and Feldman 2013 for a review). Turrigiano (2012) defines the homeostatic form of plasticity as stabilizing the activity of a neuron or neuronal circuit around some setpoint value.

Structural plasticity. I noted in Chapter 5 that dendritic spines are integrated independently and linearly, in a "great synaptic democracy" for experience-dependent structural synaptic plasticity (Yuste 2011, 2013). Nevertheless, the population of this democracy consists of cells with particular identities. The consequence of this cellular heterogeneity is a continuum of structural changes that may vary from long-range axon growth to the "twitching" of dendritic spines and synaptic receptor composition dynamics (Holtmaat, Randall, and Cane 2013). Spines come in many forms—from large, mushroom-like spines that are the most stable to small, thin spines that are the most dynamic (Holtmaat et al. 2013) (see Table 7.5). Long-term *in vivo* imaging studies in the adult (mouse) brain indicate that

> the large-scale organization of axons and dendrites is remarkably stable and some synaptic structures also persist,

> perhaps for most of the lifespan of the animal. By contrast, it is now clear that a subset of synaptic structures displays cell-type specific, experience-dependent structural plasticity: axonal boutons turn over in a cell type-specific manner. Spines grow and retract, and the dynamics of this turnover is regulated by sensory experience. (Holtmaat and Svoboda 2009, 654)

The scientific toolbox for imaging the relationship between spine growth and synapse formation includes two-photon laser scanning microscopy (and super-resolution techniques) as well as fluorescent probes (Holtmaat et al. 2013). Studies of spine structural plasticity have also been advanced by the development of a technique, called two-photon glutamate uncaging, that allows selective stimulation of a single spine while simultaneously imaging the morphology of the stimulated spine with two-photon microscopy (Nishiyama and Yasuda 2015). These studies show that biochemical computation occurs not only in a single spine, but also in a short stretch of dendrite around the spine and in the whole dendritic branch. Despite these technical advances, however, imaging studies do not reveal how and when a synapse on a new spine is recruited relative to network function. Addressing the relation between structural and functional plasticity requires a different strategy.

To address this question, Margolis et al. (2012) turned to a population-level description of plasticity of cortical areas, or "maps," in the barrel cortex of mice (see Figure 7.3). In mice, there is a one-to-one correspondence between facial whiskers and the organization of barrel columns in somatosensory cortex, and in response to sensory deprivation resulting from experimental whisker trimming, maps of the deprived sensory inputs shrink, while maps of the remaining spared inputs expand (Feldman 2009). Two methods form the neuroscience toolbox allow long-term tracking of neuronal activity following sensory deprivation: chronic imaging of genetically encoded calcium indicators (GECIs) and repeated patch-clamp physiology (Margolis, Lutcke, and Helmchen 2014). GECIs are calcium-sensitive fluorescent proteins that can be expressed in neurons and imaged over

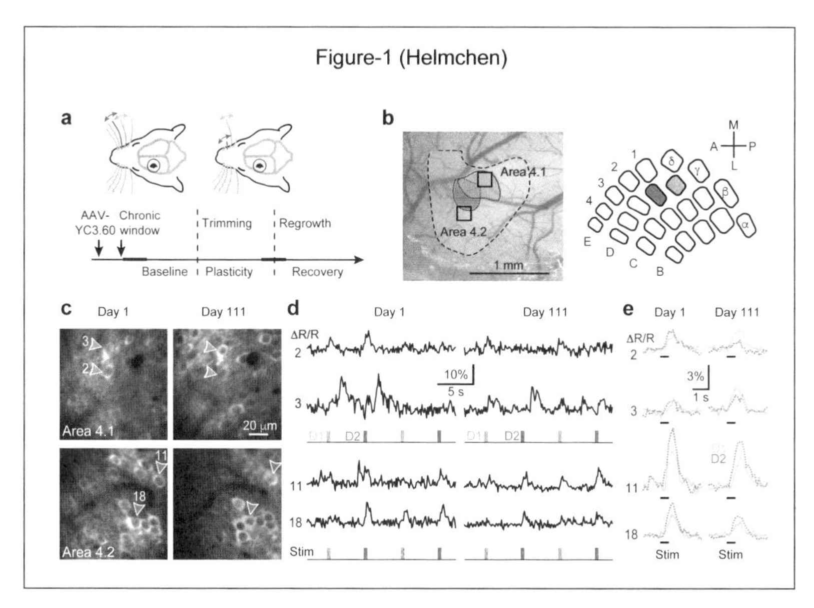

Figure 7.3 Long-term imaging of neuronal population activity in mouse barrel cortex. (a) Experimental time line. After intracortical injection of rAAV-hSYN-YC3.60 and implantation of a cranial window, mice recovered for several weeks before starting two-photon imaging sessions. Two neighboring whiskers were stimulated in repeated imaging sessions under isoflurane anesthesia spaced at least two days apart. Baseline (two–five sessions), plasticity (three–nine sessions) and recovery periods (three sessions after twelve–sixteen weeks in two animals). (b) Image of cortical surface through a chronic cranial window. YC3.60 expression area shown in yellow, intrinsic signals for D1 and D2 whisker in colors (50 percent threshold), imaging areas indicated by squares. (c) Two-photon images of two L2/3 populations as marked in panel (b), on an initial day and 111 days later. (d) Raw calcium transients (ΔR/R) for two cells from each population as marked by arrowheads in panel (c). Timing of whisker stimulation is indicated below traces (cyan, D1 whisker) and (magenta, D2 whisker). Areas are from mouse 4. (e) Averaged stimulus-evoked calcium transients ($n = 25$ per whisker) from panel (d). David J. Margolis, Henry Lütcke, Kristina Schulz, Florent Haiss, Bruno Weber, Sebastian Kügler, Mazahir T. Hasan, and Fritjof Helmchen (2012). Image reprinted with permission from Nature Publishing Group/Macmillan Publishers Ltd.

time periods of days to multiple weeks (Knopfel 2012). Margolis et al. (2012) used these chronic imaging methods to examine experience-dependent changes in barrel cortex organization. They first conducted baseline whisker stimulation with *in vivo* imaging and found stable functional networks of low-, mid-, and high-responsive neurons. They then trimmed all contralateral whiskers except one, spared, whisker of two used for the baseline stimulation. Whiskers were retrimmed and then allowed to regrow after each imaging session. Margolis et al.

found that during sensory deprivation, responses to trimmed whisker stimulation decreased overall, but responses to spared whisker stimulation increased for low-responsive neurons and decreased for the high-responsive ones. Thus, sensory-driven responsiveness controlled experience-dependent activity changes in individual neurons.

Possibilities for Plasticity following Spinal Cord Injury

Spinal cord injury. A human spinal cord injury (SCI) disrupts the axonal connections between brain and spinal cord, resulting in a devastating loss of function (Thuret, Moon, and Gage 2006). SCI creates a particularly hostile milieu for axon regeneration: the formation of a glial scar halts advancing axon growth cones (Cregg et al. 2014), and astrocytes surrounding the lesion become reactive such that severed axons are exposed to a cocktail of inhibitory extracellular matrix molecules around injury sites (Schwab and Strittmatter 2014; Tuszynski and Steward 2012). As discussed earlier, though, the glial scar is not simply an inhibitor of axon regeneration: it may also play a beneficial role in reducing inflammation and secondary tissue damage (Liddelow and Barres 2016). This and other insights have led to major breakthroughs in harnessing plasticity to restore spinal cord integrity in rats through (1) the use of neural stem cell grafts and inhibitory cocktails to both modify and *overcome the inhibitory environment* of the adult CNS (Bonner and Steward 2015; Kadoya et al. 2016; Lu et al. 2012; Wang et al. 2016) and (2) the use of combinations of neuromodulators and epidural or intraspinal electrical stimulation (Bachmann et al. 2013; Courtine et al. 2008, 2009; Harkema et al. 2011).

In rat models of incomplete spinal cord injury, there is spontaneous functional recovery of locomotor behavior (Raineteau and Schwab 2001; Bareyre et al. 2004). Bareyre et al. (2004) found in adult rats that twelve weeks after placement of a spinal lesion, hindlimb corticospinal contacts with long propriospinal neurons (PNs) formed a spinal detour circuit. The functional connectivity of the new circuit was then established by intramuscular injection of a retrograde transsynaptic tracer (Bareyre et al. 2004). Courtine et al. (2008) provide further evidence that newly formed propriospinal circuits mediate plasticity during recovery from incomplete spinal cord injury. They examined recovery of locomotor behavior in mice following various combinations of lateral spinal hemisections. One procedure involved placing two unilateral hemisections, ten weeks apart. This sequence of hemisections separated by ten weeks effectively created an entirely complete bilateral transection of all direct projections from the brain to spinal locomotor circuits but allowed time for the formation of propriospinal relay circuits (Courtine et al. 2008). After the ten-week recovery period, the mice exhibited virtually normal recruitment of leg muscles and coordinated stepping behavior.

The formation of propriospinal relay circuits as a foundation for neuronal plasticity following spinal cord injury raises the important question of the role played by sensory information during the spontaneous recovery period. One possibility is that the new relay circuits restore sensory feedback loops, normally provided by the muscle spindles. To examine this possibility, Takeoka et al. (2014) placed incomplete spinal cord injuries in a group of genetically mutant mice, called Egr3, and a group of wild-type mice with no genetic mutation. Due to early postnatal degeneration of muscle spindles, Egr3 mice exhibit two distinctive motor characteristics: they are proficient in basic locomotor tasks, but poor at tasks requiring precision, such as ladder walking. Takeoka et al. (2014) used fluorescently labeled trans-synaptic rabies virus and found the hallmark of detour circuits: the presence of new connections to ipsilateral spinal circuits. It is these two characteristics of spinal cord circuits—ascending and descending communication with supraspinal centers in the brainstem and higher areas and sensory feedback loops—that are pointing the way to new strategies in murine research on spinal cord injury.

Stem cell grafts. One experimental approach to creating neuronal relay circuits between injured long tract neurons and denervated neurons following experimental placement of partial or complete SCI in rats is the use of neural stem cell transplantation (Bonner and Steward 2015; Kadoya et al. 2016; Lu, Kadoya, and Tuszynski 2014; Wang et al. 2016). Neural stem cells (NSCs) are multipotent cells, having the potential to differentiate into both neuronal and glial lineages within the central nervous system (spinal cord and brain) (Wyatt and Keirstead 2012). One series of studies from the laboratory of Mark Tuszynski at the University of California–San Diego demonstrated that grafts of NSCs to experimentally placed spinal cord lesion sites extend many new axons into the host spinal cord for long distances (more than 20 mm, or seven spinal segments) and receive inputs from injured host axons (see Lu et al. 2014 for a review). These NSCs were derived from green fluorescent protein-expressing rat embryos, making it possible to track the process extension of the grafted cells. Even in a complete transection model, grafted neurons extended axons that grew at a rapid rate of 1–2 mm / day (Lu et al. 2012).

Neuromodulation. Another experimental procedure for establishing detour circuits following spinal cord injury involves neuromodulation (Courtine et al. 2008; Musienko et al. 2011; Wenger et al. 2014; Dominici et al. 2012). As one illustration of this paradigm, Van den Brand et al. (2012) developed a multisystem neuroprosthetic program with paralyzed rats that combined epidural electrical stimulation of lumbosacral circuits, treadmill training, and a robotic postural interface that required the animals to use their paralyzed hindlimbs to locomote bipedally toward a target. The rats not only regained voluntary locomotion, but there was formation of intraspinal detour circuits that relay supraspinal information (Van den Brand et al. 2012). However, these stimulation protocols were restricted to constant modulation patterns regardless of the current state of the lower-limb movements. To better capture the actual patterns of activation of the spinal circuitry underlying locomotion, as demonstrated, for example, by Hagglund et al. (2013), Wenger et al. (2016) developed a methodology that matched electrical neuromodulation to the natural dynamics of motoneuron activation. The foundation for this spatiotemporal neuromodulation methodology was their discovery that walking involves the activation of spatially delimited motoneuron "hot spots," alternating between a spinal region controlling limb flexion and one controlling limb extension (see Chapter 8 on "synergies"). Wenger et al. (2016) then used computer simulations to identify optimal electrode locations for spinal implants and used real-time control software to selectively modulate flexion and extension of each hindlimb. Wenger et al. (2016) found that after complete SCI, rats treated with five weeks of spatio-temporal neuromodulation exhibited gait patterns that were closer to that of intact rats than after continuous neuromodulation.

Despite these encouraging achievements with mice and rats for promoting recovery from spinal cord injury, there are fundamental differences between murine and primate nervous systems, such as the architecture of the corticospinal tract, which may limit the kinds of questions that can be addressed by murine models. In humans and nonhuman primates, axons projecting to the spinal cord originate from both the left and right motor cortices, but in mice and rats, these axons originate exclusively from the contralateral motor cortex (Rosenzweig et al. 2009). Does the bilateral projection of the primate corticospinal tract reveal a process of recovery from partial spinal cord injury that is not apparent in murine models? To examine this question, Friedli et al. (2015) first conducted a prospective study of functional recovery during the first year following incomplete spinal cord injury in 437 patients. They found that recovery of motor function was significantly correlated with laterality of the spinal cord injury: patients who presented a noticeable lateral asymmetry in motor performance two weeks after injury regained extensive bilateral motor function between six and twelve months after the injury.

Friedli et al. (2015) next modeled the human spinal cord injury by placing a C7 lateral spinal cord hemisection in monkeys and rats. Humans, monkeys, and rats were evaluated with motor tests of locomotion (such as walking along a horizontal ladder) and hand function (reaching for food in monkeys and rats and for plastic shapes in humans). A notable finding was that hand function recovery was more extensive in monkeys and humans than in rats: over time, the hemisected monkeys regained the ability to reciprocally recruit extensor and flexor digit muscles, greatly improving object retrieval, while rats never regained the ability to retrieve food with the injured paw (Friedli et al. 2015). Finally, to examine reorganization of the corticospinal tract below the injury in the rats and monkeys, corticospinal fibers were labeled with injections of anterograde tracers in the left and right cortex. Only the monkeys showed any evidence of sprouting of corticospinal detour circuits below the injury (Friedli et al. 2015).

These findings raise the question whether, at this stage in the development of neuroprosthetic devices, a nonhuman primate model may represent a crucial validation step for spinal cord injury research (Collinger et al. 2014; Nielsen et al. 2015; Schwarz et al. 2014). How might scientists implement a strategy for nonhuman primate research that both follows the highest standards of neuroethics for the animals and provides a means for developing safe neuroprosthetic devices for ameliorating the devastating effects of spinal cord injury in humans (Farah 2015)? One criterion is whether the most benefit is being gained from data collected from each animal. The California Spinal Cord Consortium is a multidisciplinary team that has implemented a "big data" strategy for understanding the relationship between plasticity and recovery in a cervical SCI model in nonhuman primates. As part of their comprehensive approach to nonhuman SCI research, the researchers use a shared methodology for assessing recovery and for pooling data resources (Nielsen et al. 2015). To illustrate, the informatics workflow for every nonhuman primate SCI experiment involves collecting open-field and chair behavior, treadmill and kinematics, and a unified database infrastructure, which allows querying and statistical analysis shared by all members of the consortium (Nielson et al. 2015).

Another way to evaluate benefit of data collection in nonhuman primates is to consider the comparability of the experiments to the way that the device will be used in humans. For example, one major challenge in neuroprosthetics studies is the longevity of the means for decoding intended movements from motor areas of the brain. Advances in using tethered and wireless recordings of large-scale brain activity are allowing researchers to study individual animals for periods as long as five consecutive years, demonstrating a degree of longevity (Schwarz et al. 2014). A further challenge is providing somatosensory feedback, especially for upper-limb neuroprostheses, to provide patients with the experience of embodiment (see, for example, Bensmaia and Miller 2014). The use of intracortical microstimulation (ICMS) through chronically electrode arrays has the potential to provide sufficient benefit of intuitive sensory feedback that balances the risks of invasive surgery. A crucial requirement for the use of ICMS is that the evoked sensory experience during microstimulation remains stable over time. Callier et al. (2015) studied the longevity of ICMS in two adult male rhesus monkeys with an electrode array implanted in the hand representation of cortical area 1. The animals performed a detection task in a two-alternative forced choice paradigm, in which they initially reported when a mechanical stimulus was presented to the skin and subsequently when ICMS was delivered through the implanted array. The major result was that the sensitivity to ICMS remained stable over several years. Both of these paradigms with nonhuman primates illustrate the way in which information valuable to developing components of neuroprosthetics—including decoding and providing somatosensory feedback for embodiment—is being obtained in a manner that balances risk to the animal and benefit for humans.

PART THREE

Understanding and Emulating Nature's Responses to Injury or Damage

A newt, whose tail is chomped by a predator, is able to regenerate the missing distal segment, even as an adult. A dog, after losing one leg in an accident, quickly learns how to use the remaining three healthy legs to form a tripod gait for locomoting. Following a cerebral stroke in an elderly man, the brain's own protective neural pathways are activated to counteract tissue damage. After her mother rearranges the furniture in their home, a toddler born at twenty-six weeks' gestational age uses a wheelchair to explore and play, as other toddlers do. All of these healing responses to injury illustrate the abundant set of solutions by which biological and social systems repair themselves. Our current technologies for neuroprosthetic devices face challenges that include how to decode the future-directed neural activity of brain networks, how to stimulate the same muscle groups in different patterns for different functions, and how to build control architectures that share biological and machine resources. As our technologies for promoting healing become increasingly sophisticated, how might we use lessons from nature to build biologically inspired devices for repairing the body and nervous system? What role should robots play in neurorehabilitation? What types of environments may best promote recovery from brain injury? Can we regenerate nervous systems following injury? Is it possible to fabricate vascularized tissue to replace damaged biological organs? How may we better protect vulnerable nervous system networks? The chapters of this section address these questions with a particular emphasis on how we may emulate nature in developing technologies for repairing and remodeling injured nervous systems.

✦ 8 ✦

Neuroprosthetics: The Embodiment of Devices

Fans of the original *Star Trek* television series may fondly recall the episode in which an alien being removes and abducts Mr. Spock's brain so that it could be incorporated into an advanced computer. Ever prepared, the ship's medical officer, Dr. McCoy, is able to temporarily control Spock's body with a 1960s-looking black box so that the biological Spock could at least be mobile, if not sentient. (Fear not. Spock's brain is eventually restored to its rightful owner by the good doctor.) The contemporary term for that black box is a brain–machine interface, as defined in the field of neuroprosthetics (see, for example, Borton et al. 2013; Raspopovic et al. 2014). As evident in the truly incredible advances in neuroprosthetics already made in these early years of the twenty-first century, we may be steps closer to indirectly or directly linking animal or human nervous systems (spinal cord and / or brain) with robotic devices (Bouton et al. 2016; Collinger et al. 2012; Courtine et al. 2013; Downey et al. 2016; Goldfarb, Lawson, and Shultz 2013; Hochberg et al. 2012).

When Spock's brain is removed, a hidden assumption of the screenwriters is that his spinal cord remains intact and that the black box somehow interfaces with the remaining anatomy in order to allow Spock to walk. One interpretation is a relatively simple black box that taps into the spinal cord, a structure that is sufficiently sophisticated to help Spock walk. We now know this assumption about the spinal cord is scientifically accurate, if incomplete: populations of spinal cord neurons provide central pattern generation for combining the muscles into functional groupings, called *synergies*, and combinations of synergies are involved in the patterning of locomotion (see, for example, Bizzi et al. 2008; Kargo and Gizster 2008; Overduin et al. 2012). Fundamentally, these functional groupings reflect a reduction of dimensionality of the central pattern-generating network that drives motoneuron bursts at distinct phases of the locomotor cycle (Levine

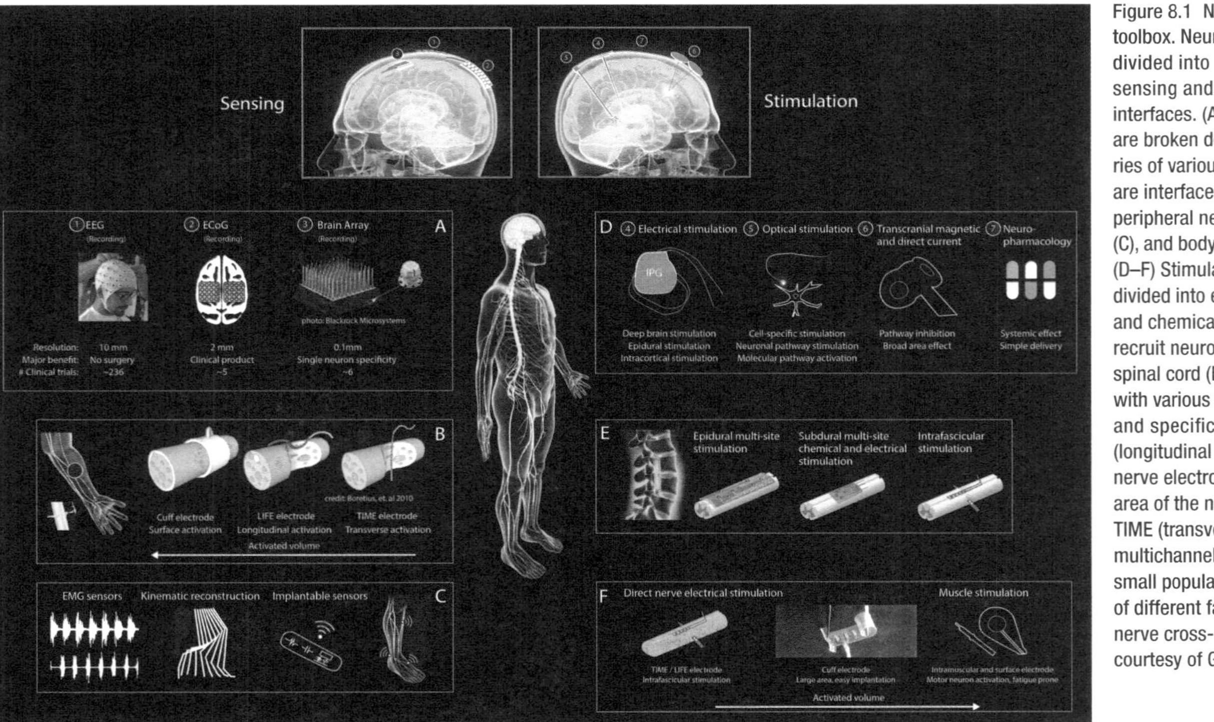

Figure 8.1 Neuroprosthetic toolbox. Neurotechnologies are divided into two categories: sensing and stimulation interfaces. (A–C) Sensing devices are broken down into subcategories of various resolutions that are interfaced with the brain (A), peripheral nerves (B), muscles (C), and body kinematics (C). (D–F) Stimulation probes are divided into electrical, optical, and chemical classes that can recruit neurons in the brain (D), spinal cord (E), and periphery with various activation volumes and specificities (F). LIFE (longitudinal intra-fascicular nerve electrode) excites a small area of the nerve cross-section; TIME (transversal intra-fascicular multichannel electrode) excites small populations of nerve fibers of different fascicles over the nerve cross-section. Figure courtesy of Gregoire Courtine.

et al. 2014). But in considering the nature of synergies, it is important to note that Spock still also has a body, characterized by an anatomical layout of tendons and fascia that add non-neural contributions to the mix during synergy formation (Bizzi and Cheung 2013; Kutsch and Valero-Cuevas 2012; Turvey and Fonseca 2014). We may speculate, then, that the signals to the black box share a *common mathematical foundation in dynamical systems, in which the nervous system, body, and machine are based upon populations of components that self-assemble, dissolve, and reassemble into functional ensembles*, as needed (see, for example, Turvey 2007; Turvey and Fonseca 2014).

We may not need to wait until the twenty-fourth century to build machine interfaces based upon a dynamical systems foundation for such synergies. But there are still numerous challenges this century, evident in Figure 8.1 from the laboratory of Gregoire Courtine at the Swiss Federal Institute of Technology (EPFL) (Borton et al. 2013). This figure presents the components of a multifaceted approach called *personalized neuroprosthetics* and identifies the complex interplay between components of neuroprosthetic systems. Both Dr. McCoy's black box and our contemporary technology for personalized neuroprosthetics raise questions about (1) how to build machines with sensor networks capable of recording signals indicating how the brain looks ahead, as we do, to prepare actions in advance of performing them; (2) where the functional boundary is between the nervous system and machine, and (3) how a projection of the unfolding of future events is continuously updated with information obtained in real time during exploratory and performatory behaviors in order to adapt to the exigencies of the environment. Here, I present a dynamical systems perspective on the relation between humans and devices, in general, and on the relation between humans and neuroprosthetic devices, in particular, that attempts to address these three questions (see Table 8.1). I attempt to pry open the black box to reveal three pillars for progress in this field.

The *first pillar for neuroprosthetics* is the future-directed neural activity of cortico-cerebellar and other neural loops (see, for example, Churchland et al. 2010, 2012; Cunningham and Yu 2014; Shenoy, Sahani, and Churchland 2013; Shenoy and Nurmikko

Table 8.1 A dynamical perspective on motor cortex to guide the design of neuroprosthetic devices

Motor cortex	Characteristics
What motor cortex does in the production of movements	During movement preparation, the motor cortex generates a pattern of activity (a time-varying vector) that is mapped to muscle activity to produce the forces that move the body for achieving an individual's goals.
The population dynamical state	The motor cortex is an extensively connected network coupled through input and feedback signals with the rest of the nervous system and so is described by its population activity.
Dimensionality reduction	In neurons within local regions, it is possible to measure the population dynamical state at a reduced dimensionality by using techniques of dimensionality reduction to map the responses of many neurons onto a small number of variables that capture the basic patterns present in those responses.
Evolution of the preparatory state	Preparation may bring the population dynamical state to an initial value from which accurate movement-related activity follows. This initial value is generated within an optimal subspace / subregion of neuronal firing rates, ensuring that movements occur with minimal reaction time.
Relationship between movement period neural responses and the movement itself	During movement, the collective activity of motor cortical neurons (a) is driven by a low-dimensional dynamical model, and (b) the lower dimensional dynamics rotate with a phase and amplitude set by the preparatory state.

Source: K. Shenoy, M. Sahani, & M. Churchland, (2013), Cortical control of arm movements: A dynamical systems perspective, *Annual Review of Neuroscience, 36,* 337–359.

2012) that must be decoded by machines to transform agent-generated goals into dynamical systems for action. A dynamical systems perspective on the motor system, proposed by Mark Churchland at Columbia and Krishna Shenoy at Stanford, shifts the emphasis from the *meaning* of neural output, its representation in the cortex, to *how neural output is generated* (Churchland and Cunningham 2014; Shenoy et al. 2013). The *second pillar for neuroprosthetics,* the dynamics of embodiment (synergy formation), refers to the *shared biological and physical resources of the functionally coupled nervous system, body, and machine.* The relative contribution of biological and machine resources to synergies is continually shifting as an individual's goals and opportunities for action unfold. The shared resources include the living flesh of human muscles and the actuators of the machine; the biological receptor fields of the visual, haptic, and auditory systems and dual sets of machine-sensor arrays; and the shared control systems of living brain tissue and electronic circuitry that together assign values and make trade-offs for using resources to perform tasks in particular ways. The *functional connectivity and neuromodulation* of the nervous system; the multiple configurations of the body's musculoskeletal system; and the possible rearrangements of biomaterials, sensors, and actuators to work together as a functional system all reflect a *shared control structure.* The third pillar provides a potential means for biological and machine systems to share their combined resources. It explores the relationship between human and machine, modeled on the task–dynamic framework introduced in Chapter 2. Before getting to these three pillars for neuroprosthetics, though, I first consider the urgent need for neuroprosthetic technology and how animal studies provide a foundation for safe devices for humans.

Opportunities and Challenges in Building Neuroprosthetic Devices

The possibility that tetraplegic individuals could use their thoughts to control a robotic prosthetic arm and hand may seem as far-fetched as the *Star Trek* script just described. Its actual achievement was highlighted in two publications that appeared in 2012. In *Nature,* a group of collaborating scientists at Brown University and Massachusetts General Hospital reported the ability of two human patients, with long-standing tetraplegia due to brainstem stroke, to use their thoughts to direct a robot to perform reaching and grasping actions (Hochberg et al. 2012). Each of the study participants was implanted with a ninety-six-channel microelectrode array for recording motor cortex activity. Over multiple sessions, the study participants had to imagine reaching and grasping objects, such as a bottle containing liquid. The electrical potentials from each of the channels of motor cortex activity were filtered and used to train a signal classifier, which eventually was able to identify intended hand states. In a widely published photo sequence, one of the study participants used her thoughts to direct the robot to grasp the bottle, bring it toward her mouth, drink coffee from the bottle through a straw, and place the bottle back on the table (Hochberg et al. 2012) (see Figure 8.2). These technologies also hold great promise for replacing lost function in amputees, a process that requires several steps (see Table 8.2).

Emulating the emergent behavioral dynamics of the human hand with a robotic prosthetic device presents an enormous scientific and engineering challenge because of the anatomical complexity of the hand (its muscles, tendons, articular surfaces, bones, and ligaments; Valero-Cuevas et al. 2007); its multiple functions (Kwok 2013; Tabot et al. 2013); and the high demand for visual and haptic guidance of the hand for reaching, grasping, and exploring (Janssen and Scherberger 2015; Turvey and Fonseca 2014). At the end of *The Force*

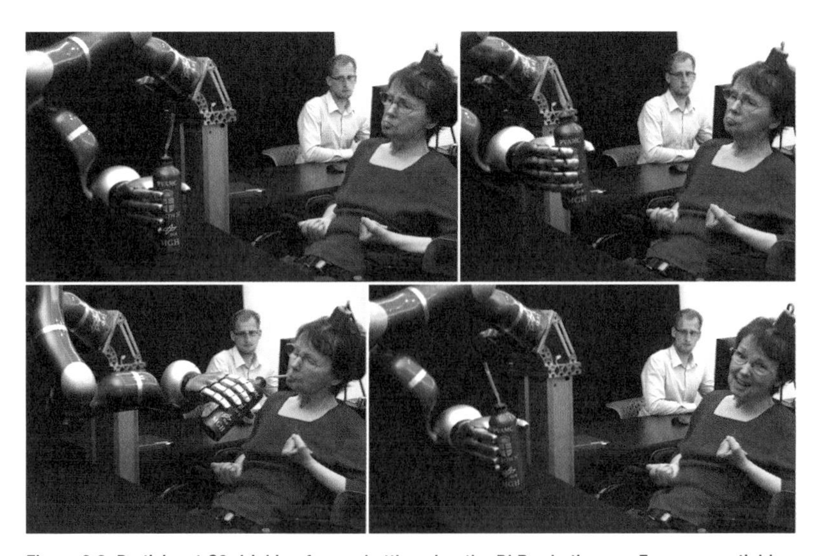

Figure 8.2 Participant S3 drinking from a bottle using the DLR robotic arm. Four sequential images from the first successful trial showing participant S3 using the robotic arm to grasp the bottle, bring it toward her mouth, drink coffee from the bottle through a straw (her standard method of drinking), and place the bottle back on the table. The researcher in the background was positioned to monitor the participant and robotic arm. Leigh R. Hochberg, Daniel Bacher, Beata Jarosiewicz, Nicolas Y. Masse, John D. Simeral, Joern Vogel, Sami Haddadin, Jie Liu, Sydney S. Cash, Patrick van der Smagt, and John P. Donoghue (2012). Photos reprinted with permission from Nature Publishing Group / Macmillan Publishers Ltd.

Table 8.2 Closing the loop: Restoring sensory feedback during use of a prosthetic hand

Procedure	Description
Targeted muscle reinnervation	To create new surface EMG signals that may be used to control a motorized prosthetic arm, the remaining arm nerves are transferred to residual chest or upper arm muscles that are no longer biomechanically functional due to loss of the limb. A sensory-neural machine interface is created by denervating a patch of target skin near the nerve redirection site to provide a receptive environment for sensory reinnervation. When the reinnervated target skin is touched, the amputee experiences that the missing limb is being touched.
Stimulation of median and ulnar nerve fascicles	Uses multichannel intrafascicular electrodes for stimulating the median and ulnar nerve fascicles. To "feel" an object, electrodes deliver electrical stimulation to the nerves that are proportional to finger sensor readouts. This sensory feedback provides near-natural sensory information to an amputee's hand prosthesis during real-time decoding of different grasping tasks.
Intracortical stimulation of somatosensory cortex	Uses intracortical stimulation of primary somatosensory cortex timed with the onset and offset of object contact during manipulation with a prosthetic hand. Stimulation provides sensory information about contact location, pressure, and timing.

Sources: T. Kuiken, P. Marasco, B. Lock, R. N. Harden, & J. DeWald (2007), Redirection of cutaneous sensation from the hand to the chest skin of human amputees with targeted reinnervation, *Proceedings of the National Academy of Sciences, 104,* 20061–20066; S. Raspopovic, M. Capogrosso, F. Petrini, M. Bonizzato, J. Rigosa, G. Di Pino, J. Carponeto, et al. (2014), Restoring natural sensory feedback in real-time bidirectional hand prostheses, *Science Translational Medicine, 6,* 1–10; and S. Flesher, J. Collinger, S. Foldes, J. Weiss, J. Downey, E. Tyler-Kabara, S. Bensmaia, A. Schwartz, M. Boninger, & R. Gaunt (2016), Intracortical microstimulation of human somatosensory cortex, *Science Translational Medicine, 8,* 1–10.

Awakens, the 2015 return of the *Star Wars* saga, we get only a brief glimpse of one of the original 1977-vintage heroes, Luke Skywalker. As you may recall, Luke loses his light-saber-wielding hand in a battle with Darth Vader, and it is replaced by a flawlessly functioning prosthetic one. Inspired, perhaps, by Luke's fictional prosthetic, as well as by the unmet needs of thousands of upper-limb amputees with lost function, and supported by generous funding from a U.S. Defense Advanced Research Project Agency (DARPA) initiative, engineer-inventor Dean Kamen and his company DEKA created the "Luke hand" and DEKA arm (Resnik, Klinger, and Etter 2014). The DEKA arm has multiple grasp options and the ability to accommodate multiple amputation levels (Gonzalez-Fernandez 2014). A second DARPA-funded effort, this one at Johns Hopkins University, began development of a modular prosthetic arm that has now been successfully used as part of a human neuroprosthetic system. In a report of the work, published in *Lancet* by a group at the University of Pittsburgh, a tetraplegic patient was implanted with two ninety-six-channel intracortical microelectrode arrays, and over a thirteen-week period after implantation, participated in thirty-four brain–machine interface training sessions (Collinger et al. 2012). During an initial phase of the calibration, the participant was instructed to carefully watch the prosthetic limb as it automatically moved to the targets. Calibration and neural decoding progressed from three-dimensional endpoint translation control (weeks 2 and 3) to four-dimensional control of translation and grasp (week 4) to seven-dimensional control of translation, orientation, and grasp (Collinger et al. 2012). Following the calibration period, the participant was given control of the prosthetic limb's capability for translation, orientation, and grasp and explored a range of arm and hand configurations. In the period of less than four months provided for training, the participant was able to move the prosthetic limb in a "smooth, coordinated, and skillful" manner (Collinger et al. 2012).

The role of extended periods of exploratory behavior for establishing a stable functional relation between intentions and the movements of a multi-degree-of-freedom device, apparent in the remarkable patients just discussed, is powerfully illustrated in another study of a tetraplegic individual, FM, as reported by Nonaka (2013). Before a diving accident at age 29, in which he dislocated vertebrae C4, FM practiced calligraphy. As a result of the accident, FM lost all voluntary control and sensation of his body below the level of the injury. Despite this horrific and tragic loss, FM courageously continued his calligraphy with a mouth-held brush and, over a period of twenty-five years, achieved master level of competence.

From the perspective of motor control, the mastery of calligraphy requires that the artist feel the contact of the brush with the paper, in order to control the dynamics of the brush tip as it interacts with the paper. How did FM achieve such mastery? What can engineers and roboticists building neuroprosthetic devices learn from this tetraplegic master calligrapher? One major lesson for engineers follows from the methodology that Nonaka (2013) used to measure the relation between the variability of body movements and the invariant functional relation between brush and paper maintained by the master calligrapher in this tetraplegic individual. Due to his high cervical injury, FM was limited to head and brush movements relative to the cervical spine, with little movement below the neck. Of interest was whether, to produce his beautiful calligraphic strokes, FM minimized the variability of the relation between brush and paper (a task variable) by using a compensatory coupling of head rotations relative to the cervical spine. Nonaka (2013) addressed this question by analyzing kinematic data from motion capture and determining the sources of joint angle variation that contributed to the relation between brush and paper. Nonaka (2013) reports that, as hypothesized, the movement of the head and that of the cervical spine were coupled in a compensatory manner to stabilize the upright posture of the head relative to the contact with the paper, as the head flexed down and up, and rotated, to produce strokes. In other words, FM established new organizational patterns involving the muscle groups of his head and neck that remained functional in order to achieve what he had previously done with his hand.

There are many wrenching scientific and ethical challenges in the field of neuroprosthetics (Belmonte et al. 2015; Farah 2015). It is clear that the use of animals in research must be justified in terms of the value of the research in ameliorating devastating human disease. How will it be possible to evaluate the safety and efficacy of newly developed neuroprosthetic devices and therapeutic interventions for humans, without endangering human lives?

A First Pillar for Neuroprosthetics: Interfaces, Decoding, and Embodiment

Brain-machine interfaces. Electrical signals of the mammalian brain and spinal cord lie beneath layers of skin, bone, dura mata, and other tissue. For this reason, signals at the surface of the skull are notoriously weak, and neuroscientists seeking robust signals for recording brain activity implant chronic sensor electrode arrays through the skull into brain tissue. However, due to their volume and material composition, these implanted electrodes tend to degrade over long periods and cause inflammation and scar formation (Saxena and Bellamkonda 2015). An ideal implantable neural probe should possess (1) stiffness similar to brain tissue in order to minimize mechanically induced scarring; (2) a degree of porosity and cellular / subcellular scale features to allow interpenetration of, and integration of neurons with, the electronics, and (3) a means of implanting the resulting highly flexible structure and an input / output system to allow multiplexed recording (Xie et al. 2015).

J. Liu et al. (2015) have been able to incorporate many of these essential characteristics of implantable neural probe into a syringe-implantable flexible mesh. The mesh has a "parallelogram" shape that allows it to roll up longitudinally. By rolling up the mesh, it is able to pass through a syringe for injection into the brain of mice, without damaging tissue. The mesh design provides a remarkable degree of biocompatibility because these probes are similar in flexibility to brain tissue and because the feature sizes of the electronics are less than or equal to those of individual cells (J. Liu et al. 2015). The mesh was then injected into the lateral ventricle and hippocampus of live mice, and confocal microscopy showed that (1) the mesh expanded to integrate within the local extracellular matrix; (2) cells form tight junctions with the mesh; and (3) neural cells migrate along the mesh structure, indicating that mesh electronics have the potential to monitor neural cells following brain injury, as well as to record and stimulate brain structures (J. Liu et al. 2015).

Another major advance in electrode implants for neuroprosthetics is the "electronic dura mater," or "e-dura" developed by Minev et al. (2015). The e-dura has mechanical properties matching the viscoelastic properties of the mammalian dura mater, the thick protective sheath surrounding the brain and spinal cord. Matching viscoelasticity in this way makes it possible to implant the device below the dura mater and may enable long-term biointegration and functionality with the brain and spinal cord, a major step toward implanted neuroprosthetic devices providing lasting therapeutic benefit. The device architecture integrates a 120-μm-thick substrate, 35-nm-thick stretchable gold interconnects, a compliant microfluidic channel, and soft electrodes coated with a platinum-silicone composite (Minev et al. 2015). This architecture is designed to provide multimodal stimulation as well as electrophysiological recording: the microfluidic channel delivers drugs locally, while interconnects and electrodes transmit electrical excitation and signal recording. A first test successfully demonstrated that e-dura implanted into the subdural space of rat spinal cord conformed along the entire extent of the lumbosacral segments. Next, Minev et al. (2015) compared the kinematics of locomotor behavior of groups of rats treated with either e-dura, a stiff implant, or no implant (sham operation only). Even after six weeks, the behavior of the rats with soft implants was indistinguishable from the sham-operated animals. By contrast, by one to two weeks after implantation, the rats with stiff implants displayed significant locomotor deficits (for example, altered foot control during basic walking and a higher percentage of missed steps on a horizontal ladder). Finally, a partial spinal cord injury (sparing less than 10 percent of spinal tissues) was placed in a group of adult rats, leading to permanent paralysis of both legs. An e-dura covering lumbosacral segments was implanted to engage spinal locomotor circuits below the injury. Using the functionality of the e-dura, the animals received multimodal therapy consisting of chemotrode injection of 5HT agonists and continuous electrical stimulation. This concurrent electrical and chemical stimulation enabled the paralyzed rats to walk.

The initial published work on optogenetics revealed how a microbial opsin could be expressed in neurons to make their activity controllable by light (Boyden et al. 2005). Optogenetics has now become a mature technology (for perspectives, see Boyden 2015; Deisseroth 2015). In addition to its other applications, optogenetics is being developed for implantable devices in neuroprosthetics (see, for example, Buzsaki et al. 2015; Gerits et al. 2012; Packer et al. 2015; Park et al. 2015; Wu et al. 2015). Here, I consider two illustrations of advances in optogenetics with potential applications for neuroprosthetics. The first is the use of soft, fully implantable, miniaturized optoelectronic system for modulating peripheral and spinal pain circuitry (Park et al. 2015). A remarkable feature of this device is its use of energy-harvesting technology to allow drastic reduction in size. A stretchable antenna harvests radio-frequency (RF) power, with a design that absorbs energy most effectively. The flexible antenna and soft interface are sufficiently small to be inserted into the spinal epidural space for optogenetic modulation of the spinal terminals of the peripheral nerves.

The second illustration is the use of optogenetic devices for closed-loop control of animal behavior in what Grosenick, Marshel, and Deisseroth (2015) call "closed-loop optogenetics," a procedure that modulates the light signal on the basis of desired and measured outputs. A major challenge for closed-loop optogenetics is the development of online algorithms for making light stimulation conditional on the detection of neural activity or behavior. This is a more general problem for closed-loop neuroprosthetic devices, and it involves calculation of an estimate of neural state as input to algorithms that compute the necessary input for control to achieve some target activity level, as discussed in the next section.

Decoding cortical signals. Another daunting challenge for multidisciplinary teams of neuroscientists and engineers building neuroprosthetic devices for patients with injuries to the nervous system (for example, spinal cord injury, stroke, Parkinson's disease, amyotropic lateral sclerosis) has been to elucidate the role of the nervous system in generating intentions. This is called the "decoding problem" because it assumes that the initial step of a neuroprosthetic interface—for example, implanted or surface electrodes—is to record neural electrical impulses as a basis for inferring impending behavior. Implicit in this step is that during the period prior to performing a behavior, such as visually guided reaching, there is a preparation for action. In order to participate in the work of the nervous system, neuroprosthetic devices are designed to "decode" intent beginning during the preparatory period and continuing throughout performance. A decoder is a set of algorithms that converts neural activity into desired prosthetic actuator movements (Shenoy and Carmena 2014). There has been considerable success in decoding the neural activity that occurs in preparation for action in the primary motor cortex (Brodmann's area 4, most typically called M1) and in the posterior parietal cortex and frontal cortical areas in nonhuman primates (Andersen and Cui 2009; Hwang, Bailey, and Andersen 2013).

At the heart of the decoding problem is how to understand the vast information contained in neural signals, the challenge of what is being called "big data" (Lichtman, Pfister, and Shavit 2014; Sejnowski, Churchland, and Movshon 2014). For neuroscientists, big data is a means for exploring populations of neurons to discover the macroscopic signatures of dynamical systems, rather than attempting to make sense of the activity of individual neurons (Engert 2014). Two surprising results from numerous experiments recording from neurons in different brain regions have revealed a wonderful secret of nature about the relation between the number of neurons recorded and their dimensionality (the number of principal components required to explain a fixed percentage of variance). First, the dimensionality of the neural data is much smaller than the number of recorded neurons. Second, when dimensionality procedures are used to extract neuronal state dynamics, the resulting low-dimensional neural trajectories reveal portraits of the behavior of a dynamical system (Gao and Ganguli 2015). This means that it may not be necessary to record from many more neurons within a brain region in order to accurately recover its internal state-space dynamics (Gao and Ganguli 2015).

The value of dimensionality reduction may be illustrated during recording of the population response structure of the motor system during monkey arm reaches (Churchland et al. 2012, Cunningham and Yu 2014; also discussed later). In this work, it was initially very difficult to interpret the individual responses of hundreds of neurons in the motor cortex. As an alternative, Churchland et al. (2012) applied dimensionality reduction techniques so that they could examine the behavior of an ensemble of neurons during arm reaches. They found a coherent mechanism at the level of the population that was not apparent in single-neuron responses alone. This use of dimensionality reduction has profound implications for neuroprosthetics because it may potentially solve the problems of (1) decoding intention at the level of populations of cortical neurons, (2) discovering robotic controllers that couple synergies of the body's muscles and tendons to the dynamic movement primitives of robotic components, and (3) revealing the layout of environmental attractors that may introduce disequilibria that shift the body's current postural state to a new one in order to guide behavior.

Spock redux. In our *Star Trek* scene at the outset of the chapter, we might imagine that Dr. McCoy was able to animate Spock's limbs by using a recording he somehow obtained of motor cortex signals to the muscles during specific functional activities. This was, indeed, the approach developed in a recent study by Bouton et al. (2016) of a study participant who was paralyzed due to a cervical spinal cord injury. The 24-year-old man underwent chronic implantation of an intracortical microelectrode array. After implantation, he participated in a fifteen-month training session during which he learned to use a custom-built, high-resolution, neuromuscular electrical stimulator, with 130 electrodes embedded in a sleeve wrapped around his right forearm. During training, cortical activity was continuously decoded by machine-learning algorithms while the participant viewed selected movements of an animated virtual hand. Signals from six simultaneously running decoders were then used to activate patterns of electrical stimulation of the muscles in his right forearm, allowing the participant to grasp, manipulate, and release objects. Ajiboye et al. (2017) report a further advance in neuroprosthetics for tetraplegic patients: they combined intracortical signals from an implanted recording microelectrode array with percutaneous electrodes for functional electrical stimulation of coordinated reaching and grasping movements. Remarkably, the formerly paralyzed tetraplegic patient was able to feed himself, including drinking a cup of coffee! A significant challenge for such a system, though, is devices that may not require cortical implants for a brain–machine interface, and instead use noninvasive recording equipment to acquire signals from the brain. Toward that end, Soekadar et al. (2016) have demonstrated a hybrid noninvasive EEG / EOG system that uses brain electrical activity and eye gaze electrooculography signals along with a wireless tablet computer to control a hand exoskeleton. Such systems promote a further consideration of the body's functional systems that contribute to the activities of daily living to be recovered through rehabilitation.

Another approach to Dr. McCoy's conundrum is to consider more broadly the coordination of microscopic components of brain and body: not just muscles, but also the mechanical components of the body required for human performance in the arts, sports and leisure, and play. During my graduate school education at the University of Connecticut, Michael Turvey, a former track and field star, experimental psychologist, and dynamicist, introduced us to the writings of the Russian physiologist and movement scientist Nikolai Bernstein (Whiting and Whiting 1983; see also Latash and Turvey 1996). Bernstein's major work, *The Coordination and Regulation of Movement,* was not yet widely known in the United States when I first read it in 1980. Watch as a ballet dancer performs flawless pirouettes; cheer as an outfielder leaps to catch a ball and rob the hitter of a home run; try hula-hooping! This latter may seem like an anachronism, but the scientific challenge of understanding coordination of body parts involved in keeping an unstable hula hoop spinning around the waist was recognized by an Ig Nobel award to Michael Turvey for his long-standing work on coordination dynamics (see, for example, Balasubramaniam and Turvey 2004). From the standpoint of physics, the task involves conservation of angular momentum: carefully timed impulses are applied to the small portion of the hoop interior that makes contact with the body to oppose the force of gravity (Balasubramaniam and Turvey 2004). As a problem in dynamics, hula-hooping is the constraint of a high degree-of-freedom system so that it behaves as a lower-dimensional one: through a mathematical decomposition technique, hula-hooping was revealed as a coordination of the high degree-of-freedom system into a vertical suspension mode and an oscillatory fore–aft mode. These two modes may, therefore, be a means for reducing making the task of control one of stabilizing the vertical and horizontal components of the hoop's angular momentum (Balasubramaniam and Turvey 2014).

A dynamical systems perspective on neuroprosthetic systems offers a compelling approach to discovering decoding algorithms that will serve as a "Rosetta stone" between nervous system and machine. In a dynamical systems approach, a fundamental initial step in decoding involves addressing the question of how many neurons need to be measured (Gao and Ganguli 2015). Mark Churchland and Krishna Shenoy propose that preparatory neural activity "functions as the initial state of a dynamical system and may not explicitly represent movement parameters," as had long been assumed (Churchland et al. 2010, 387). The neuronal preparation for action involves the specification of a goal space for making choices—for example, in activations of the motor cortex (Churchland, Afshar, and Shenoy 2006; Churchland et al. 2010). Their research focus is on the period of preparation prior to movement and how the motor cortex, as a dynamical machine, is able to generate a pattern of neural activity that maps onto muscle activity to produce forces that move the body through an environmental landscape of attractors (Shenoy et al. 2013). Preparation before movement is believed to bring the population dynamical state to some initial value from which accurate movement-related activity follows (Shenoy et al. 2013).

The experimental paradigm with monkeys used by Churchland and Shenoy to study cortical activity during movement preparation involves an instructed-delay task. A task trial begins when the monkey fixates and touches a central target, triggering a peripheral target (see, for example, Churchland et al. 2006; Shenoy et al. 2013) and directing the animal to where, after a randomized delay period and go signal, the movement is to be made. Most notable in this experiment is that neurons in many cortical areas, including the parietal reach region, systematically modulate their activity during the delay. The neural activity of each animal may be plotted in state-space, with the firing rate of each neuron contributing an axis. As the movement is triggered, the population dynamic state departs from the initial state and follows a trajectory through state-space.

It is the state-space trajectory, as determined by the initial state, neural dynamics, and any feedback, that governs movement (Shenoy et al. 2013). The movement period neural responses are quite remarkable in their simplicity: the neural trajectory simply rotates with a phase and amplitude set by the preparatory state, tracing a circular trajectory! The transition from a state of motor preparation to overt action poses another challenge for this dynamical approach to the preparation of action. How does the animal hold still while planning a course of action until it is time to act (Sanger and Kalaska 2014)? Kaufman et al. (2014) propose the output-null hypothesis: neural activity patterns during the preparatory period are structured so that they are "output null"—that is, they cannot cause muscle contractions.

These future-oriented dynamics of the motor and parietal cortices are part of a multiregional control system for perceiving and acting, which includes a set of closed loops with the cerebellum (Witter and De Zeeuw 2015). This large, and still mysterious, brain structure may play key roles in (1) *predicting the state trajectories of dynamical systems* and (2) using available sensory feedback to *update the dynamical system based upon deviations from these predictions* (Bastian 2006, 2011; Brooks, Carriot, and Cullen 2015; Therrien and Bastien 2015). The cerebellum may use estimates of body properties, such as inertia and damping, to predict how applied forces would move the body (Bhanpuri, Okamura, and Bastian 2012, 2013). And through cortico-cerebellar loops, the cerebellum may recalibrate cortical networks to the changing properties of the encountered body and world (Baumann et al. 2015; see also Chapter 9). One strategy for revealing how the cerebellum is involved in motor control has been to study patients with cerebellar damage. These patients exhibit uncoordinated, variable, and dysmetric movements, called ataxia. For example, Bhanpuri, Okamura, and Bastian (2014) studied ataxic patients making single-joint reaching movements between two targets. They proposed that the injured cerebellum makes a biased estimation of limb inertia (that is, either under or over) and confirmed this by using a robotic arm to alter the inertial properties of the patient's limbs. Another research strategy has been to present patients with objects undergoing changing visual dynamics. Deluca et al. (2014), for example, presented cerebellar patients with decelerating targets and asked them to judge the distance required for each to come to a complete stop. They attributed observed deficits of the cerebellar patients to the process of calibrating estimates of the visual object dynamics.

The recognition that *multiple* brain regions may be involved in generating future-oriented neural population dynamics has motivated an approach to developing brain–machine interfaces in which neurons are recorded not only in motor cortex, but also in posterior parietal cortex (PPC) (Andersen et al. 2014). PPC neurons are both sensory and motor, and are involved in sensorimotor transformations. The parietal reach region (PRR) is involved in decision making and, thus, has the potential to specify a reach for a decoder more quickly than motor cortex because of its rich source of information about movement trajectories and about the final goal of a movement (Andersen et al. 2014). The goal of embodying decoders for brain–machine interfaces used for reaching so that the control strategy is similar to that of natural arm movement has resulted in a decoder design referred to as a *biomimetic decoder* (Fan et al. 2014; Shenoy and Carmena 2014). This design is being combined with a capability for decoders to update in response to neural adaptation—changes in neural activity due to recovery from injury or to the process of learning to control a prosthetic device (Shenoy and Carmena 2014).

A challenge for brain–machine interfaces is that decoding occurs on single noisy trials. One way to improve decoding is to incorporate the dynamics of the way neural populations modulate themselves over time into the decoding algorithms, as in the work of Kao et al. (2015). As an illustration of this approach, consider the parabolic behavior of a cannonball tracked by low-resolution video. Due to low resolution (analogous to recording of neural populations), the video image is noisy. However, knowledge of Newtonian mechanics makes it possible to better estimate the trajectory of the cannonball, improving the resolution of the video image. The trajectory of a neural state inferred from noisy single trials may, similarly, be very noisy and may be improved by taking into account the known neural state dynamics. Kao et al. (2015) did just that by linearly weighting the predicted neural state based upon its dynamics, creating a neural dynamical filter (NDF). The NDF filter was able to achieve between 31 percent and

83 percent higher performance than current decoding algorithms, such as Kalman filters.

Embodiment. A growing consensus in motor control is that the design of biological motor systems should be based on natural principles, rather than on Newtonian and Euclidean formalisms (as in robotics) (Kalaska 2009). In hierarchical models from robotics, premotor cortex initiates a high-level motor plan that is transformed by the primary motor cortex (M1) into a low-level plan to be executed (see, for example, Scott and Kalaska 1997). But we have just seen, in the work of Churchland and Shenoy, that the population dynamics of M1 are governed by natural principles of population dynamics rather than formal representations. This implies that brain oscillations involved in motor control are embodied: they somehow have a natural affinity for the dynamical characteristics of limb biomechanics.

It is known, for example, that neurons in monkey M1 are most active for certain directions of hand movement. To examine what underlies this distribution, Lillicrap and Scott (2013) trained a simple network that controlled a model of the primate upper limb. The model was optimized to (1) make reaches and maintain postures under static loads while (2) keeping neural and muscle activities and synaptic weights small. They examined distributions of preferred movement patterns as a function of manipulations of multiple features of limb mechanics, including limb geometry, intersegmental dynamics, and muscle mechanics, and found that the best fit between model and empirical observations occurs when all of these are included (Lillicrap and Scott 2013). They also examined predictions about how preferred movement patterns should change as a function of limb posture and anatomy. They found, for example, that when the model limb performed center-outreaching in the right half of the workspace, then the optimal preferred movement patterns rotated substantially in the clockwise direction. This is consistent with the finding that M1 neural populations systematically rotate their preferred directions to different parts of the workspace (Sergio and Kalaska 2003). Thus, the model illustrates (1) how movement biases depend specifically upon properties of the skeleton-muscular system and (2) that these distributions closely parallel those observed in M1 neurons of nonhuman primates performing similar tasks.

A Second Pillar for Neuroprosthetics: Perceiving, Synergies, and Coupling

Perceiving. Imagine a soft, rubber glove sitting on a flat, rigid tabletop. As I look at the glove, my eyes in my mobile head actively explore the glove's contours and the light reflected from its surfaces. As I continue to look at the glove and move my fingers over these surfaces and along the contours, I can see and feel that it is shaped like my hand and is about the same size, so it affords fitting my hand inside it. I can no longer see my hand after I place it inside the glove, but I can look at the glove and feel my hand. So long as there is a direct correspondence between the information obtained by the skin and joint receptors of my hand and from my eyes as I move my hand and explore how it looks and feels, I continue to experience my hand inside the glove.

There is now substantial evidence that sensory information from a body part is crucial for the experience of psychological ownership of that body part. However, it remains a difficult technical challenge to incorporate what we know about the process of information pickup by active perceptual organs—such as the skin surface, visual system, and auditory system—into profoundly embodied synthetic devices such as prosthetic limbs. The insights of James and Eleanor Gibson into the way that the body's perceptual systems detect the same amodal information through multiple means via a process of active exploration (see, for example, Gibson 1986; Goldfield 1995) provides one source of biological inspiration for addressing this challenge. As I described in Chapter 3, each perceptual system transduces patterned information revealed through active movements of the perceptual organs with respect to environmental structure. It is the correlation between information obtained from the multiple sensory modalities that may strengthen the synaptic connectivity of neuronal networks and provide a basis for the emergence of the psychological state of body ownership [see, for example, the perspective of Steinberg et al. (2015) on the relation between neural activation states and the emergence of psychological states].

In a series of experiments, Ehrsson (2012, for a review) has investigated sensory contributions to the experience of body part ownership, such as the hand. In the "rubber hand" paradigm (preferable, I think to calling it an illusion, as in the literature), the participant's own hand is placed behind a screen so that he or she cannot see it. A life-sized rubber hand is placed in front of the participant, and the experimenter uses two small paintbrushes to simultaneously stroke the rubber hand and the hidden hand. After between 10 and 30 seconds, most participants in these experiments have the experience that the rubber hand is their own and that it is the rubber hand that senses the touch of the paintbrush (Ehrsson 2012).

The role of sensory information for body ownership has important implications for the embodiment of prosthetic devices. Petkova and Ehrsson (2008) examined the experience that amputees have of their prosthetic hand during the "rubber hand" experiment. A "map" of referred phantom sensations on the arm stump was first established. After a period of synchronized brushing of specific spots on the map of phantom sensations, as well as on the index finger of a prosthetic hand, amputees reported an increased ownership of the prosthetic hand. In this early experiment, there was no physical connection between the amputee and prosthetic hand, but more recent work with patients undergoing a surgical procedure called targeted muscle reinnervation (TMR) (Kuiken et al. 2007) has used this technique to demonstrate a further step toward at least moderate embodiment of a prosthetic hand. One goal in TMR is to create new surface EMG signals that may be used to control a prosthetic limb (Kuiken et al. 2007, 2009; Hargrove et al. 2013; Marasco et al. 2011). For example, following arm amputation, the remaining arm nerves are transferred to residual chest or upper arm muscles that are no longer biomechanically functional due to limb loss (Kuiken et al. 2007, 2009). Following amputation at the knee, TMR has been used to reinnervate residual thigh (hamstring) muscles (Hargrove et al. 2013). The EMG signals from the natively innervated and surgically reinnervated muscles were then used for control of a prosthetic robotic leg.

To create an "artificial sense of touch" for a prosthetic hand, Marasco et al. (2011) used Ehrsson's multisensory rubber hand paradigm to attempt to improve the experience of "hand ownership" of a prosthetic hand with two amputees who underwent TMR surgery. During the TMR procedure, a sensory-neural machine interface is created by denervating a patch of target skin near the nerve redirection site. When the reinnervated target skin is touched, the amputee experiences that their missing limb is touched. Marasco et al. (2011) used a miniature haptic robot, called a "tactor," to convert a signal from a load cell to the skin interface site and provide direct physiologically and anatomically appropriate sensory feedback to the amputee. Each of the two amputee participants watched the investigator touch the prosthetic hand and load cell, while the tactor simultaneously pressed into the surgically reinnervated target skin of the residual limb (Marasco et al. 2011). The results were similar to the previous findings by Ehrsson et al. (2008): compared to a control condition, the amputees reported an increased experience of ownership of the prosthetic hand. Another advance in restoring natural sensory function to amputees has been the use of "bidirectional" hand prostheses (Raspopovic et al. 2014). The bidirectional flow of information is achieved through multilayer decoding of surface EMG signals and simultaneous electrical stimulation via transversal intrafascicular multichannel electrodes (TIMEs) implanted into the median and ulnar nerves. The participant was able to control the prosthesis through voluntary contractions of the remaining muscles on the stump and was also able to identify the stiffness and shape of different objects.

There have also been advances in using intracortical microstimulation of the somatosensory cortex to restore tactile sensation on locations of the hand of patients with long-term spinal cord injury (Flesher et al. 2016). Restoring somatosensory feedback is crucial not just for the perceptual experience, but also for providing feedback in motor control (Raspopovic et al. 2014). In the Flesher et al. (2016) study, electrode arrays were chronically implanted in a 28-year-old male with tetraplegia due to a spinal cord injury sustained ten years earlier. Over a six-month period after electrode array implantation, they evaluated (1) quality of evoked sensations, (2) projected locations on the body of these sensations from (3) sensitivity to each intracortical microstimulation. With the possibility of verbal report, the participant was asked what the microstimulation felt like. He responded that it felt possibly natural, that is was both at and below the skin surface, that there was no experience of pain, and that the somatosensory quality was mostly of pressure. The goal of future work is to expand the range (for example, shape, motion, and texture) and realism of sensory experience by emulating natural brain signals.

Synergies. Late-night movie trivia buffs may smile slyly when asked about films featuring a disembodied hand. The Peter Lorre and Robert Alda film from 1946, entitled *The Beast with Five Fingers*, had a third star, a hand that could ambulate across a room in order to climb up the leg of, and strangle, a poor, horrified soul. This cultural icon of the 1940s was satirized in the 1960s television series, and subsequent films, *The Addams Family* (and its sequel), with an equally competent, if more gentle, "Thing." The incongruity in these depictions, of course, is that a disembodied hand has no nervous system to impel its muscles. However, clever experiments by Francisco Valero-Cuevas, at the University of Southern California, demonstrate that the topology of the tendon network of the hand generates its own non-neural "switching functions," capable of performing a kind of logical computation (see, for example, Valero-Cuevas et al. 2007; Kutch and Valero-Cuevas 2012). Moreover, the tendon architecture of the hand has characteristics of a tensegrity, redistributing perturbing forces throughout the structure (Rieffel, Valero-Cuevas, and Lipson 2010). The muscle, connective tissue, and skeletal system may also be the medium for rapid, long-distance, zero-delay mechanical propagation of information (Turvey and Fonseca 2014; Wang, Tytell, and Ingber 2009). So, a disembodied hand may be smart. But what does it mean for a body organ—a hand, a vocal tract, a leg—to be smart? One possibility is that the rich anatomical architecture of the hand allows it to become so many different types of devices.

In the decades since the English translation of Bernstein's (1967) work became available, his concept of "synergies"—functional units of motor behavior—has become a cornerstone of research in motor control. During this time, work in separate subdisciplines of motor control—dynamics, neuroscience, and biomechanics—has focused on different aspects of synergies. Dynamicists have emphasized how synergies are formed, how their composition is selected for achieving particular functions, and how their composition changes according to task demands. For example, from the dynamical systems perspective developed by Turvey (2007) and his collaborators, the concept of synergy highlights a collection of independent degrees of freedom behaving as a single functional unit by means of adjusting together to any perturbations.

Classic studies of speech production by Turvey's colleagues at Haskins Laboratories in New Haven, Connecticut, have provided experimental evidence for this adjustment of an entire ensemble of articulators to a perturbation of just one of them. A demonstration of this synergic response to perturbation comes from a study by Kelso et al. (1984). They applied brief perturbations to the jaw of a speaker repeating a single syllable utterance. In the case of the syllable / bab / , the jaw perturbation was applied as it moved upward to close the vocal tract for producing the final phoneme / b / . Kelso et al. (1984) found that within 15–30 msec, there was also an adjustment of the upper and lower lips, indicating that a collection of articulators forming a synergy with the jaw, and not the jaw alone, responded to the perturbation. This kind of experiment reveals the signature characteristic of a synergy—its cooperative organization of parts. But how are individual synergies organized into larger ensembles?

Soon after my graduate school introduction and subsequent tutorials on dynamical systems, I had the opportunity as a postdoc to learn about an influential neurophysiological perspective on synergies, in lectures given in my Harvard / MIT neuroanatomy class by Emilio Bizzi. Research by Bizzi and his collaborators since that time has established the neurophysiological foundation for discrete basis elements from which synergies are constructed (see, for example, Bizzi et al. 2008). The work has focused on the spinal cord of animals, especially the frog (Kargo and Giszter 2008) and rat (Tresch, Saltiel, and Bizzi 1999) and the relation among spinal interneurons, patterns of muscle activations, and what Bizzi's group calls "force fields," defined as "collections of isometric muscle forces generated at the limb's endpoint over different locations of the workspace" (Bizzi and Cheung 2013, 1). Their assertion, then, is that synergies have a neural origin in the motor cortex and spinal cord.

One of the most dramatic demonstrations that muscle synergies are encoded in the nervous system comes from a series of studies by Overduin et al. (2012, 2014) that applied long-duration electrical intracortical microstimulation (ICMS) to motor areas in two awake monkeys and simultaneously recorded muscle EMG and finger movements. Overduin et al. (2012) found that ICMS drives the hand and digits toward particular postures at each stimulation site. Moreover, the evoked EMG patterns resembled muscle coactivations seen in temporally complex behaviors, such as reach and grasp (Graziano, Aflalo, and Cooke 2005; Graziano and Aflalo 2007). As further evidence for cortical encoding of synergies, Overduin et al. (2014) examined whether the muscle synergies evoked during ICMS were similarly encoded during purely voluntary behaviors. At each microstimulation site, they determined that the synergy most strongly evoked among those extracted from muscle patterns were matched by the synergy most strongly encoded during voluntary behavior more often than expected by chance.

Comparing the definitions of synergy from the physical-dynamical tradition of Turvey, the neurophysiology of Bizzi, and the biomechanics of Valero-Cuevas makes clear that each has captured the essence of dimensionality reduction and flexibility at a particular level of analysis. As noted by Santello, Baud-Bovy, and Jorntell (2013), kinematic-level synergies are covariation patterns among joints, kinetic synergies denote coordination patterns among digit forces that are thought to minimize given cost functions, and neural synergies consist of common divergent inputs to multiple neurons.

An integration of these three perspectives on synergies—dynamical, neurophysiological, and biomechanical—may also be fundamental to the design of neuroprosthetic devices. A key to such an integration is that *the circuitry of the spinal cord as well as the mechanical properties of the body function are considered as dynamical systems with multiple stable states*. From a dynamical systems perspective, invariant positions on the body—such as the location of the joints—are the attractor nodes of the body-referenced control system (see, for example, Saltzman and Kelso 1987; Saltzman and Munhall 1992; Saltzman et al. 2006). Here, neurophysiology and dynamical systems are complementary perspectives on the same process of creating disequilibria that drive the spatial organization of parts into new configurations. This process of driving the body toward attractors and away from repellors may be understood in the context of perception-action (or P-A) coupling. P-A coupling refers to the reciprocal processes of obtaining information and using information to tune each postural adjustment to future and current states of the environment (Warren 2006). During exploratory behavior, agent and environment are coupled *informationally* through energy fields that are structured by the environment and *mechanically* through forces exerted by the agent (Warren 2006). *The role of the environment for contributing to the neuromechanical disequilibria that move the body toward a goal is determined by a layout of environmental attractors*.

How might motor cortex activity influence the dynamical systems spontaneously generated by spinal circuitry? Santello et al. (2013) propose that the intrinsic dynamics of spinal circuits may be characterized as a set of potential functions (see Chapter 2) of premotor neurons. In this view, *motor cortex influences spinal function by controlling the shape of each of a set of potential functions*. The potential field, a composite shape of multiple interacting potential functions, results in the emergence of a stable state (as illustrated by a ball rolling into the valley of a potential well), which establishes a synergy between the alpha-motoneurons belonging to specific motor nuclei. The heuristic power of identifying potential functions in neural circuits as the basis for synergies is that the size and shape of their peaks and valleys is an indication of the many possible solutions that maintain system stability. How might we, then, relate these neural potential functions to force production in behavior? One possibility is with respect to the visualization of covariation of solutions that maintain stability via the mathematics of the component subspaces of what is called the uncontrolled manifold (UCM). The mathematical technique for revealing the emergence of subspaces of variables at the level of behavior (forces) returns us full circle to Churchland's view of the motor cortex as a dynamical system that generates an optimal subspace for action. Following Santello's perspective on premotor circuits, each subspace may influence the shape of spinal potential functions, which, in turn, sets threshold levels for muscle activation in particular muscle groups (Feldman 2009).

Coupling. Pushing a young child on a swing is always great fun and not very strenuous work. A "child on swing" system dissipates kinetic energy on the downswing, but gains back potential energy on the upswing, so that the swing will remain in motion for many cycles, even without a push (or child's own kicking attempts) (see Goldfield, Kay, and Warren 1993 for analysis of a similar case—infants bouncing while suspended in a harness attached to a spring). When the swing begins to come to rest, timing the next push so that it occurs just at the moment when upswing motion stops and downswing motion begins is all that is required to maintain the swing's oscillation. (The illustration, of course, controls for cries to a parent to "make it go higher.") It is this possibility of injecting small pulses of energy into pendulum systems that has, for centuries, made the pendulum a primary timing mechanism for clocks. In 1665, the Dutch scientist Christiaan Huygens serendipitously discovered the "sympathy of two clocks": two pendulum clocks hanging from a common support kept in pace such that the two pendula always swung together (in opposite direction) (Peña Ramirez et al. 2014). We now know that a fundamental requirement for the phenomenon of sympathy, referred to in contemporary physics as *synchronization,* is due to the mechanical energy transferred by the support bar between the two pendula. Just a few decades before Huygens's discovery, Galileo turned his attention to swaying lamps in the cathedral at Pisa, eventually conducting his now-famous experiments on the relationship between period and pendulum length. In his *Dialogue on the Two Principal World Systems,* Galileo proposed that for small-amplitude oscillations, the square of the period was proportional to the length of the pendulum (Dugas 1958).

Returning to the case of the child on a swing, the pendulum has a regular source of energy, gated by an observer using the behavior of the swing to determine when to add energy to the system to sustain its oscillation. Information about the behavior of the system thus regulates the timing of the energy pulse, creating a feedback loop that keeps the pendulum system swinging as a self-sustained, limit-cycle oscillatory system. The system oscillates because gravity pulls the swinging child back to a resting position (that is, damps the motion) while the force acting in the same direction as the motion imparts energy. Information (in this case, the observed motion of the swing) balances the friction of the swing with the excitation of the push. Thus, feedback is a powerful means by which nature may selectively assemble and dissolve a variety of oscillatory systems. Do humans explore the relation between their behavior (for example, pumping the legs) in relation to the behavior of an oscillatory system to which they are attached?

The "child-on-swing" system may be put in the more general form of shared control. The toolbox of the dynamicist holds a set of modeling "building blocks" for characterizing shared control between humans or between human and machine. The building blocks have been given different names, including "motor primitive" (Flash and Hochner 2005; Giszter 2015), "dynamic primitive" (Hogan and Sternad 2012), and "dynamic movement primitive" (Ijspeert et al. 2013), the one I adopt here. Dynamical movement primitives are a set of building blocks coupled together to generate complex movements. They take the mathematical form of a set of equations with coupling terms for closed-loop P-A systems, which generate complex attractor dynamics. The inclusion of the adult as part of the behavior of the child-on-swing pendulum system is made possible by considering the forces generated by the adult as a forcing term, f, acting on the equation of motion that captures the attractor dynamics of the pendulum. By choosing a forcing term that is periodic, capturing the regular push given by the adult to the swing, the equation of motion is now defined over both the child and adult. With periodic forcing, the child–pendulum system exhibits the behavior of a nonlinear oscillator.

A Third Pillar for Neuroprosthetics

Shared control. The abstract nature of the primitives comprising the equation of motion for the child-on-swing system suggests the possibility of a more general approach to modeling the dynamics of shared control of oscillatory and other behavior in neuroprosthetic human–machine systems. Toward this end, I now return to the task dynamic framework introduced in Chapter 2 and a task involving two interacting humans (agents) in close proximity of each other. Unlike the child pushed by an adult, this interacting pair must avoid colliding while they each move a screen-displayed object between two target locations (Richardson et al. 2015). Following Richardson et al. (2015), there are three modeling steps. The first step is to define the task space for an individual, identifying the goals or terminal movement objectives. More generally, a task space includes the minimal number of relevant dimensions (task variables or task-space axes) and a minimal set of task-dynamic equations of motion (Saltzman and Kelso 1987). Here, the task was modeled on a two-dimensional plane, with an oscillating point mass on the x-axis and point-attractor dynamics (a damped mass-spring equation) indicating deviations from the primary motion axis. The dynamics in this functionally defined task space was then modeled as a set of equations of motion along each axis.

The next step models the second agent with an identical system of equations and, critically, introduces a set of linear, dissipative coupling functions to link the point mass movements of the two (Richardson et al. 2015). More generally, coupling functions are a powerful means for modeling how agent and environment may influence each other. The third step in the Richardson et al. (2015) task was meant to model the particular task instruction that subjects avoid bumping into or colliding with each other. This modeling step involved including a repelling coupling force. With this model in hand, Richardson et al. (2015)

were able to explain the major finding for pairs of subjects (agents): two agents in each pair cooperatively performed the task by virtue of one of them spontaneously producing a more straight line trajectory between the targets, while the other adopted a more elliptical trajectory. Model-based simulations revealed that these asymmetries emerged through modifications of the strength of the repeller coupling dynamics. In other words, a "behavioral dynamics" emerged through the combined influences of the coupled equations of motion under environmental constraints.

How might this strategy of task-dynamic modeling help to inform the interactions between a human and a neuroprosthetic robotic device? Here, I begin by considering the simpler case of a human wearing a robotic exoskeleton and a task in which the individual is instructed to flex and extend the arm at the elbow in order to match a visually displayed sinusoidal oscillation (Ronsse et al. 2011). As an illustrative exercise, I apply the three general steps of task-dynamic modeling outlined earlier—defining the task space and primitives (equations of motion), selecting coupling functions between human and robot, and identifying constraints on the permissible dynamics of the robot (for example, to avoid obstacles)—from which behavioral dynamics on this task emerge.

The particular exoskeleton worn on the arm of a healthy subject for this task is meant to assist rather than replace an intended limb movement: repeatedly flexing and extending the arm at the elbow. Therefore, the selected primitive is what Righetti, Buchli, and Ijspeert (2006) and Ronsse et al. (2011) call an *adaptive oscillator*, and the task space involves an oscillatory motion along a single axis. The adaptive oscillator is designed to modify its output on the basis of sensory input from motion of the arm of a human following a sinusoidal pattern on a screen. This primitive "is expressed as a dynamical system characterized by a limit cycle whose features (phase, frequency, amplitude, etc.) are changed in adaptation to an external input" (Ronsse et al. 2011, 1002). The oscillator serves as an estimator, a kind of filter that anticipates in real time the state evolution of the elbow joint undergoing (necessarily sinusoidal) oscillatory rotations. The equation of motion defining the system includes joint dynamics, based on a simple pendulum model, and an adaptive oscillator, a system of differential equations based on a Hopf oscillator (see Chapter 2), which learns the parameters of sinusoidal input by using the difference between elbow position and the output of the estimator (Ronsse et al. 2011).

The second step of task-dynamic modeling requires further consideration of the coupling functions between a human and robot. One way for a robot to become useful as a limb is to provide it not only with primitives, but also with coupling functions that guide the behavior of the limb. An example is a task space that models point-to-point reaching and obstacle avoidance, similar to the one used by Richardson et al. (2015) to model coordination behavior of pairs of individuals and used by Ijspeert et al. (2013) to model robot reaching while avoiding an obstacle. In both cases, reaching involves online reactive behavior because the obstacle appears suddenly and in an unforeseen way.

Smart perceptual instruments. In addition to congenital blindness, two age-related retinal disorders leading to profound vision loss are retinitis pigmentosa and macular degeneration (Merabet 2011). Modern techniques already make it possible to implant electronic devices, such as a synthetic retina (Zrenner 2013), in order to restore some visual function. However, providing useful vision to individuals with profound vision loss requires not only sophisticated technologies for retinal implants, but also an understanding of how to embody signals from the implant within the natural movement patterns of the eye (Merabet 2011). The mammalian retina is an exquisitely complex organ that (1) decomposes outputs of the photoreceptor cells into parallel information streams; (2) connects these streams to bipolar cells that transmit a "cafeteria" of stimulus properties selected by retinal ganglion cells, whose arbors cover the retina completely and evenly; and (3) creates a mosaic of independent tilings simultaneously superimposed on each other (Masland 2012).

One approach to restoring vision is to replace the function of the damaged neuronal components that comprise the visual pathway (Merabet 2011). Among the earliest visual devices was a cortical prosthesis for individuals with nonfunctioning retinae or optic nerves (see, for example, Brindley and Lewin 1968). It involved implanting intracortical electrodes into the visual cortex and using electrical stimulation to generate isolated spots of light, called "phosphenes." With increasing sophistication in surgical techniques and microelectronics over the ensuing decades, device development has expanded to include retinal prostheses (Ong and Cruz 2012). These take patterned light input (for example, from a camera) and transduce that information into electrical patterns. Epiretinal devices (such as Argus II and EpiRetIII) include a component that lies on, and stimulates, the inner surface of the retina (Humayun et al. 2016). Subretinal prostheses involve implantation of a microphotodiode-electrode array device between the central visual field of the retina (for example, the bipolar layer) and the retinal pigment epithelium (Zrenner 2013). Devices such as the Retina Implant Alpha IMS detect light through the eye's own optics and then stimulate retinal bipolar cells to project visual signals to the brain (Hafed et al. 2016).

Congenital cataracts, a leading cause of childhood blindness, are currently treated by surgically replacing the natural lens with an artificial intraocular lens (Daniels 2016). However, because the eyes of children are still growing, these surgeries provide only limited success in providing good vision (Daniels 2016). A new strategy for surgical cataract removal, inspired by the process of embryogenesis discussed in Chapters 2 and 6, is to regenerate a naturally transparent lens inside the body (H. Lin et al. 2016). H. Lin et al. (2016) developed a surgical means of cataract removal that preserves endogenous lens epithelial stem / progenitor cells (LECs). These epithelial cells express genes PAX6 and SOX2, which make possible differentiation into the lens fiber cells that form a new lens. Following successful use of this procedure for lens regeneration in rabbits and macaques, H. Lin et al. (2016) succeeded in regenerating a functional lens with refractive and accommodative capabilities in both eyes of twelve infants with congenital cataracts.

About 5 percent of the world's population, or 360 million people, suffer from disabling hearing impairment (Olusanya, Neumann, and Saunders 2014). There has been considerable success in restoring hearing with surgically implanting synthetic cochleas (Tan et al. 2013).

However, the wide spread of current around each electrode leads to channel cross-talk so that less than ten frequency channels may be part of the design (Friesen et al. 2001). The result is that individuals with cochlear implants suffer from poor comprehension of speech in noisy environments and an inability to appreciate music. One attempt to eliminate these limitations in current cochlear implants combines optogenetic manipulation of spiral ganglion neurons in the cochlea with innovative optical stimulation technology: light emitted from microscale LEDs are focused by lenses or emitted from waveguide arrays (Moser 2015). This approach offers the promise of spatially confined activation of spiral ganglion neurons that encode sound in the cochlea; however, a remaining challenge is to increase maximal firing rates (Moser 2015).

A major challenge for regeneration strategies in treating deafness is that cochlear hair cells, the sensory cells that transmit signals to the auditory nerve when stimulated by acoustic energy, are susceptible to damage, but do not regenerate (Fujioka, Okano, and Edge 2015). The discovery that the epithelial cell protein Lgr5+ is expressed in the cochlear supporting cells surrounding hair cells, and is a progenitor of cochlear cells (Bramhall et al. 2014), has led to a new procedure for expanding the number of hair cells in mice (McLean et al. 2017). McLean et al. (2017) used a cocktail of drugs and growth factors to clonally expand supporting Lgr5+ cells from newborn as well as adult tissue. From a single mouse cochlea, they obtained over 11,500 hair cells (compared to less than 200 in a control condition) from the cochlea supporting cells. In future work, the Lgr5+ will need to be targeted *in situ*, because cochlear mechanics are dependent upon its precise anatomy.

✦ 9 ✦

Neurorehabilitation for Remodeling and Repairing Injured Nervous Systems

Children on crutches or in wheelchairs are a frequent sight at the Boston Children's Hospital. These children and their families make regular visits to the Cerebral Palsy (CP) Clinic at the hospital for diagnosis and treatment. Parents bring with them the hopes that medical treatments, rehabilitation, and new technologies may improve the lives of their children by helping them learn to walk, use their hands, and speak with some assistance, or even independently. Multistage orthopedic surgeries; dorsal rhizotomies; chemical treatments to ameliorate the stiffness and dysfunction of the muscles; and intensive physical, occupational, and speech therapies are valiant efforts made, from early childhood onward, to provide each pediatric patient with expanded opportunities for more typical sensorimotor function. CP illustrates some of the central challenges for using advanced neuroimaging, medical treatments, neurorehabilitation, and new devices to remodel injured nervous systems: (1) it is not currently possible to predict with great assurance whether an injury to the nervous system around the time of birth will result in cerebral palsy, (2) current emergent behavior reflects a developmental history of earlier sensorimotor achievements and (3) the developing parts of a complex system have uneven rates of growth and recovery from injury, and interventions not properly timed may result in maladaptive plasticity.

At collaborative sites around the world, including the Massachusetts Mental Health Center and in my home Department of Psychiatry at Boston Children's Hospital, there is new hope for preventing schizophrenia through early intervention with children at clinical high risk (CHR) (Seidman and Nordentoft 2015; C. H. Liu et al. 2015). Although psychotic episodes (delusions, hallucinations) characteristic of schizophrenia have their full onset in adulthood, these episodes are often preceded by the emergence of more subtle changes in mood / anxiety, cognition, and sleep disturbances during the CHR and prodromal periods of development (Cannon 2015; Fisher et al. 2013). One framework for understanding the developmental processes underlying schizophrenia is based upon the concept of critical period, especially as developed by Takao Hensch (for example, Do, Cuenod, and Hensch 2015; see also Chapters 6 and 7). The potential value of such an approach is that scientists are closer to identifying biomarkers of schizophrenia in nervous system functioning during the CHR period and to targeting oxidative stress and neuro-inflammation to prevent or repair these dysfunctions well before they progress to psychotic episodes.

In the nearby Neonatal Intensive Care Unit at Beth Israel Deaconess Medical Center (BIDMC), a premature infant, born at 28 weeks' gestational age, and at a birth weight of 820 grams, is asleep, and will be hungry when she awakens. Premature infants often continue to not only have breathing and feeding problems, but also visual impairments such as retinopathy of prematurity and neuromotor disabilities, including enhanced risk of cerebral palsy and autism (Kuban et al. 2009; Stoll et al. 2010). During the period soon after birth, a routine head ultrasound scan showed that this infant had a germinal matrix hemorrhage with some cysts, placing her at risk for later neuromotor disability (see Chapter 6). Even here in Boston, with extraordinary medical facilities and staff to care for her, the prognosis for this child's early neuromotor development is uncertain (Einspieler and Prechtl 2005). As medical technology continues to improve around the globe, infants born at extremely low birth weights will survive but may face lifelong disability. Can we develop neurodiagnostic strategies during the infancy period that can reduce this burden? One possibility, discussed later, is to turn to advances in the mathematics of complex systems to reveal biomarkers that are better predictors of the trajectories of neurobehavioral development.

At the Spalding Rehabilitation Hospital in another part of the city, an adult stroke patient walks on a treadmill with the assistance of a rehabilitation robot. There have been significant advances in the capabilities of rehabilitation robots, including body weight–supported treadmill training. Nevertheless, these machines still lack the insight and skills of physical and occupational therapists, and machine control algorithms do not fully incorporate what we know about learning (Reinkensmeyer et al. 2016). Indeed, randomized clinical trials of gait rehabilitation training with machines show that more conventional opportunities for home-based exercise in patients poststroke produce similar results in measures of strength, walking speed, distance, quality of life, and dependence on assistive aids (Dobkin and Duncan 2012). There is also increasing recognition that the use of robots for neurorehabilitation must be mindful of the particular vulnerabilities to injury of the mammalian nervous system, its limitations for spontaneous repair, and the mechano-chemical environment of injured neurons, as discussed in Chapters 6 and 7, respectively.

Neurorehabilitation of the human nervous system may be most successful by employing multiple technologies—including pharmacological manipulation of the internal milieu at the injury site, electrical stimulation, and the assistance of adaptive robotic systems—in order to improve the individual's functional relationship with the environment. For example, by sequentially promoting growth of descending corticospinal tract fibers through pharmacological blockade of the neurite growth-inhibitory protein NogoA (also known as RTN4), followed by (rather than simultaneously with) motor training of grasping, lost motor function following experimental stroke was restored in rats (Wahl et al. 2014). Cerebral palsy, schizophrenia, stroke, blindness, and deafness continue to place a substantial burden on families—and on our economy.

Major Challenges of Cerebral Palsy, Schizophrenia, and Stroke

Cerebral palsy. CP is a developmental disorder affecting four children per 1000 in the United States, with a lifetime cost for all CP patients in the United States of $11.5 billion (Christensen et al. 2014). CP was originally reported by Little in 1861, but its definition has remained elusive because it includes a "heterogeneity of disorders" (Rosenbaum 2007). Critical emphases are currently placed on the *developmental* and *nonprogressive* nature of CP. So, for example, the motor impairments of children who are eventually diagnosed with CP, as a result of perinatal pathophysiology, typically begin to be noticed before 18 months of age and are clinically defined at age 4 to 5 years (Krigger 2006). The "encephalopathy of prematurity" of very preterm infants, which includes cerebral white matter injury and neuronal, axonal, synaptic disturbances (Volpe 2012) as well as neuro-inflammation (Leviton et al. 2015), puts infants at high risk for cerebral palsy. Additionally, though, while these pathophysiological mechanisms leading to CP arise from an *initial* series of perinatal events, there may be persistent cascades of neuro-inflammation in cerebral palsy, which Dammann and Leviton (2014) call intermittent or sustained systemic inflammation (ISSI). In this view, secondary and tertiary systemic inflammation, not just the initial perinatal insult, may contribute to the adverse neurodevelopmental outcomes in children born prematurely. Tertiary forms of brain damage may worsen outcome, promote further injury, or prevent repair or regeneration after an initial brain insult (Fleiss and Gressens 2011). Two potential sources of prolonged systemic inflammation, for example, are assisted ventilation and bacteremia (Dammann and Leviton 2014).

The effects of the early brain pathophysiology on action and perception in individuals with CP is apparent at both the neuroanatomical level of descending corticomotor and ascending sensorimotor tracts in the brain (Scheck, Boyd, and Rose 2012), and in sensorimotor behaviors, such as the reaching and grasping (precision grip) activities of children with CP (Bleyenheuft and Gordon 2013; Robert et al. 2013). This implies a multifaceted strategy in which biomarkers and agents targeting perinatal neuro-inflammation are used in infants for *prevention* of brain injury, while early neurobehavioral diagnostics, neurotherapeutics for tertiary inflammation, and neurorehabilitation are used to promote more typical development once brain injury has occurred (see Fleiss and Gressens 2011). One approach in using neurotherapeutics for treating tertiary inflammation is to target activated microglia and astrocytes (for example, Kannan et al. 2012), as discussed further later.

Schizophrenia. The onset of psychosis has a devastating effect on the lives of an individual with schizophrenia, as evident, for example, in the 1975 portrayal of a psychotic man by Jack Nicholson in *One*

Flew Over the Cuckoo's Nest. More than four decades later, treatments administered during psychosis may ameliorate symptoms, but they do not reverse circuit dysfunctions (Uhlhaas and Singer 2015). New findings on brain region synchronization (Uhlhaas and Singer 2010, 2011, 2015) and connectomics (Fornito, Zalesky, and Breakspear 2015; Fornito and Bullmore 2015; van den Heuvel and Fornito 2014) have reenergized clinical research in schizophrenia because it may now be possible to identify and interventionally target developing circuits during the developmental period preceding onset of psychosis (Cannon 2015; Fisher et al. 2013). So, for example, disturbances in excitation / inhibition balance that give rise to abnormal network dynamics, such as thalamic gating of communication between cortical areas (Uhlhaas and Singer 2015), may lead to deficits in sensory processing, attention, and executive functioning, from which the devastating psychotic symptoms of schizophrenia emerge (see, for example, Wolf et al. 2015). Therefore, drugs that target the nexus of oxidative stress (neuroinflammation, NMDAR hypofunction) during early development, such as Omega 3, sulforaphane, and N-acetylcysteine (NAC), are potential candidates for repairing these developmental anomalies (Seidman and Nordentoft 2015).

Stroke. Stroke is a leading cause of brain injury resulting from lack of blood flow to the brain (George and Steinberg 2015) (see Table 9.1). Approximately 15 million people worldwide suffer a stroke every year (Starkey and Schwab 2014). Stroke is the second leading cause of death worldwide and the leading cause of acquired neurological disabilities in the developed world (Corbett et al. 2015). Up to 85 percent of stroke survivors experience hemiparesis, resulting in an impairment of one upper extremity immediately after stroke (Levin, Kleim, and Wolf 2009). In the clinical literature on stroke rehabilitation, there has been a major concern with the distinction between *motor recovery*, the reappearance of behavior patterns that were present prior to central nervous system injury and *motor compensation*, in which former functions are taken over, replaced, or substituted by different body parts (Levin et al. 2009). Recovery and compensation occur simultaneously at the neuronal, performance, and functional level. Motor recovery involves restoration of the neural tissue surrounding the primary lesion (neuronal), return of the ability to perform a movement in the same manner as before injury (performance), and successful task accomplishment using end effectors typically used for the task (function). By contrast, motor compensation involves neural tissue acquiring a function it did not have prior to injury (neuronal), use of alternative movement patterns (performance), and successful task accomplishment using alternative end effectors, such as opening a bag of chips with one hand and the mouth rather than with two hands (function) (Levin et al. 2009) (see Table 9.2).

There are two characteristics of brain structure and function that may facilitate stroke recovery: diffuse and redundant connectivity and remapping between related cortical regions. First, due to the brain's network organization, the poststroke recovery of any brain region must be understood with respect to interactions with other parts of the network (Corbetta 2010; Carter, Shulman, and Corbetta 2012). Second, after stroke, cortical remapping is both activity-dependent and

Table 9.1 Pathophysiology of Stroke

Mechanism	Description
Excitotoxicity	Lack of blood flow (ischemia) results in a deficiency of glucose and oxygen and excessive glutamate release and influx of calcium. This triggers cell death pathways.
Mitochondrial response	Rapid influx of calcium leads to excess accumulation in the mitochondria, causing dysfunction in cell energy homeostasis.
Free radical release	The influx of calcium triggers nitric oxide production that leads to injury through the formation of oxygen-free radicals and further oxidative stress. Free radicals not only contribute to initial toxicity, but also prevent recovery, making them an important poststroke therapeutic target.
Protein misfolding	Ischemic injury induces stress on the endoplasmic reticulum, an organelle that regulates protein synthesis. This leads to an accumulation of misfolded proteins and halts new protein synthesis.
Inflammatory changes	An inflammatory response initially contributes to cellular injury through release of cytokines and harmful radicals but eventually helps to remove damaged tissue to enable remodeling.

Source: P. George & G. Steinberg (2015), Novel stroke therapeutics: Unraveling stroke pathophysiology and its impact on clinical treatments, *Neuron, 87,* 297–309.

Table 9.2 Rehabilitation robots: Control strategies and artificial intelligence

Control strategy		Description
Cooperative assistance	Impedance-based assistance	While a patient moves along a trajectory (e.g., within the walls of a virtual tunnel), the robot does not intervene. Guiding force (mechanical impedance) is applied when the person deviates from this trajectory.
	EMG-based assistance	Augments the activity of individual muscles, e.g., when an exoskeleton applies torque to a joint proportionally to the EMG of the muscle.
	EEG-triggered assistance	When the rehabilitation robot detects increased EEG activity in the brain's motor planning areas, it activates assistance along a predefined trajectory.
Challenge-based control	Resistive control	The robot generates forces that constantly resist any movements the patient attempts, forcing them to apply a larger force (to improve strength).
	Error augmentation	Aims to improve coordination by using artificial intelligence to identify a patient's deviations from a desired trajectory and then uses the robot to magnify those deviations to reveal errors.
Control based on longer-term success	Performance-based adaptation	The task is made more difficult if the patient is performing well over time but is simplified if the patient is largely unsuccessful.
	Bio-cooperative control	The patient's physiological responses (e.g., heart rate) are measured to determine physical and mental workload. These may be inferred by means of machine learning.

Source: D. Novak & R. Riener (2015), Control strategies and artificial intelligence in rehabilitation robotics, *AI Magazine, 36,* 23–33.

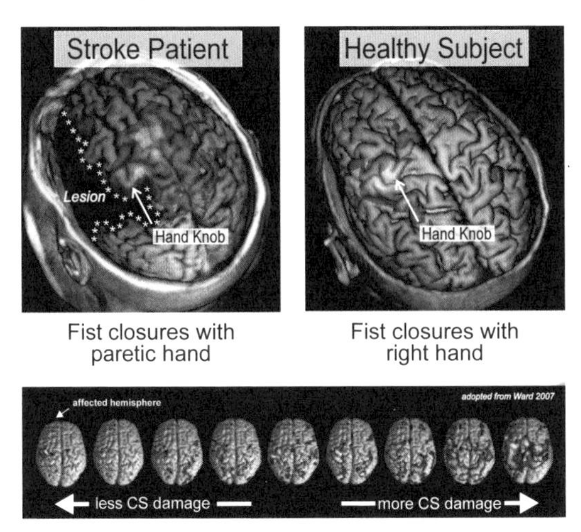

Figure 9.1 Representative functional MRI blood oxygenation level-dependent activity reconstructions after stroke in the motor system. *Top:* Patient with persistent hemiparesis more than ten years after stroke (*left*) and an age-matched healthy control (*right*) during a fist closure task. In the healthy control, activity is strongly lateralized to the left hemisphere with movements of the right hand. By contrast, fist closures with the paretic hand were associated with enhanced and more extended neural activity in both hemispheres. Overactivity is particularly seen in the supplementary motor area region (see activity maximum along the interhemispheric fissure) and in the intact cortex adjacent to the lesion, which has spared the motor hand area formation (white arrows; asterisks delineate lesion region). Furthermore, additional clusters of activity are evident in prefrontal cortex, which might show higher cognitive control when doing this relatively simple task. *Bottom:* Different appearances of functional MRI activity in patients who had chronic stroke with corticospinal tract damage and different lesion volumes. Increased damage was associated with a greater amount of overactivity. CS = corticospinal. Christian Grefkes and Gereon R. Fink (2014). Images reprinted with permission from Elsevier.

based on competition for available cortical map territory (Grefkes and Fink 2014) (see Figure 9.1). Stroke-affected circuits in the peri-infarct region gain an advantage in rewiring due to proteins that encourage growth-related processes, including sprouting of new axons and the increased elaboration of dendrites and dendritic spines (Murphy and Corbett 2009). As an endogenous response to focal cerebral ischemia, the injured brain exhibits a "reemergence" of ontogenetic organizational patterns: there is intense neuronal sprouting and brain capillary sprouting, and glial cells create a favorable cerebral environment for neuronal growth and plasticity (Hermann and Chopp 2012). One implication of the apparent reemergence of ontogeny-like growth processes in the immediate poststroke period is that therapeutic intervention must be conducted with an appreciation for appropriate time windows of endogenous recovery.

A major discovery in understanding the regulation of plasticity has been that certain proteins, including those in the Nogo family, determine the degree to which environmentally induced activity drives patterns of synaptic change (Kempf and Schwab 2013; Schwab 2010; Schwab and Strittmatter 2014). For example, by blocking NogoA or its receptor, it has been possible to functionally restore skilled forearm reaching in rats with large unilateral strokes (Tsai et al. 2011), and there is both recovery of hand function and cortical map shifts in macaques (Wyss et al. 2013). This implies that it might be possible to combine immunotherapy against neurite-inhibitory proteins, such as Nogo, along with environmentally based rehabilitation strategies to promote poststroke synaptic connectivity and restoration of function. A study by Wahl et al. (2014) using a rat stroke model did just that, combining four different therapy and rehabilitation schedules. The most successful of the four strategies, both in terms of growth of corticospinal fibers and functional recovery of skilled use of the forelimbs, was to first promote nerve fiber growth (a process akin to the proliferation period in nervous system ontogeny) and then use rehabilitative training to induce selection and stabilization of connectivity.

Neurodiagnostics

Biomarkers. The underlying etiologies for developmental neuropsychiatric diseases, such as schizophrenia, have been notoriously elusive. The process of untangling the interdependencies of multiple factors over development have been aided by the emergence of patterns in meta-analyses (Seidman and Nordentoft 2015) as well as by large datasets that include genetic, neurophysiological, imaging, and behavioral data on individuals at high clinical risk (Cannon 2015). After decades of work, an understanding of the fundamental neurobiology of schizophrenia may be at hand (Dhindsa and Goldstein 2016). The crucial link for risk association in schizophrenia appears to involve a cascade of developmental processes which includes (1) *mutations in loci near gene 4* of the major histocompatibility complex (MHC), a region on chromosome 6 involved in acquired immunity (Sekar et al. 2016); (2) *greater activation of microglia and neuroinflammation during brain development;* (3) *excessive synaptic pruning resulting in loss of connectivity;* and (4) *loss of cortical gray matter and connectivity in regions involved in monitoring predictions about beliefs relative to experiences* (Cannon 2015). This cascade of processes not only fits the genetic, neurophysiological, imaging, and behavioral data on individuals at clinical high risk for schizophrenia, but also helps to explain the characteristic onset of psychotic symptoms between ages 18 and 24 years.

Complexity in time series. The Margaret and H. A. Rey Institute of Nonlinear Dynamics in Physiology and Medicine at Beth Israel Deaconess Medical Center, in Boston, is named for the authors of the "Curious George" books, including my favorite, *Curious George Goes to the Hospital* (1966). H. A. Rey, who came to America to escape Hitler's atrocities against the Jews, also became famous as author of a beginner's guide to the night sky, called *The Stars*, published in 1952. Rey was a brilliant nonconformist, and his little book reconceptualized the way that illustrations of the constellations actually represented the animal or person in the constellation's name. *The Stars* was a favorite of Albert Einstein, who appreciated Rey's unconventional thinking and skill at revealing simplicity in presenting the constellations (Ashmore 2013). The director of the Rey Institute, Ary Goldberger, shares Rey's passion for using unconventional thinking to reveal hidden patterns. Ary is a cardiologist and has innovated techniques for visualizing and revealing clinically meaningful patterns in complex physiological signals, such as the multifractal fluctuations apparent in cardiac interbeat intervals (Goldberger et al. 2002; Burykin et al. 2014).

The body's transportation and communication networks exhibit a fractal geometry that is a structural signature of physiologic complexity. In fractal geometry, there is a branching of subunits in which there is self-similarity, a resemblance at all scales (Peitgen, Jurgens, and Saupe

1992). For example, cardiovascular structures, such as arterial and venous trees, serve a fundamental physiologic function: rapid and efficient transport of fluid and signals over complex distributed networks (Goldberger et al. 2002). These and other bodily systems may degrade with aging or disease, and mathematical tools for fractal analysis demonstrate a loss of complexity and an anticipation of system failure. Illustrations include the altered fractal dynamics in manual behavior and postural control in aging (Kelty-Stephen and Dixon 2013).

Neuroimaging. *In vivo* neuroimaging of mice and connectome analyses (see Chapter 5) have provided new insights into the mechanisms underlying recovery of function following stroke. It is clear from neuroimaging of mouse brain that within minutes, the effects of stroke on the brain connectome are unavoidable. Silasi and Murphy (2014) offer a compact description of the effects of experimental stroke and reperfusion on structural connections in the mouse brain:

> Within 60–120 s of the induction of forebrain ischemia, energy supplies rapidly dwindle and ischemic depolarization occurs. This massive ischemic depolarization leads to the loss of membrane potential from neurons and spreading waves of damage. Individual dendrites were rapidly swollen and beaded within 3 min. In addition to regular dendritic swellings, there was a loss of dendritic spines, as well as swelling of glial elements. (p. 1357)

Because of the vascular contribution to stroke, a valuable complement to the neuronal connectome is a high-resolution, capillary-level reconstruction in rodents of what Silasi and Murphy (2014) call the *angiome*. There are now remarkably complete brain-wide capillary data for predicting local vulnerabilities to vascular disruptions.

Computational Approaches to Neurorehabilitation

Motor learning. Current approaches to neurorehabilitation, especially the use of robotic systems and exoskeletons, are based upon our knowledge about the nature of motor learning. According to Shmuelof and Krakauer (2011), motor learning is refers to any practice-related change or improvement in motor performance. Motor learning may be further differentiated into (1) a model-based system in which improvements in motor performance, such as the reduction in error in adaptation paradigms, are presumed to be based upon an internal forward model of the environment updated on the basis of prediction errors, and (2) a model-free system in which learning occurs directly at the level of the controller (Haith and Krakauer 2013). In model-based learning, it is assumed that the motor system compensates for systematic perturbations by first identifying the dynamics of the system being controlled through a forward model and then transforming this knowledge into a control policy in (Haith and Krakauer 2013, 12). In an uncertain environment, a model-based learning strategy is the most powerful and flexible and allows for faster and more precise estimation of the state of the body and environment, but it requires "unwieldy computations" (Haith and Krakauer 2013). Direct, model-free approaches require only simple computations but involve extensive exploratory behavior. In model-free learning, reward prediction errors directly drive changes to a control policy.

A widely used experimental paradigm to study motor learning is *adaptation*. At the heart of the adaptation paradigm is the use of a robotic system to generate precisely timed force fields that act upon the body. A classic example of adaptation is a task in which subjects are asked to reach for a visually displayed target while holding the handle of a robotic arm (Shadmehr and Mussa-Ivaldi 1994). The motors of the robot generate specific force fields, such as a viscous curl perpendicular to the direction and proportional to the velocity of the hand movement. The force-field adaptation paradigm is described by Krakauer (2006). Without the motors turned on, the mechanical properties of the robotic arm maintain the subject's reach trajectory in a horizontal plane, and subjects make smooth and straight trajectories. On initial trials with the motors turned on, subjects perform skewed trajectories but, with practice, are able to adapt to the force field and again are able to make smooth and nearly straight trajectories. Then, while subjects are in this "adapted state," the force field is turned off and subjects exhibit "after-effects": trajectories are now skewed in in the direction opposite to that seen in the initial adaptation period. What is the significance of the after-effect for probing the healthy nervous system and for rehabilitation following brain injury? As summarized by Krakauer:

> The presence of after-effects is strong evidence that the central nervous system can alter motor commands to the arm to predict the effects of the force field and form a new mapping between limb state and muscle forces (internal model). . . . The importance of the concept of internal model to rehabilitation is that the model can be updated as the state of the limb changes. This rehabilitation needs to emphasize techniques that promote formation of appropriate internal models and not just repetitions of movements. (2006, 85)

Variability. In motor adaptation paradigms, the goal of an agent is both to *exploit* current information in choosing the highest valued option and to *explore* the many local suboptimal alternatives available in ever-changing local environments in order to maintain accurately updated values (Louie 2013; Sutton and Barto 1998). This exploration–exploitation process in human motor learning is illustrated in a study by H. G. Wu et al. (2014). They wanted to know whether intrinsic motor system variability is harnessed during trial-and-error learning to improve performance as subjects traced a curved guide line with a manipulandum. During a baseline period, shape-tracing movements were not rewarded, but subjects were given feedback about their movement speed. This provided an estimate of intrinsic task-relevant variability. Then, during training, each subject was assigned a performance score based on deviation from his or her mean baseline trajectory. H. G. Wu et al. (2014) found a strong positive relationship between intrinsic variability during baseline performance and the rate of learning during training. Individuals with higher task-relevant variability at baseline learned faster than those with lower baseline variability.

H. G. Wu et al. (2014) then examined whether the motor system may improve learning by modulating the structure of variability along the task-relevant dimension of the motor output space. To do so, they measured motor variability before and after a training procedure that repeatedly exposed subjects to either velocity-dependent or position-dependent force-field perturbations. They found that motor variability *increased* with respect to the task-relevant component (that is, either position or velocity). As summarized by Wu et al.:

> The ability to produce finely sculpted changes in the temporal structure of motor variability thus allows the motor system to improve the efficiency of learning by guiding exploration to the relevant parts of motor output space. . . . The human motor system does not simply exploit what it currently knows, but instead actively engages in motor exploration, possibly sacrificing accurate performance in lieu of facilitating learning. (2014, 8–9)

The intrinsic variability of trial-and-error learning is consistent with a control system comprised of "good enough" ways of discovering local minima in the task space (Loeb 2012). The spinal cord may be a "regulator" of trial-and-error learning, a structure whose outputs depend on multiple sources of feedback from the lower level according to control signals established by a higher level (Raphael, Tsianos, and Loeb 2012). Such feedback is a general and flexible means to govern system-level behavior (Roth, Sponberg, and Cowan 2014).

The role that variability plays in motor learning also has important implications for neurorehabilitation, as illustrated by a series of studies in spinal mice (Cai et al. 2006) and rats (Shah et al. 2012) comparing fixed-trajectory assistance with assistance as needed that provided step-to-step variability. Spinal-transected animals were able to more effectively step bipedally when the mode of forward step training allowed for some critical level of variability in behavior and, by inference, in spinal locomotor circuitry (Cai et al. 2006; Ziegler et al. 2010). Further, use of step training that included variations in the direction of stepping (that is, sideward and backward) exhibited greater forward stepping consistency and more highly coordinated interlimb behavior than controls trained to step forward (Shah et al. 2012).

Mice and other rodents occupy a niche at the light / dark boundary between earth and sky, nesting in subterranean burrows during the day and emerging at dusk to explore for food and other resources. Their visual systems are able to detect the shadows of flying predators looming in the sky and promote an adaptive escape or freeze response (Yilmaz and Meister 2013), but murine retinas lack foveas and, therefore, provide limited capacity to lock the eyes onto a target (de Jeu and de Zeeuw 2012). For detecting the layout of the environment and its affordances, such as things that are edible or places for hiding, rodents rhythmically sweep their whiskers in an exploratory fashion, a behavior called "whisking" (Diamond et al. 2008).

Whisking is particularly intriguing with regard to the acquisition and remodeling of behavior during learning for at least four reasons. First, rodent vibrissae are smart instruments, consisting of muscle-actuated whiskers that locate and identify objects by means of phase-sensitive detection of vibrissa deflection angle, with neural coding of deflection amplitude at a slow scale and phase of motion at a fast scale (Hill et al. 2011; Kleinfeld and Deschenes 2011). Second, the rhythmic pattern of whisking becomes precisely phase-locked with respiration-driven sniffing, suggesting that at the brain stem level, the breathing cycle binds other rhythms into an integrated pattern of exploratory behavior, analogous to thalamocortical gamma serving as a reference oscillation to align sensory input (Kleinfeld et al. 2014; Moore et al. 2013; see also Chapter 5). Third, whisking behavior emerges developmentally during a period when premotor circuitry begins to provide new input to vibrissal facial motor neurons (Takatoh et al. 2013). And fourth, while learning to lick during whisking, ensembles of layer 2/3 pyramidal neurons are remodeled by means of enhanced dendritic spine growth (Holtmaat and Svoboda 2009; Huber et al. 2012; Kuhlman et al. 2014). Together, these four findings identify how a smart instrument for exploratory behavior is assembled and remodeled at the neural and behavioral levels.

Whisking provides a means for finding food. A mouse emerging from a hole in the ground to sniff, taste, and consume a fallen nut displays both a range of locomotor maneuvers and skilled use of the forelimbs, highlighting the organization of the spinal cord, brainstem, and higher centers for sensorimotor behavior. The mouse may run at different speeds, sometimes using a gait pattern of left-right alternation, and at other times, a galloping pattern. Upon reaching the food morsel, the mouse may grasp it with both hands, quickly take a few nibbles, and then run back to the hole. The advent of novel molecular and genetic tools is providing new opportunities for discoveries about the neuronal circuitry of the spinal cord, how descending projections may regulate spinal networks, and how neuromodulatory remodeling and sensory information may drive these networks through multistable dynamical modes (see Arber 2012).

Rehabilitation robotics. The use of interactive clinical devices for remodeling injured nervous systems has made enormous strides since the 1990s, with a new focus on *mechanisms of recovery* and *evidence-based treatment* (Krebs et al. 2006; Krebs, Volpe, and Hogan 2009). Chapter 7 has already reviewed the former, and here I focus on clinical trials that have evaluated the use of robotic devices for upper-limb (Maciejasz et al. 2014) and lower-body rehabilitation in cerebral palsy (Dobkin and Duncan 2012; Fasoli et al. 2008) and stroke (Klamroth-Marganska et al. 2014; Pennycott et al. 2012).

In rehabilitation of upper-limb cerebral palsy, there is a consensus that the most effective therapies include intensive, structured, task repetition; progressive incremental increases in task difficulty; and enhancement of the motivation and engagement of individuals in therapy (Reid, Rose, and Boyd 2015). Novak and Riener (2015) distinguish assistance strategies in robot rehabilitation (see Table 9.3). One exoskeleton robot, ARMin, combines several of these by supporting physiological movements of the shoulder and arm, providing intensive and task-specific training strategies for the arm that are particularly effective for promoting motor function, and using a "teach-and-repeat" therapist-guided procedure (Nef, Guidali, and Reiner 2009).

Table 9.3 Motor control principles used in virtual reality training of upper-body movements

Principle	Description	Virtual environment (VE)
Task demands	Task difficulty considers speed and precision of intended movement, as well as object location and distance.	Virtual objects are adjustable so that task difficulty is graded according to speed and precision; objects are placed at different workspace locations.
Task environment	Organization of movement is related to the quality of the viewing environment.	Viewing environment includes context for three-dimensional viewing (e.g., perspective lines).
Affordances	Hand orientation for grasping is related to object properties.	Objects in VE should be of various shapes, sizes, and locations.
Goals	Organization of reach-to-grasp movements depends upon what the individual intends to do with the object.	Tasks involving grasping should have purposeful goals.
Feedback	Salient feedback about quality of movement and joint ranges used is essential to improve motor behavior.	Auditory, visual, and tactile feedback are incorporated into the VE.

Source: M. Levin, P. Weiss, & E. Keshner (2015), Emergence of virtual reality as a tool for upper limb rehabilitation: Incorporation of motor control and motor learning principles, *Physical Therapy, 95,* 415–425.

Klamroth-Marganska et al. (2014) report a randomized clinical trial of task-specific robot therapy of the arm using ARMin, compared with conventional therapy after stroke. Their results show a small but statistically significant difference in the ARMin exoskeleton compared to conventional therapy. However, the study does not reveal what subjects actually learn as they improve performance.

It is well known that hemiparetic stroke patients learn to adapt to, or compensate for, their motor deficits by using their trunk and affected arm muscles differently (Jones 2017), but it is not clear in the trial the extent to which the improvement involves compensatory strategies (Kwakkel and Meskers 2014). Some clarification on this point comes from the EXPLICIT-stroke Program (van Kordelaar et al. 2013). They report that a dissociation of shoulder and elbow movements during reach-to-grasp movements occurs mainly during the early (first five weeks) poststroke period. After that, recovery includes pathological compensatory strategies. It is notable that even without exoskeleton training following stroke, therapeutic constraint of the arm on the nonaffected side of the body results in clinically relevant improvements in arm function that persist for at least a year, as demonstrated in the EXCITE trial (Wolf et al. 2006). Together, these results question the efficacy of exoskeleton rehabilitation. However, a somewhat different picture comes from studies that use both robots and virtual reality for rehabilitation.

Virtual reality. Our hands enable us to do so much that we take for granted: touching the soft skin of a beloved child, reaching and grasping for a cup of water to slake our thirst, opening a car door and driving to work, and writing notes with a pen. The upper-limb impairment resulting from stroke frequently prevents or interferes with all of these functional activities. Rehabilitation that enhances experience-dependent neuroplasticity and provides opportunities for relearning upper-limb function is challenging because it requires intensity, variability, specificity, motivation, and interactivity of practice (Levin, Weiss, and Keshner 2015; Weiss, Keshner, and Levin 2014). Virtual environments made possible using virtual reality (VR) technology have the potential for promoting such relearning because they can be systematically manipulated to engage learning processes (Levin et al. 2015) (see Table 9.4).

A crucial capability of effective VR rehabilitation systems is that they integrate visual information with haptic feedback that allows a user to touch, feel, and manipulate objects during VR simulations (Magdalon et al. 2011; Merians and Fluet 2014). (It is noteworthy that this capability is similar to the advanced capabilities of neuroprosthetic devices with haptic feedback.) One approach to using VR systems with stroke patients has been to combine two- or three-dimensional virtual environments with a force-reflecting upper-extremity exoskeleton, the *CyberGlove*, in order to improve task-based activity scaling, such as using individual fingers to press targeted piano keys (Fluet et al. 2014; Merians et al. 2011). In pressing some combination of keys, the fingers may play different roles in managing

Table 9.4 Approaches to stroke therapeutics

Approach	Description
Restoration of blood flow (acute)	Intravenous administration of tissue plasminogen activator (tPA), if given within 3–4.5 hours of symptom onset, is the mainstay of acute stroke therapy.
Neuroprotection (acute)	Minimizing the damage to the peri-infact region, which contains factors promoting growth and sprouting, is a major goal of neuroprotective strategies, such as hypothermia.
Stem cell replacement (recovery)	Stem cells are self-perpetuating pluripotent or multipotent cells with the ability to transform into multiple cell types. As part of the brain's normal response to injury, endogeneous stem cells, called neural progenitor cells (NPC), move to injured areas of the brain. Therapeutic approaches are aimed at augmenting this normal endogenous reaction to injury.
Modulation of circuits (recovery)	Following ischemia, there is a shift in the excitatory-inhibitory (E-I) balance across the brain. Potential ways to restore E-I balance include noninvasive repetitive transcranial magnetic stimulation (rTMS), transcranial direct current stimulation (tDCS), and invasive implantable epidural electrodes. Optogenetics to selectively stimulate ipsilesional primary motor cortex neurons following stroke may improve functional outcomes.
Brain–machine interface	In patients whose primary cortical areas have been damaged, plasticity in alternative areas may be promoted by means of closed-loop control by robotic systems.

Source: P. George & G. Steinberg (2015), Novel stroke therapeutics: Unraveling stroke pathophysiology and its impact on clinical treatments, *Neuron, 87,* 297–309.

overall force production, and so the differentiated production, or fractionation, of forces is important for skilled performance (Kapur et al. 2010).

Computational neurorehabilitation. The increased availability of kinematic data obtained from robotic devices during neurorehabilitation has promoted a computational approach that (1) provides a quantitative description of the patient's sensorimotor experience, (2) mathematically models the motor learning processes and plasticity underlying the rehabilitation process itself, and (3) describes quantitatively the patient's behavioral outcomes (Reinkensmeyer et al. 2016). An influential approach to learning with implications for mathematically modeling the learning process in computational neurorehabilitation is reinforcement learning (RL) theory (Sutton and Barto 1998). RL theory is centered on a "control policy" that maps world states to actions that an individual should take to maximize future rewards. This is unsupervised learning so that the learner actively explores the environment via stochastic search to obtain information about future rewards (see also Gibson's view on obtaining information in Chapter 3). As one illustration of the computational neurorehabilitation approach, Reinkensmeyer et al. (2012) consider recovery of upper-limb strength following stroke, motivated by the view that strength predicts extremity functional activity. A first assumption is that wrist force is produced by the summed effect of corticospinal cells targeting motor neuronal pools and that the basis for plasticity after stroke is a reinforcement learning process, in which repetitive movement experiences modify corticospinal cell activations (Reinkensmeyer et al. 2016). Next, they assume that reinforcement learning via stochastic search may explain a broad range of stroke recovery phenomena, including exponential-like recovery curves that do not asymptote but, instead, exhibit a residual capacity for further recovery with further movement practice. And finally, they conclude that if RL via stochastic search can explain recovery of upper limb strength, then the RL mechanism should be targeted during rehabilitation training to facilitate the search process (Reinkensmeyer et al. 2016). One possible way of targeting RL during rehabilitation is to add variability and opportunities for exploration to training regimens.

The Bernstein Tradition and Dynamical Systems

The uncontrolled manifold. Lessons learned, thus far, from robotics and virtual reality suggest that in order for neurorehabilitation to be most effective, it should target the way that the motor control system continuously dissolves and reassembles itself to discover an organization of parts that maintains a stable *functional* relationship between the body and environment (for example, locomoting, reaching, speaking). This functional relationship changes with an individual's goals, capabilities, and discovery of new opportunities for action in the environment through exploration of affordances (see Table 9.3). Studies from the Bernstein tradition in the field of motor control have adopted coupled oscillator models of behavior to reveal the organizational principles by which multiple synergies form, combine, dissolve, and reform for particular modes of behavior, such as locomotion (Couzin-Fuchs et al. 2015; Holmes et al. 2006). One approach, for example, has been to model animal and human gait patterns as a symmetry group of coupled oscillators (Golubitsky et al. 1998, 1999; Pinto and Golubitsky 2006). The Bernstein tradition also emphasizes the *flexibility and stability* of synergies:

> Synergy is a neural organization of a multi-element system that (1) organizes sharing of a task among a set of elemental variables; and (2) ensures co-variation among elemental variables with the purpose to stabilize performance variables. (Latash, Scholz, and Schöner 2007, 279)

A technique that is becoming widely used to measure the process by which body parts are organized for particular functions is called the uncontrolled manifold (UCM) analysis (Scholz and Schöner 1999; Schöner and Scholz 2007). The UCM analysis characterizes the structure of variability of a collection of elemental variables (for example, body joint angles) in relation to the functional demands of a task by statistically quantifying the extent to which variability of a collection of variables tends to stabilize or destabilize the functional parameter of a task (Nonaka 2013). Within the UCM, a collection of variables is free to vary without disrupting task performance. Identifying the structure of variability of motor elements specific to task demands provides a means of scaling current actions to environmental properties.

As an illustration of the procedure, consider the act of goal-directed reaching in order to touch an object with an index finger of one hand (van der Steen and Bongers 2011). Repetitions of reaching to touch an object exhibit trial-to-trial variability of joint angles of the shoulder, elbow, and wrist. The variance of joint angles during movement may be divided into two parts: (1) joint angle variability that *does not affect* the index finger position and (2) joint angle variability that changes the index finger position. The UCM method decomposes the overall joint angle variability into one of these two types of variability. This is done by formally obtaining a model that relates joint motion to movement of the index finger in space. The null space of the equation relating the task space (for example, the two- or three-dimensional space where the position of the index finger is defined) to the space of joint angles provides a linear estimate of UCM—all combinations of the joint angles that do not affect the value of the position of the index finger in space at that point in a movement trajectory. The null space is computed around the mean value of the joint angles at each point in the movement trajectory. Values of the joint angles for each movement repetition are projected onto the null space (a linear estimate of the UCM) and onto the subspace orthogonal to the UCM (where different combinations of the joint angles lead to a different position of the index finger in space). The UCM method makes it possible to determine whether joint angles vary more within the UCM (variability that does not affect the finger position) than within its orthogonal counterpart (variability that does affect the index finger position). If the joint angles of the arm vary more in the direction parallel to the UCM than in the direction orthogonal to it, this would mean that the stable behavior of the arm is coordinated in such a way that it is controlling the index finger's position.

The UCM analysis illustrates how synergies share a task among a set of elemental variables. Consider a person grasping an object in order to produce a net force of 20 Newtons with two fingers. Sharing means that there is a co-variation (flexibility) that leaves the net force invariant (stable), such as 5 and 15 Newtons, 10 and 10 Newtons, and 15 and 5 Newtons. Covariation that preserves invariance is indicated by the different ways that the forces may be distributed by the two fingers while still producing a net force of 20 Newtons. The uncontrolled manifold analysis is a quantitative technique that attempts to capture covariation that preserves invariance by partitioning the variance of the elemental variables (forces) into two components: one that affects and one that does not affect the value of a particular performance variable (Latash et al. 2007). For tasks involving multifinger force production, the UCM analysis has been used to understand the structure of force variability that preserves invariant net force. Scholz et al. (2002), for example, instructed adult subjects to press and release two fingers on a force transducer at a metronome paced frequency (112 beats per minute). A UCM analysis revealed that the variance of individual finger forces was structured in specific ways to achieve control of total force, using a range of covariations. The results of this and many other experiments on finger force production, suggest to Latash (2012) that the abundance of motor solutions used to maintain an invariant relation between body and environment is not a curse but is, rather, "bliss" for motor control systems.

Stochastic resonance. Stochastic resonance (SR) is a nonlinear phenomenon in which the addition of "noise" may enhance the information content, and detection, of a signal by improving the output signal-to-noise ratio (Moss, Ward, and Sannita 2004). The "classical" SR paradigm begins with a weak, undetectable periodic input (a "subthreshold" signal) to a nonlinear dynamical system. A power spectral density analysis of the response is used to determine how added noise allows the input signal to be detected (Collins, Imhoff, and Grigg 1996; McDonnell and Ward 2011). What role might SR play in neural systems? One clue comes from research by Levin and Miller (1996) on the role of SR in the way that the cricket's abdominal cercal system detects low-frequency air disturbances of approaching predators. They made intercellular recordings of interneurons excited by cercal mechanosensory afferents in response to experimental presentation of air current stimulation with or without added noise. During the attack of a predatory wasp, which generates air displacements in the frequency range of 5–50 Hz, Levin and Miller (1996) found that added noise resulted in significant improvement in the signal-to-noise ratio. An implication of this and other experimental studies—including shark sensory cells (Braun et al. 1994) and human muscle spindles (Cordo et al. 1996)—is that intrinsic and extrinsic modulation of noise by neural mechanisms may govern a selection process among computational modes (McDonnell and Ward 2011).

MIT and Wyss biomedical engineering professor Jim Collins has been conducting a research program that investigates how SR may be used to enhance somatosensory input from the skin surface during balance control. Human skin is innervated by somatosensory neurons projecting to the spinal cord, including thickly myelinated Aβ fibers (for example, around hair shafts) of hairy skin and Merkel cells, Meissner corpuscles, and Pacinian corpuscles of glabrous (nonhairy) skin (Abraira and Ginty 2013; Lumpkin and Caterina 2007; Zimmerman, Bai, and Ginty 2014). During balance control, changing pressure patterns along the plantar surface of the foot generates a gradient of mechanical cutaneous stimulation. Priplata et al. (2002) hypothesized that application of subthreshold mechanical noise to the soles of the feet would enhance sensory feedback about pressure changes along the gradient as the body sways. In a comparison of elderly and younger subjects, they found that sway decreased with application of noise. Here, SR is presented as an exogenous source of stimulation, within the context of endogenous fluctuations at multiple scales of each individual's behavioral organization (Kelty-Stephen and Dixon 2013; Palatinus, Dixon, and Kelty-Stephen 2012). The effect that SR has in reducing sway, therefore, is likely a function of the way that SR interacts with intrinsic body fluctuations. Indeed, when Kelty-Stephen and Dixon (2013) reanalyzed the Priplata et al. (2002) data, they found that subthreshold vibratory effects were moderated by (interacted with) endogenous fluctuations in postural sway. This context-specific influence on how mechanical stimulation influences behavior has important implications for device design.

Developmental Neurorehabilitation from a Task-Dynamic Perspective

Thelen and Ulrich. Esther Thelen was a developmental psychologist who recognized that the earliest motor behaviors of newborns, such as stepping (Thelen and Ulrich 1991) and reaching (Thelen et al.

1993) revealed the hidden dynamics of underlying self-organizing processes for the emergence of new forms of behavior (see, for example, Thelen 1995; Thelen and Smith 1994). A classic illustration of these hidden dynamics is a study of treadmill stepping in infants as young as 1 month: when the children were held supported so that the soles of their feet rested on the belts of a small motorized treadmill, they performed coordinated alternating stepping patterns characteristic of mature walking (Thelen and Ulrich 1991). The hidden pattern was revealed in advance of its typical appearance in development by providing both postural support and perceptual information for guiding the phases of the gait cycle. In looking back on this study, Thelen (1995) emphasized that treadmill stepping is not a simple reflex. Instead, this behavior is complex, perceptual-action cycle in which the dynamic stretch of the legs provides both energetic and informational components that allow the complex pattern to emerge.

Subsequent work on infant treadmill stepping by Beverly Ulrich at the University of Michigan has demonstrated that even though the kinematics of stepping (consistency of producing steps, reduction in leg flexion, and improved foot placement) improves with development, the activity of the four major muscles of the thigh and shank contributing to these limb trajectories is quite inconsistent (Teulier et al. 2012; Teulier, Lee, and Ulrich 2015). Teulier et al. (2012) conclude that during the first year, observable stepping parameters were created by a wide variety of muscle combinations and considerable variability in the timing of muscle activations. From the standpoint of underlying motor control primitives, as discussed in Chapter 8, one implication of this work is that the *neural activation of the muscles is not prescriptive:* spinal activation of the muscles alone does not determine the behavioral dynamics of the leg during locomotion. Instead, as part of a perception-action cycle, stepping may provide sensory information for *a posteriori selection* of particular dynamics of the leg by actively modulating and stabilizing the composition and timing of activation of muscle groups. In other words, *it is the variability of behavior during developmental experience that provides an opportunity to modify the neural circuits modulating spinal primitives.*

Dynamics of exploratory behavior. There is ongoing interest in using kinematic data and analyses from the dynamical systems toolbox to better understand sensorimotor development of children born early and how to use principles of typical sensorimotor development to inform early intervention (see, for example, Morgan et al. 2016). Recording and analysis of infant kicking using multicamera motion capture has made it possible to identify specific differences between typically developing and prematurely born children with white matter disease (Fetters et al. 2004) and to examine the process of exploratory behavior that typically developing infants use to learn how the body behaves in a gravitational field (Sargent et al. 2015). Some of my own work from a dynamical systems perspective has examined how infants explore the naturally occurring variability in their self-produced arm-waving and kicking behaviors (Goldfield et al. 2012; Stephen et al. 2012) and how infants begin to use the arms and legs for different functions.

A hallmark of a dynamical systems approach is that nervous, musculo-tendon, and link-segment systems are all used in flexible combinations to assemble different behavioral functions. This raises the question of how infants may learn that the same body components may be used in different ways. One possibility is that developmental changes in postural control and in coordination fluctuate over time and converge at certain periods to promote certain opportunities for learning specific functions, such as reaching or locomoting. So, for example, changes in the ability to stabilize the arms at the shoulders in order to hold the hands away from the face and, at the same time, hold the head so the eyes look straight ahead may promote opportunities for reaching, grasping, and exploring objects within arm's reach at the midline. By contrast, the ability to stabilize the torso while the hips and knees are flexed brings the feet close to the mouth to explore the relation between touching the feet with the tongue and lips at body midline. In this posture, the spring-like muscles of the legs

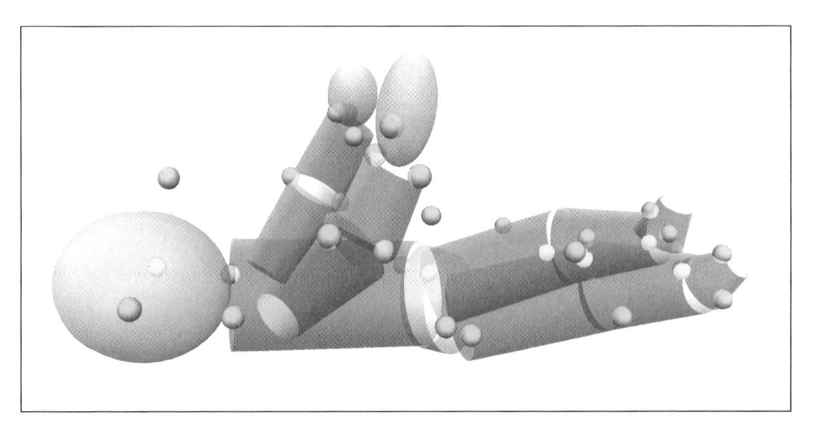

Figure 9.2 Infant exploratory behavior: arms and legs. Three-dimensional model based on motion capture data of a six-month-old supine infant exploring the different functions of the arms and legs when presented with an overhead mobile.

are maximally flexed, and the infant may excitedly extend the legs from this midline posture.

The infant's experience of the arms and legs are thus quite different. Thus, the convergence of posture and coordinated flexion-extension around the joints may be the nexus for differentiating using the hands for exploring and the legs for propulsion (Goldfield 1989). To examine this possibility, Hsu et al. (2014) conducted a longitudinal kinematic study of spontaneous arm and leg motions in four supine infants presented with an overhead mobile at 3, 4.5, and 6 months of age (see Figure 9.2). Motion capture and three-dimensional modeling allowed us to simultaneously measure joint rotations and the spatial locations of the hip, knee, and ankle joints as well as the shoulder, elbow, and wrist joints. We found that during the period at around 6 months of age, when the joint rotations of the arms were moving more independently of each other than at 3 months, infants were more likely to hold their hands away from the body in preparation for making contact with the mobile. However, at the same age, infants brought their feet close to the body to maximize propulsive kicks. Thus, the convergence of postural development and coordination of joint rotations appears to be related to the emergence of different arm and leg functions.

A project at the Wyss Institute and Boston Children's Hospital, called the "second skin," has been inspired by these and other findings that we and others have made over the years about infant sensorimotor development. The second skin project has been developing soft sensors, pneumatic actuators, and decentralized control for a soft, bio-inspired wearable robot for active control at the knees and ankles (Goldfield, Park, et al. 2012; Park, Chen, et al. 2014; Park, Santos, et al. 2014; Wehner et al. 2012). One architectural highlight of the second skin is its decentralized modular functionality, making it configurable for a diverse range of tasks. This functionality was achieved by means of a decentralized network of self-configuring nodes that manage the collection of sensor data and the delivery of actuator feedback (Park et al. 2012). Given our findings about the emergence of different functions of the arms and legs during infancy, this decentralized control network will allow the second-skin actuators to be used in different combinations for different functions.

The lessons from the work of Esther Thelen, Bev Ulrich, and Linda Fetters, as well as my own research, motivate an expanded developmental approach to neurorehabilitation that (1) builds biologically inspired robotic control systems comprised of *adaptive primitives* that promote initial exploration of intrinsic variability as a starting point for assisting learning and subsequent opportunities for introduction of combinatorial complexity; (2) uses *graph dynamics* as the basis for *adaptive control architectures* that couple control networks to the biomechanics of the body and the physical plant of a robotic device; and (3) provides enriched *task-dynamic spaces* that allow the architecture of an adaptive control system to remodel itself in response to the ongoing behavioral dynamics of the individual and adults providing assistance during neurorehabilitation. It is here, in the rich context of socially assisted learning, that the adaptive architecture of a robotic control system may become profoundly embodied with a neurologically injured individual.

Adaptive primitives. One strategy of biologically inspired engineers whose goal is to build rehabilitation robots to assist human gait has been to intensely study and model the pattern-generating neurons of animals, such as lamprey and salamanders, in order to identify and emulate the central pattern-generating neurons underlying rhythmic locomotor behavioral patterns (see, for example, Bicanski et al. 2013b; Floreano, Ijspeert, and Schaal 2014; Ijspeert 2008, 2014). An outcome of this work has been the development of what are called adaptive oscillators (AOs), mathematical tools that extract the periodic characteristics of locomotion and, thereby, emulate spinal neural oscillators generating locomotor behavior.

Early studies with human subjects coupled AOs to user movements to synchronize with and learn the gait pattern in order apply virtual impedance fields, but the systems had difficulty in transitioning between locomotor modes—for example, from walking to ascending stairs (Ronsse et al. 2011). This approach has evolved by now basing the control system for an assistive device on combinations of AOs and "dynamic motor primitives" (Garate et al. 2016) and algorithms for intention detection. The controller for a robotic assistive device combines a set of adaptive oscillators to produce virtual muscle stimulations. This is done by means of an inverse model based on available kinematic and dynamic data. A musculoskeletal model, consisting of a set of muscle-tendon units, transforms these stimulations into desired joint torques (Garate et al. 2016). The AOs are used to synchronize the control primitives to the actual gait. Then the musculoskeletal model is used to introduce a local force feedback in the muscle-tendon units, based on muscle length and velocity. Muscle forces are converted into muscle torques through geometric relationships, and the muscle model computes the joint torques as a function of the torque provided by each muscle (Garate et al. 2016). The intention detection algorithms monitor behavior to select among different locomotor modes—for example, standing, walking, and ascending / descending stairs. But remaining challenges include how intention detection algorithms modify the selection of particular modes of an individual who is learning, developing, and / or recovering from injury and how the controller is able to select primitives for assistance when the same body parts are used for different functions—for example, when the legs are used to kick a ball rather than walk.

Graph dynamics. One answer to the question of how a controller might address these challenges for a primitives-based controller rests with an approach in dynamical systems developed by Elliot Saltzman, called *graph dynamics*. Saltzman and Munhall (1992) and Saltzman et al. (2006) propose that the dynamical system primitives modeling a particular behavior have an architecture, or graph structure, and that change in behavior is reflected in the way that the parts of the graph are connected. The parts of the graph are its state variables, the dependent variables of the set of autonomous differential equations of motion used to describe the system. As an illustration, Saltzman et al. (2006) use graph structures to model the relative timing of speech gestures, where a gesture is defined as "an equivalence class of goal-directed movements" by a set of articulators in the vocal tract (Saltzman and Munhall 1989). Gestures are primitives for speech, as in the production of the bilabial gestures for/p/,/b/, and/m/. These bilabial gestures

> are produced by a family of functionally equivalent movement patterns of the upper lip, lower lip, and jaw that are actively controlled to attain the speech-relevant goal of closing the lips. Here the upper lip, lower lip, and jaw comprise the *lips* organ system, or effector system, and the gap or aperture between the lips comprises the controlled variable, of the organ / effector system. (Goldstein, Byrd, and Saltzman 2006, 218)

Gestural primitives for speech exhibit an underlying point-attractor dynamic: they act similarly to a damped mass-spring system. Point-attractor dynamics successfully model the creation and release of constrictions of the end-effectors being controlled in at least two ways. First, regardless of their initial position, and despite any unexpected perturbations, articulatory gestures with point-attractor dynamics are able to reach their target successfully. Second, the activation of a gesture's point-attractor dynamics attains the constriction goal in a flexible and adaptive way (Goldstein et al. 2006). The activation of the trajectories of multiple gestural primitives, called a *gestural score*, occurs at a second organizational level of the graph architecture, called the intergestural level. At this level is a higher-order set of primitives—activation variables—that reflects the strength with which the associated gesture (for example, lip closure) "attempts to shape vocal tract movements at any given point in time" (Goldstein et al. 2006). The intergestural level models the dynamics of planning:

> It determines the patterns of relative timing among the activation waves of gestures participating in an utterance as well as the shapes and durations of the individual gestural activation waves. Each gesture's activation wave acts to insert the gesture's parameter set into the interarticulator dynamical system defined by the set of tract-variables and model articulator coordinates. (Goldstein et al. 2006, 220)

The higher-order primitives at the intergestural level are called *planning oscillators,* and they are organized into what are called *coupling graphs* to achieve a dynamics of planning (Byrd and Saltzman 2003; Goldstein et al. 2006; Saltzman and Byrd 2000; Saltzman et al. 2008). The coupling graph for the utterance "*spot*" determines the coupling between gestures: nodes indicate gestures, and internode links are intergestural coupling functions (Saltzman et al. 2008). Each coupling graph sets the parameters of the equations of motion for the planning oscillators, which are "numerically integrated until the system reaches a steady-state pattern of interoscillator relative phasing" (Saltzman et al. 2008, 2). The relative phasing pattern is the basis for the activation waves of a gestural score used to engage the associated constriction gestures.

While most fully developed for speech, coupling graphs have the potential to be applicable to all of the domains of perceiving and acting (see, for example, Ijspeert et al. 2013) and for identifying the relation between abstract models of behavior and a physical level of description (Saltzman and Holt 2014). Consider the foot as a body end-effector serially making and breaking contact with the support surface during the locomotor gait cycle of walking. At a physical level of description, each foot and other parts of the leg alternately generate propulsive forces at the support surface in order to move the body's center of mass through the environment (see, for example, Holt et al. 2006). At the same time, at an abstract functional level, the end-effector is any agent-related property involved in creating "functionally appropriate, task-specific motions and force patterns" (Saltzman and Holt 2014).

A way to further explicate the relation between physical and functional in task dynamics is to use a mathematical tool of the dynamicist, a graph, to identify how body topology and kinematics maps onto a layout of environmental attractors. A physical graph consists of topological features, the physical linkages between body segments in contact with the environment and other segments, forming what Saltzman and Holt (2014) call "loops." During the gait cycle, changes in the relation between body and environment may be portrayed in the loops of the physical graph. At the abstract level of description, the nodes of the graph form a layout of environmental attractors—an abstract navigation space—in which an end-effector becomes coupled to a task-relevant "attracting" target location via perceptual information. Here, at the functional level, there is informational rather than physical coupling that creates the possibility of a closed loop for control (Warren 2006).

In Chapter 6, I highlighted the process of activity-dependent pruning of initially exuberant synaptic network connections in the cerebellum (Hashimoto and Kano 2013), in the hippocampus (Paolicelli et al. 2011), and at the neuromuscular junction (Turney and Lichtman 2012). This process, called *apoptosis*, involves the selective strengthening of some synaptic connections and the weakening and dying off of others, depending upon both species-specific regulatory gene expression and individual experience (see, for example, Buss, Sun, and Oppenheim 2006). Here, I present a vision of adaptive control systems for developmental robots that emulates this process. First, like apoptosis, the coupling of both the network architectures and electromechanical components of assistive wearable robots may be strengthened and maintained, or weakened and dissolved, depending upon which components are most active. A further characteristic of this envisioned adaptive control is that the coupling strengths of the system components may be more or less difficult to modify, depending on their significance for maintaining system integrity and safety. So, for example, a developmental adaptive controller for a wearable assistive device for walking would always maintain some synergic patterns of assistive muscles to provide support against perturbations that could disrupt balance and promote falling.

Organizational changes to a system over time occur at the timescales of performance (real time), learning (over minutes or longer), development (over a single life span), and evolution (over eons). As presented in this book, *changes in coupling graphs over time may accommodate unfolding dynamics at multiple time scales through differences in coupling strength:* (1) performance in real time is assigned the weakest coupling to allow rapid assembly and dissolution of the parts of the system's equations of motion, and (2) coupling strength increases during development and evolution to conserve some parts of graph structures, but not others. For example, developing systems do not simply grow in an additive fashion, but rather may establish an overabundance of connections between the parts and then prune some while strengthening others (see Chapters 5, 6, and 7). In graph dynamics as applied to developing systems, therefore, I propose that the architecture of a graph of may contain some *early primitives that are conserved,* while *others may be remodeled or removed,* as the basis for behavioral change. This would explain the findings of apparent "disappearance" of some early patterns of behavior and the dominance of others reported by Dominici et al. (2011) and in the early work of Thelen and Fisher (1983). An important implication of this conceptualization of developmental changes in graph structure is that injury to mature systems in later childhood or adulthood may leverage parts of the graph that have been conserved over development as the basis for remodeling (that is, repairing) it. This opens the possibility of a developmental neurorehabilitation that uses opportunities for exploring primitive dynamical systems as a way to reconstruct lost behaviors.

A second extension of graph dynamics proposed here is that the *primitives comprising the graph may be contributed by more than one individual* (Richardson et al. 2015) *or by an individual and a machine* (Ijspeert et al. 2013). The basis for shared control by human and machine, as proposed here and in Chapter 10, is a mathematical structure called a *manifold,* a topological space that contains a layout of attractors or, in other words, the solution sets of equations of motion (Huys, Perdikis, and Jirsa 2014). Consider as an example the complex functional relationships of the body and environment in a task such as reaching while standing (Saltzman and Kelso 1987; Saltzman and Holt 2014). At a physical level of description, in order for the hand to reach for a target while a person is standing, the body center of mass (COM) must stay within the base of support. At an abstract functional level, the task space takes the form of a manifold containing two regions. One region, called the control manifold, is a set of solutions that create task-specific patterns of motion of the end-effectors in task space (Saltzman and Holt 2014). The second, orthogonal region, called the uncontrolled manifold, identifies those solutions in joint space that cause no motion of any of the end-effectors in task space (see Latash, Scholz, and Schöner 2007 for a further description of the methodology). The graph of solutions, a task-dynamic control structure, thus provides a way for generating solutions that not only create task-specific patterns of the end-effectors along the control manifold (that is, for reaching), but also allow contextually responsive postural adjustments (that is, for standing) that "do no harm" to the attainment of the task-specific goal (Saltzman and Holt 2014).

Task dynamic spaces and socially assisted learning. Task dynamics, introduced earlier, provides a framework for socially assisted learning based upon adults assembling a layout of attractors as a task space for a child's exploratory behavior (Saltzman and Kelso 1987; Richardson et al. 2015; Saltzman et al. 2006; Warren 2006). I build this framework here with the assumption that learning begins with *an opportunity to explore the behavioral dynamics of the body in task-specific environments* (see, for example, Saltzman and Holt 2014). On this assumption, all learning is governed by the same underlying attractors, with functionally specific behavioral dynamics emerging through repeated cycles of perceiving and acting in a structured environment (Warren 2006). The same abstract attractor dynamics govern each functionally distinctive mode, including locomoting, manipulating, eating and drinking, and communicating by gesture and speaking (Goldfield 1995).

Each task environment consists of a high density of local attractors and repellors, initially promoting an opportunity for trial-and-error learning through variable performance (see, for example, Loeb 2012) and, only then, allowing optimization of performance to increase stability and minimize energy expenditure. These dense nestings of attractors include not only environmental surfaces such floors and walls, and physical objects such as furniture and tools, but also the dynamic postures of other people and of animals. Adults may guide early development within the context of these task dynamics, as follows:

1. Adults may organize the environment so as to increase the density of attractors (food, attractive toys) within a particular task space.
2. If there are sufficient numbers of good enough attractor minima, then the child's actions will be more likely to find them quickly, regardless of where a particular action begins.
3. The cumulative actions produced by adult and child over repeated opportunities for learning in the task space establish a graph of the shared task dynamics.
4. With multiple opportunities for repetition of actions in the task space, some pathways of child action in the graph become more robust to perturbation than others; these are more likely to be responded to by the adult and thus become more likely to persist over time.
5. As certain action pathways become more robust, adults modify the task to introduce other attractors around which new learning may occur.

Longitudinal studies of the context of learning in the natural environment of the home, and in laboratory simulations of home learning environments, illustrate the ways in which adults play a crucial role in arranging the environment to provide opportunities for safe exploration and in modifying task content and complexity to promote learning of complex sequences, with optional combinations. An illustration of the dynamics of socially assisted learning is the process by which adults prepare the environment to promote the child's use of a feeding implement, such as a spoon, for scooping and transporting food from a bowl or plate and bringing it to the mouth.

Tetsushi Nonaka is a Japanese developmental psychologist who brings insights from his studies of stone napping during human evolution (Nonaka, Bril, and Rein 2010; Rein, Nonaka, and Bril 2014), as well as an appreciation of Gibson's concept of affordance, to advance our understanding of the development of human tool use. During a year-long visit to the Wyss Institute, Tetsushi conducted a detailed analysis of longitudinal video recordings from the Akachan database (Nonaka and Sasaki 2009), which considers how a caregiver activities during meals may introduce specific changes in the density and layout of the ingredients of a meal for a young child (such as rice, vegetables, and broth, in Japan). Nonaka and Goldfield (2017) propose that these caregiver activities may reflect a process by which the caregiver initially promotes the child's simple exploration of ways to use the spoon to scoop, transport, and remove a *particular component of the meal,* such as rice. By placing the bowl of rice at the child's body midline, any food that drops falls back into the bowl and keeps the child's attention focused on hand, spoon, rice, and bowl. At this period of learning, the caregiver completes the meal by either feeding the child or by placing an additional food item in the rice bowl. Then, as the child becomes more proficient at using the spoon to eat rice, the caregiver modifies the layout of bowls of rice, vegetable, and broth to enrich the possibilities for the child to independently combine different ingredients of a more complete meal. Here, then, the adult's understanding of the combinatorial complexity of the mature skill establishes a pathway for the child's learning over an extended period. This process of allowing variability of an initial repetitive pattern, and then adding combinatorial complexity, is similar in many ways to the role of the adult in human language acquisition and, perhaps, the role of bird tutor to the acquisition of birdsong (see Chapter 7).

• 10 •

Toward Devices That Are Seamless Parts of Collective, Adaptive, and Emergent Systems

Neuron and glia. Bacteria and host. Womb and fetus. Nature does not build *in vacuo.* Nature builds its devices as interwoven parts of complex ecosystems, as *collective structures within support networks.* An implication of this ground truth of modern biology is that to build devices we may trust with our lives, we will need to build them so that they function seamlessly as partners within hybrid biologic-synthetic-social ecosystems. But biological organs are also *organized across multiple spatial and temporal scales;* are *adaptive,* shrinking or growing in composition during development and learning, as well as in response to damage; and exhibit *emergent* behavior. Can developmentally oriented bioinspired technologies emulate nature's collective, distributed, and adaptive systems and exhibit emergent behavior?

Some new technologies guided by a hybrid ecosystems vision of biological inspiration include microscale robot swarms that function cooperatively within natural ecosystems (Wood, Nagpal, and Wei 2013), cells grown and nourished within microfluidic systems (Benam et al. 2015, 2016; H. J. Kim et al. 2016; Millet and Gillette 2012), and cells and microbes programmed to anticipate and remedy disease in the body's ecosystems (Khalil and Collins 2010; Kotula et al. 2014; Lienert et al. 2014; Litcofsky et al. 2012). Some of the manufacturing techniques include bioprinting threads of living cells in specialized inks to form vascularized tissue (Kolesky et al. 2016). None of this is science fiction. Instead, it is technology inspired by biological systems.

Bioinspired hybrid ecosystems may also address certain challenges posed by the American BRAIN initiative and other international efforts toward deeper understanding of the human brain (see Part II of the book), as well as provide new directions for fulfilling the urgent need for translational strategies to build neuroprosthetic devices and machines for neurorehabilitation (see Chapters 8 and 9). For example, mutualistic relationships, existing between squid and bacteria, may provide a heuristic framework for energy sharing between neuroprosthetics and body. The embodied nervous system architecture of colonial animals, such as *Volvox*, may inform the design of decentralized control and communications networks for soft wearable devices used for neurorehabilitation. And an appreciation of our place within microbial communities (Alivisatos et al. 2013; Charbonneau et al. 2016; Hays et al. 2015; McFall-Ngai et al. 2013; Sommer and Backhed 2013) may provide a model for the way that manufactured devices living inside us may engage in cellular repair. All of these possibilities raise intriguing questions, including

- Can we develop technologies based on the kind of mutualism seen in microbiota?
- Can we inject into the body engineered micromachines as drug delivery and repair devices?
- Can we regenerate neurons in the central nervous system as a means for remodeling following injury?

Mutualism

Subcellular collective behavior. Nature's actuators are evident at multiple scales, including subcellular scale actin-myosin molecular motors at 10^{-9} m (Feinberg 2015) (see Table 10.1). Myosin motors do their work by progressively moving along actin filaments to form contractile elements, the basis for muscle tissue at the macroscopic scale (Feinberg 2015). Does nature use the collective motion of actin filaments at the subcellular scale to progressively build more complex cellular functions? One way to begin to address this question is to examine whether the collective motion of actin filaments may exhibit emergent patterns, including transition to ordered phases. To examine this possibility, Schaller et al. (2010) conducted experiments in which they placed a mixture of unlabeled actin filaments, fluorescently labeled F-actin, and ATP (as fuel) on a microscope slide and examined under the microscope the outcome of manipulating actin

Table 10.1 Molecular-, cellular-, and tissue-scale engineered biological machines

Scale of force generation	Function	Illustrations
Biomolecular motors (1pN-45pN forces)	Transport and assembly using molecular motors (e.g., DNA, kinesin, myosin, F1-ATPase)	Nanomotors, DNA origami, DNA walkers, DNA transporters
Biological actuators using cardiac and skeletal cells or clusters (80nN-3.5μN contractile force)	Flagellar pumping, bacterial transport, cardiomyocyte microcantilever action	Flagellar motors, cardiomyocytes, skeletal muscle cells
Tissue, i.e., sheets of cells (25μN-1.18mN contractile forces)	Micropumping and propulsion by sheets of cardiomyocytes, and propulsion by skeletal muscle	Medusoid and Ray

Source: A. Feinberg (2015), Biological soft robotics, *Annual Review of Biomedical Engineering, 17,* 243–265.

filament density as a control parameter (see Chapter 2). Below a critical density, there was a disordered phase in which the filaments performed random walks without any specific directional preference. However, by raising the filament density above the critical value, there was a transition to an ordered phase with coherently moving, wave-like structures. How might a cellular locomotor function emerge from molecular motors interacting with other molecular structures, such as microtubules?

The core structure of eukaryotic cilia and flagella, called an axoneme, is assembled from a molecular motor, dynein, which converts energy from ATP into linear movement along a microtubule track (Chan, Asada, and Bashir 2014). The collective activity of thousands of these molecular motors within each axoneme results in self-sustained oscillatory beating patterns, an emergent behavior that is qualitatively different from the linearly moving motors themselves. To examine whether cilia-like beating emerges as a consequence of the collective self-organizing interactions of microtubules and dynein motors, Sanchez et al. (2011) performed *in vitro* experiments with mixtures of biotin-labeled kinesin motors bound into clusters, microtubules, and polyethylene glycol (PEG) (to induce attractive interactions between microtubules and create microtubule "bundles").

Those bundles that were trapped under air bubbles began to exhibit uniform, large-scale beating patterns. The major difference between these synthetic bundles and biological cilia and flagella is the significantly slower beating frequency in the former, probably due to the higher density of motors in the biological system (Sanchez et al. 2011). This and other experiments, therefore, support the hypothesis that the highly coordinated locomotor behavior of biological cilia is not controlled by a chemical signaling pathway, but rather emerges from the hydrodynamic coupling between neighboring cilia. Is it possible that more complex biological structures, such as the spindle that controls separation of chromosomes during cell division, may similarly emerge from microtubule interactions?

The metaphase spindle is a collection of microtubules, molecular motors, and associated proteins that segregates chromosomes during cell division (Karsenti 2004). To examine how the collective behaviors of these interacting molecular components may self-organize to form the spindle, Brugués and Needleman (2014) used confocal microscopy movies to compute spatiotemporal correlations in the fluctuations of microtubule interactions. Their hypothesis that spindles form due to local interactions of its component parts overwhelming other fluctuations was confirmed: microtubules close to each other tended to orient in the same direction, while those that were far apart were less well aligned (Brugués and Needleman 2014). Moreover, the ellipsoid shape of the spindle was found to be due to the distributions of microtubule density throughout, similar to the process by which a droplet of liquid crystal forms (Brugués and Needleman 2014).

Emergence of the hybrid self. A scenario for the emergence of "self" in multicellular life is presented by Newman, Forgacs, and Muller (2006; see also Newman and Bhat 2008; Newman 2012). When ancient single-celled organisms began to adhere to each other and form multicellular aggregates, they became sufficiently large (about 100 μm) for genes to be able to harness the physical processes that characterize mesoscale soft and excitable matter. Excitable soft matter adhering across their cellular boundaries made possible information flow at multiple temporal and spatial scales. Interior functions, including nervous systems, emerged from the formation and dissolution of positive and negative feedback loops capable of anticipating, and persisting, beyond events in "real time." In each multicellular animal, there was a blurring of the boundary between its inside and outside and an aggregation of microscopic parts that became a macroscopic individual. An example is the volvocine green algae whose species characteristics range from its initial unicellular form to its multicellular form, *Volvox*.

Volvox illustrates an evolutionary transition in individuality in which the unit of selection changes from a single cell to a cooperating group of cells (Herron et al. 2009; Kirk 2005; Srivastara et al. 2010). All volvocine algae stick together as they replicate by means of cytoplasmic bridges that link cells together. In later evolving forms, these algae break down as cellular differentiation begins and cell walls transform into the extracellular matrix (ECM) (Kirk 2005). ECM assembly is made possible by an initial scaffolding protein, called ISG, upon which the remainder of the ECM expands 10,000-fold in volume. The ECM holds the organism together as a single entity, while at the same time, asymmetric cell divisions regulated by two coexisting pathways generate both somatic (swimming) and reproductive (gonidia) cells. The adult form of *Volvox*, therefore, is comprised of differentiated internal cells that relinquish their individuality to support the success of the colony within the ECM boundary (Srivastava et al. 2010).

Endosymbiosis as profound embodiment. Endosymbiosis refers to one organism living within another, including harmful (parasitic) and beneficial (mutualistic) relationships (Wernegreen 2012). There is a growing consensus that mutualistic interactions, such as those between animals and microbes, are not exceptions, but rather are fundamentally important nested ecological interactions (McFall-Ngai et al. 2013). One illustration of these nested interactions is the relation between an animal and its bacterial microbiota, such as the homeostatic role played by bacteria in the gut and oral cavity and on the skin via communications among themselves and with the animal's organ systems (McFall-Ngai et al. 2013). Another is the relation between the giant tubeworm, *Riftia pachyptila,* and its endosymbiont sulfur-oxidizing bacteria, *Endoriftia* (Bright, Klose, and Nussbaumer 2012; Dubilier, Bergin, and Lott 2008). In these illustrations, the incorporation of bacteria is the basis for establishing chemosynthetic symbioses. However, nature also uses incorporation as a means of leveraging parts existing both inside and outside the body to build new organs.

Unicellular organisms, such as warnowiid dinoflagellates, are unicellular plankton. They are notable among microorganisms because of their surprisingly complex, eye-like visual organs, called ocelloids. Three structures of a visual organ found in many eyes are a lens, a cornea, and a retinal body (Richards and Gomes 2015). In warnowiids, these structures are actually built from organelles (mitochondria and plastids) obtained through endosymbiosis (Gavelis et al. 2015). By employing a combination of electron microscopy, tomography, and genomics, Gavelis et al. (2015) found that the retinal body is assembled from an ancient endosymbiosis with a red alga and that the cornea is made of a layer of mitochondria. In symbiotic relationships such as these, nature has discovered a surprising solution that requires us to *place the problem of device design into a broader perspective that includes the incorporation of parts from outside the body into a new whole, with emergent function, and benefits that accrue to all participants in the relationship.*

Of known symbioses with invertebrates, one that has received particularly intensive study is the partnership between the nocturnal Hawaiian bobtail squid, *Euprymna scolopes,* and the marine bacterium, *Vibrio fischeri* (Nyholm and Graf 2012). Like the emergent function of the ocelloid, the symbiosis between squid and bacteria is characterized by the emergence of an organ that provides a solution to an adaptive problem posed by the animal's ecological niche. Unlike many other symbiotic associations in which the microbial partner provides nutrients to the host, the bioluminescent *V. fischeri* provides an essential means for the squid to initiate a developmental process that results in realization of a functioning light organ (McFall-Ngai et al. 2012). When *V. fischeri* become incorporated as part of the light organ, the ventral surface of the squid's body emits light as a means of

Figure 10.1 Bobtail squid. Bacteria living within the Hawaiian bobtail squid produce a bioluminescent glow on the squid's underside that mimics the moonlight shining above, masking the animal from hungry fish below. The bacteria, in turn, get a nutrient-rich home. Photo courtesy of Margaret McFall-Ngai.

counterillumination, matching the light of the night sky (moon and stars) and obscuring its shadow on the sea bottom as a defense against predators (Nyholm and McFall-Ngai 2014) (see Figure 10.1).

The fascinating process by which host and bacterial partners are brought together involves active harvesting, winnowing, and induction. Active harvesting, a permissive process in which host epithelial cilia sweep particles into pores leading to the light organ, begins almost immediately after hatching and lasts about 30 minutes. Winnowing is a restrictive process in which the mucus lining of the light organ promotes aggregation of *V. fischeri* so that they proliferate and outcompete other bacteria (Nyholm and McFall-Ngai 2014). Induction of the light organ occurs in response to the bioluminescence of the bacteria: host tissues must detect the light for induction to occur (Heath-Heckman et al. 2013). The population growth of the bioluminescent bacteria within the light organ is regulated by the squid's response to the day / night cycle. The light organ is useful to the squid at night, when it emerges from the sand to hunt for food, but not during daylight hours, when it again burrows itself beneath the sand. Remarkably, each day at dawn, the squid expels about 95 percent of the light organ bacteria into the surrounding water. While it is buried in the sand, the remaining 5 percent of *V. fischeri* remaining within the light organ proliferate so that by mid-afternoon, the light organ is again full of bacteria (McFall-Ngai 2014).

There are several lessons to be taken from the host–symbiont partnership for developing profoundly embodied neuroprosthetic devices. First, forming a profoundly embodied relationship with a synthetic device may require a *process of induction that initiates the development of some functional capability between host and device.* Excitation by a synthetic source of pattern generation may serve a function similar to early-maturing neurons that induce function in nascent cell populations. Second, profound embodiment may require *information flow from the device to multiple levels of the neuromechanical system,* including feedback not only to the prime movers, but also to muscle groups participating in a particular functional activity, proprioceptive feedback to the spinal cord, and sensory input to the brain. Third, learning to use the limb may require an initial *permissive process of exploring* many ways to perform the same activity with synthetic muscle groups and a *restrictive process,* in which some muscle groups, and not others, are recruited for performing a task.

Holy guano, Batman. In the comics-inspired *Batman* films, the beleaguered Police Commissioner Gordon summons Gotham City's superhero crime fighter using a beacon with the symbol of a bat to light up the sky. Unbeknownst, perhaps, to Batman's artistic creators, certain pitcher plants also have a means for attracting bats: specialized organs that increase the reflectivity of bat echolocation and, thus, draw attention to themselves against the background of a cluttered environment. Why do pitcher plants need to attract bats? In Borneo, the carnivorous pitcher plant species *Nepenthes hemsleyana* has evolved a mutualistic relationship with *Kerivoula hardwickii,* an ultrasound-emitting echolocating bat (Jones 2015). While Batman is apparently driven by altruism, the benefit for *K. hardwickii* is a parasite-free roost. And for *N. hemsleyana?* Well, it receives needed nitrogen from bat guano (Jones 2015). What is the evidence that bats are attracted to structures that signal the identity and location of a particular plant species?

Schoner et al. (2015) conducted experiments in which they first demonstrated that pitcher plants do indeed have ultrasound reflecting organs: they built a "biomimetic" sonar head and speaker to broadcast 40- to 160-kHz frequencies from different angles around the pitcher plant's orifice. They then measured the echo reflectance of *N. hemsleyana* compared to another pitcher species. Schoner et al. (2015) found a clear difference in the spectral content of the echoes from the two different species, with the *N. hemsleyana* species particularly well suited to acoustically stand out in the cluttered neotropical surroundings of Borneo. The call signature of the *K. hardwickii* bats was characterized by extremely high frequencies (up to 292 kHz), resulting in high directionality to facilitate finding targets in cluttered surroundings. Schoner et al. (2015) additionally conducted behavioral experiments in which the pitcher plant reflector was left unmodified, enlarged, or completely removed. An intriguing result was that the bats approached experimentally enlarged pitcher organs significantly more often than expected by chance (Schoner et al. 2015). Thus, bat and plant have evolved a mutualistic relationship in which the former is attracted to a roost and the latter benefits from higher nitrogen intakes.

Mutualism as Biological Inspiration for Bio-Hybrid Robots

The medusoid. Jellyfish are umbrella-shaped zooplanktons that are radially symmetric around a single feeding-elimination body opening (Katsuki and Greenspan 2013). The nervous system of jellyfish consists of the rhopalium (a sensory system with photosensitive structures, gravity receptors, and pacemaker neurons for swimming), a motor nerve net that directly activates muscle contractions in response to the swim pacemaker neurons, and a diffuse nerve net that modulates pacemaker activities (Katsuki and Greenspan 2013). How are these component systems organized in a way that makes it possible for jellyfish to use their highly symmetrical (radial or bilateral) bodies for behaviors that include sun compass navigation, precise control of swimming for obstacle avoidance, escape from predators, and formation of aggregates (Katsuki and Greenspan 2013)? A radial body plan implies that all regions of the bell are more or less equally responsive to the environment and that locomotion is achieved by a ring of muscles that contract to function like a pump. Satterlie (2011, 2015) argues that for animals with a radial body, these behaviors are achieved through interactions of the diffuse nerve nets with integrative centers that compare favorably to the central nervous systems of bilateral animals.

One way to better understand how integration of diffuse nerve nets could be the basis for propulsion in biological jellyfish is to use a biologically inspired approach to engineering a synthetic jellyfish. Wyss faculty member Kit Parker and Carnegie Mellon bioengineer Adam Feinberg (Nawroth et al. 2012; Feinberg 2015) have done just that. First, they considered how, at an early developmental stage, called ephyrae, jellyfish use both muscular forces and elastic recoil of their body tissues for propulsion. At this stage, the jellyfish body is an array of eight lobes arranged radially to a central disc, rather than a closed bell shape. They then leveraged a new technology developed in their lab, called a "muscular thin film," engineered muscle tissue (cultured rat ventricular cardiomyocytes) on a free-standing, flexible thin film (polydimethylsiloxane, or PDMS) (Feinberg et al. 2007; Shim et al. 2012).

These two components, muscle tissue and elastomer, provide a means for emulating the way that jellyfish propel themselves in water: fast muscle contractions of the rat cardiomyocytes for producing an active propulsive impulse and the inherent elastic properties of PDMS for slow elastic recoil. Their experiments on the kinematics of the motion produced by this synthetic jellyfish, what they call a "medusoid," are extremely valuable for understanding what may be required to create a synthetic emulation of even a seemingly simple biological body plan. The actual body morphology of ephyrae is a heterogeneous substrate with stiffened ribs that act like springs, soft folds for compression, and pivots that allow the surface to form a bell without wrinkling (Nawroth et al. 2012). Initial medusoid designs did not incorporate these body architectural features, and the result was that the PDMS did not deform, and there was no propulsion at all. They, therefore, developed a new "lobed" design allowing each "arm" to freely bend around its base, forming a quasi-closed bell at maximum contraction. When the medusoids were placed in a water bath and an electrical field was applied to the bath, their kinematics and thrust patterns were remarkably similar to their biological counterparts.

A tissue-engineered bio-hybrid ray. Although the medusoid is able to emulate the swimming movements of a jellyfish, this bio-hybrid jellyfish lacks a crucial functionality of swimming animals in their natural habitats: it is unable to respond to sensory information in order to modulate its behavior—that is, maneuver itself—relative to information obtained from the environment. A next step in Parker's research program on bio-hybrids, therefore, has been to combine the tissue engineering technology of the medusoid with optogenetics in order to emulate the way that a batoid fish, such as a stingray or skate, controls its body undulations in order to exchange momentum with vortices generated during swimming (Park et al. 2016). The source of biological actuation of the bio-hybrid ray was a layer of approximately 200,000 live cardiomyocytes, electrically coupled with gap junctions. These were used to propel an elastomeric body that was 16.3 mm long and weighed about 0.43 mg. These myocytes were additionally engineered to express a light-sensitive ion channel (channelrhodopsin-2, or ChR2), developed in collaboration with Karl Deisseroth's lab (see Chapter 5). Due to the selective responsivity of myocytes expressing ChR2, point light stimulation at the front of the ray with 1.5-Hz frequency triggered propagation of an action potential. A body architecture emulating the serpentine architecture of biological batoids also critically contributed to the swimming pattern. Maneuverability of the bio-hybrid ray was achieved by pacing left and right optical stimulation at different frequencies. Like live rays, the bioinspired bio-hybrid was able to generate positive and negative vortices. By means of phototactic steering, the bio-hybrid ray was also able to steer through an obstacle course that required complex coordination and maneuvering. Among the challenges ahead for this research program are developing the capability for autonomous and adaptive behaviors, not yet possible in bio-hybrid robots.

The University of Illinois lab of Rashid Bashir has developed another type of hybrid muscle-powered soft robot, called a biobot (Cvetkovic et al. 2014; Raman et al. 2016, 2017). Each biobot is powered by skeletal (rather than cardiac) muscle, and is coupled to 3D-printed flexible beams that act like an articulating joint. As in the bio-hybrid ray, the muscles of the latest generation of biobots are driven optogenetically, providing light-stimulation control of muscle contraction (Raman et al. 2016; Raman, Cvetkovic, and Bashir 2017). Moreover, the use of skeletal muscle offers the possibility of higher-level control when coupled to neural networks by means of neuromuscular junctions (Raman et al. 2017).

Limitations of current robotics. Adaptive processes in nature solve problems posed by the environment. For example, nature's solutions to making closer sensory inspections of objects at a distance include approach via locomotion, and / or bringing an object closer to the body with an appendage. Robots, even those sent to other planets, are now quite adept at programmed (albeit slow) locomotion to find objects of interest. At a distance beyond the reach of its arm, a planetary rover may zap a rock with a laser. Once within reach, it emulates the extension and grasping capabilities of the human arm and hand and makes physical contact with the rock to further analyze it. How might we advance the capabilities of future planetary rovers?

Over the past two decades, an influential approach for studying how the brain controls body appendages has been the field of computational neuroscience (see, for example, Shadmehr and Krakauer 2008; see also Chapter 9). As presented in a themed issue of a publication widely read by engineers in computational neuroscience, *Proceedings of the IEEE,* one of the primary goals is to understand how electronic activity in brain cells and networks enables biological intelligence (McDonnell et al. 2014). An overarching goal of computational neuroscience is to design engineered systems by "reverse engineering" (see Chapter 1) a range of brain functions, including sensing and motion, computation and learning, devices and communication, and reliability and energy (McDonnell et al. 2014). An example of reverse engineering from the computational perspective is to build small insect robots that use biologically inspired visual circuits (Franceschini 2014) or that evolve their own control architectures over generations of interactions between changing body morphologies and the environment (Bongard and Lipson 2014).

One significant area of biological inspiration long neglected by computational neuroscience is the role of the body's mechanical properties interacting with the environment for motor control. In an important article, whose title includes the statement that "spikes alone do not behavior make," Tytell, Holmes, and Cohen (2011) argue that any sufficient analysis of the embodied nervous system also requires an understanding of body mechanics and interactions with the environment. The ways in which the activity of neural circuits is influenced by the biomechanics of the body interacting with the environment may be illustrated by the example of swimming in fish, such as lamprey. During lamprey swimming, neural circuits produce head-to-tail activity that activates muscles in a wave and a corresponding change in body curvature (Tytell et al. 2011). Significantly, however, the speed of the *mechanical* wave does not travel at the same speed as the *wave of neural activity,* but rather depends upon the fluid forces acting on the body (Tytell et al. 2010). Conversely, the body's mechanical properties may be used to advantage by neural circuits, such as by tuning the mechanical properties (for example, stiffness) of muscles to improve the accuracy of human goal-directed arm movements (Selen, Beek, and Dieën 2005).

In bioinspired approaches that build locomoting machines by emulating an animal's stable behaviors in response to perturbations, scientists and engineers develop models capable of making the machine behave like the animal (see, for example, Ijspeert 2014). This is a difficult challenge because of the high dimensionality of the variables describing interactions between an animal's nervous system, body, and environment. There have been a number of modeling approaches based upon abstract dynamical systems, which are then embodied by real-world constraints. For example, Full and Koditschek (1999) propose a modeling approach consisting of "templates" and "anchors." A template is an abstract dynamical model that reduces the dimensionality of complex sets of variables into a generative dynamical system (for example, a mass-spring pendulum system). However, templates do not provide causal explanations of the detailed neural and musculoskeletal mechanisms required for locomotor control. Therefore, in a dynamical approach to modeling locomotion, templates are complemented by "anchors." An anchor is a neuromechanical model that identifies the role of joint, musculoskeletal, and nervous system function, as well as contributions of the environment to the emergence of behavior, such as locomotion. In other words, an anchor grounds the abstract dynamical system within constraints imposed by the morphology and physiology of the animal's body (Revzen, Koditschek, and Full 2009). Together, templates and anchors constitute a more complete approach to modeling complex behaviors: templates identify generative functional units and their relationships and set policies for neuromechanical control at a high level, while anchors incorporate the details of bodily and environmental constraints on dynamics for the control of behavior within the context of the environment.

Making computation more biological. A contemporary digital computer behaves as a precise arrangement of reliable parts, and the techniques for performing computations are dependent upon this precision and reliability (Abelson, Beal, and Sussman 2009). An alternative approach to computing, one that respects principles of complex systems, is called amorphous computing. It is based upon the cooperation of large numbers of unreliable parts that are arranged in unknown, irregular, and time-varying ways. A fundamental aspect of amorphous computing is that computation is achieved by a collection of "computational particles," each of which has no *a priori* knowledge of its position or orientation (Abelson, Beal, and Sussman 2009). Additionally, (1) the particles are programmed identically, (2) they can communicate with a few nearby neighbors, (3) communication is unreliable, and (4) the maximum distance over which two particles can communicate effectively is relatively small compared to the size of the entire amorphous computer (Abelson, Beal, and Sussman 2009).

The early development of multiagent systems based on amorphous computing, by Wyss faculty member Radhika Nagpal (for example, Werfel and Nagpal 2008), has been the foundation for my own work on decentralized wearable robotic systems, discussed further in Chapter 9. A core idea is to use global-to-local compilation of goals described at a high level with programming languages based on generative rules and then map the generative rules to local programs for individual agents. How do agents locally interact with each other? Nagpal proposes a set of primitives derived from developmental biology as the basis for local programs, including morphogen gradients, chemotaxis, and various types of local competition and cooperation. For example, Yu and Nagpal (2009, 2011) developed a tissue-like modular surface, comprised of a network of connected autonomous robotic agents organized with decentralized control. The overall shape of the surface is controlled locally by the collective action of the physically connected robots and guided by a set of global constraints. The goal of the collective is to adapt in order to achieve *global homeostasis,* defined as a set of regional constraints that must be maintained, even as the environment changes. One example of such homeostasis is a surface with a particular shape, such as a planar surface (for example, a bridge) maintained at a particular orientation. For robotic agents with a limited local view of the physically connected structure, maintaining homeostasis requires cooperation. Nagpal and colleagues accomplished this by providing each agent with a simple control rule: detect the error between a measured sensor value and the desired local constraint for each neighbor, then compute an actuation change in the direction that decreases the average local error. In other words, each agent, like a cell, computes a response based upon its local sensed environment. More recent work has further investigated biologically inspired primitives for amorphous computing in other types of multiagent systems, including swarms of microrobots (Rubenstein, Cornejo, and Nagpal 2014).

Amorphous computing offers the potential for scalable control of soft morphologies, which include not only the geometries of body forms (for example, organs and articulators), but also material properties such as friction coefficients or parameters describing compliance (Correll et al. 2014; Hauser et al. 2013). The body's soft morphologies, compliance, nonlinearities, and high dimensionality generate an intrinsic mechanical dynamics capable of performing computations. Therefore, control need not be the exclusive domain of a central executive. Instead, some computation may be outsourced—to be performed by the body's intrinsic dynamics. With amorphous computing and soft morphologies, control becomes more a matter of orchestration such that the body is not controlled at every instant, but rather is occasionally guided by a conductor. For control to be "good enough" for the body to achieve particular behavior dynamics in a structured environmental energy field (Loeb 2012), the conductor may introduce energy into the system to move it to one particular region of the field and not another (Warren 2006).

Radhika and her colleagues have been developing programming languages, based upon amorphous computing, for control of robot swarms (see, for example, Wood, Nagpal, and Wei 2013), as well as construction robots (for example, Werfel, Petersen, and Nagpal 2014). In one deterministic programming language, called Karma, for example, she begins with a global authority, the hive, that issues instructions that get translated into individual programs (Dantu et al. 2012). The global instructions are organized as a flowchart of tasks to be achieved by the colony as a whole, with indications of the local conditions that trigger new tasks. As the collective behavior of the swarm members unfolds, information from individuals returning to the hive is used to adjust the priority of tasks. A second programming language, called optRAD (optimizing Reaction-Advection-Diffusion) takes a stochastic approach to swarm control. OptRAD treats the behavior of the microrobots as a fluid diffusing through the environment. The hive provides a macroscopic continuous model with a set of parameters that are optimized for particular objectives. Before each sortie, the parameters are transmitted to the microrobots to guide their movement velocities and local behavior (for example, flying and hovering at a flower).

Kilobots are swarms of low-cost open source robots capable of collective behavior while moving along a ground plane (Rubenstein et al. 2014). An individual kilobot strikes a balance between cost (about $14

US each) and functionality (being able to move forward, rotate, communicate with, and measure distance to nearby neighbors; measure ambient light levels; and allow easy debugging and scalable operations). In an extension of this work, Rubenstein, Cornejo, and Nagpal (2014) designed a low-cost 1024-robot, decentralized swarm of kilobots, whose collective behaviors are achieved solely through local interactions between neighboring robots. Each of the 1024 robots is programmed with only three "primitive" capabilities: (1) edge following, where a robot can move along the edge of a group by measuring distances from robots on the edge; (2) gradient formation, in which a source robot generates a message that increments as it progresses through the swarm, providing a kind of geodesic distance from the source; and (3) localization, where robots can form a local coordinate system using communication with, and measured distances to, neighbors (Rubenstein et al. 2014).

TERMES. The swarm of microbots from the film *Big Hero Six*, introduced in Chapter 1, were capable of assembling, dissolving, and reforming themselves into a range of large-scale patterns and physical objects, reminiscent of the desert constructions of termites (see, for example, Turner 2010). However, there is a fundamental difference between these fictional microrobots and termite architects: the fictional machines were directed by an executive, a human (our benevolent hero, or the movie's villain, depending on the plot's ups and downs) wearing a kind of brain–machine interface. By contrast, insect swarms have no single agent directing their behavior. Instead, social insects use limited local sensory information during peer interactions to make decisions, and their local behavior propagates via chemical, mechanical, and visual communications networks to form global patterns (Ramdya et al. 2015).

Mound-building termites of the genus *Macrotermes* build some of the largest and most sophisticated structures in nature, without any architectural plan or leader (Turner 2010). For insect-like robots to truly emulate the feats of termite construction would require that the robots coordinate their construction activities indirectly through manipulation and sensing of a shared environment, or *stigmergy* (derived

Figure 10.2 Artificial collective construction. Physical implementation of collective construction system, with independent climbing robots that build using specialized bricks. Photo courtesy of Justin Werfel.

from the Greek "stigma," meaning sign, and "ergon," meaning work or action; Dehmelt and Bastiaens 2010). Werfel, Petersen, and Nagpal (2014; see also Petersen, Nagpal, and Werfel 2011) have demonstrated that a decentralized system of simple independent robots with limited capabilities based upon stigmergy are capable of building large-scale structures from prefabricated "bricks." *Stigmergy*, in this work, is characterized by actions triggered by detection of distinct arrangements of bricks. In this decentralized system, called TERMES, each robot is provided with a high-level representation of the desired structure, identifying those sites meant to be occupied by bricks, and a set of "traffic laws," or directional constraints. This approach provides a true test of nature's strategy for decentralized construction by large numbers of animals with limited sensing capabilities: each robot is provided with a goal, but not a plan; a set of real-world constraints on how to achieve the goal; algorithms for using arrangements of bricks to determine what to do next; and brick properties scaled to their action capabilities (see Figure 10.2). The TERMES movies (Werfel et al. 2014) in *Science*, demonstrating the successful construction of large-scale structures, are as much fun to watch as *Big Hero Six*. But this is not science fantasy. This is the robotics of the twenty-first century.

The New AI. Alan Turing, the 1936 creator of the "universal machine" (a forerunner of the modern digital computer), ends his classic

1950 paper on "computing machines and intelligence" (Turing 2004) by indicating the challenges ahead. The 1950 paper is widely cited because it contains his "imitation game," now called the Turing test, for determining whether a machine possesses human-like intelligence. Since publication of that paper, the field of artificial intelligence (AI) has become a dominant force in industry, spawning fast algorithms for (not-so-smart) computers. However, there has been slower progress in building machines that exhibit, say, the intelligence of a swarm of insects or the cleverness of a crow.

With the 2012 centenary of Turing's birth and the release of the 2015 film *The Imitation Game,* based on Turing's astonishing insights into breaking the German Enigma Machine code during World War II, the AI community has been reenergized in its goal of building machines with human-like intelligence (Chouard and Venema 2015). The "new AI" is driven not so much by a computer answering questions (with apologies to "Siri," the IBM "Watson," and "Hal 9000"), but rather is centered on artificial beings (robots, androids) in the physical world using sensory input as the basis for adaptive intelligent behavior (Chouard and Venema 2015). Embodying AI in the physical world in this way once again addresses, perhaps, the remaining challenges anticipated by Turing in 1950. In the twenty-first century, robots are beginning to be able to respond to injury to themselves and others (Cully et al. 2015); are becoming part of human social relationships by eschewing complete autonomy in favor of "symbiotic autonomy"—that is, by asking other machines or humans for help (Rosenthal, Veloso, and Dey 2012a,b); and are becoming learning companions for early human language development (Kory, Jeong, and Brazeal 2013).

Evolutionary algorithms are methods for machine (including robot) optimization by means of Darwinian selection (Floreano and Keller 2010; Maesani, Fernando, and Floreano 2014). The methodology in evolutionary robotics, for example, involves four major steps: (1) generate a population with different genomes, such that each genome defines the strength of synaptic connections of an artificial neural network; (2) with sensor input and actuation output, evaluate by experiment with real robots, or in simulation, the fitness *f* of each robot—that is, performance in the task assigned to them; (3) selectively choose the genomes of robots with the highest fitness to produce a new generation of robots; (4) pair the selected genomes to allow crossovers and mutations; and (5) repeat the process over multiple generations (Floreano and Keller 2010).

Evolutionary algorithms emulating natural selection have resulted in dramatic advances in machine learning systems—for example, for vision-like image processing and for robot guidance systems (see Floreano, Ijspeert, and Schaal 2014; Jordan and Mitchell 2015). Evolutionary computing is at the threshold of enabling automated fabrication of "smart objects" (Eiben and Smith 2015). And the next generation of flying robot drones will use evolutionary algorithms for path planning of "reactive"—and then "cognitive"—autonomous flight (Floreano and Wood 2015).

Robots that adapt to their own body damage. The Mars rover *Curiosity* is a wheeled hexapod that began exploring the red planet in 2012. During its journeys over the sometimes-rugged terrain of Mars, *Curiosity* has encountered regions of sharp, embedded rocks that have torn many holes in its wheels. The rover's cameras have enabled NASA scientists and engineers back on Earth to monitor damage progression and come up with an adaptive strategy. To minimize further damage, the mission control team has directed the rover to take potentially longer routes with fewer wheel-damaging rocks. At this stage in the evolution of planetary rovers, *Curiosity* is not designed to decide on its own on this type of strategy that adapts to damage. Future rovers and other mobile robots will, undoubtedly, have this adaptive capability.

A hexapod robot developed by Cully et al. (2015) provides a preview of how such robots may work. The adaptive behavior of their hexapod robot is inspired by the way that animals compensate for body damage when they are injured. A quadrupedal animal—such as a dog, for example—quickly adapts to the loss of a limb by adopting a tripod gait (Fuchs et al. 2014; Jarvis et al. 2013). Cully et al. (2015) propose that the animal's behavior is based upon trial-and-error learning, guided by prior experience. They have designed a robot controller that uses a similar strategy: it is given the dimensions of the space of possible behaviors (for example, characteristics of walking gaits) and a performance measure (for example, speed) (see Figure 10.3). [I note here that Cully et al. (2015) missed an opportunity for incorporating a dynamical systems insight into the description of the space of possible behaviors—namely, the use of gait symmetries—as from Turvey et al. (2012).] The robot's knowledge base, then, is a map of the behavior performance space. The damaged robot searches this space to find different types of behaviors that are predicted to perform well, performs tests on these behaviors, and updates its estimates: a process called "intelligent trial and error" (Cully et al. 2015).

Whither Baymax? These closing sections of the book bring us back to the beginning of Chapter 1 and my fascination with the prosocial robot, and part-time superhero, Baymax. Could we, today, build a helping robot that responds to, or even anticipates, injury and disease? One kind of answer to this question centers on our progress in building macroscopic-scale humanoid robots. A source of inspiration for me in considering this question is a beloved character from the 1960s television show, *The Jetsons:* a humanoid robot maid, named Rosie. We can, perhaps, forgive the animators for portraying Rosie's body as a wheeled, aproned torso with arms and a head. After all, she talks conversationally with George, Jane, and company. Indeed, for viewers, child and adult alike, what makes Rosie so likable is her humanity, her empathy, and that she is an accepted member of the family—not the degree to which she bears a resemblance to us. Now, more than fifty years later, advances in robotics and artificial intelligence have brought us to the threshold of including social robots in our own families.

Building humanoid social robots, such as Rosie, remains a work in progress. Among the notable breakthroughs include robots that ask for help, creating a social symbiosis in which robots and humans help each other, in the work of Manuela Veloso and her students at Carnegie Mellon University. Rosenthal, Veloso, and Dey (2012a,b) define a symbiotic relationship with a robot as one in which robots help us,

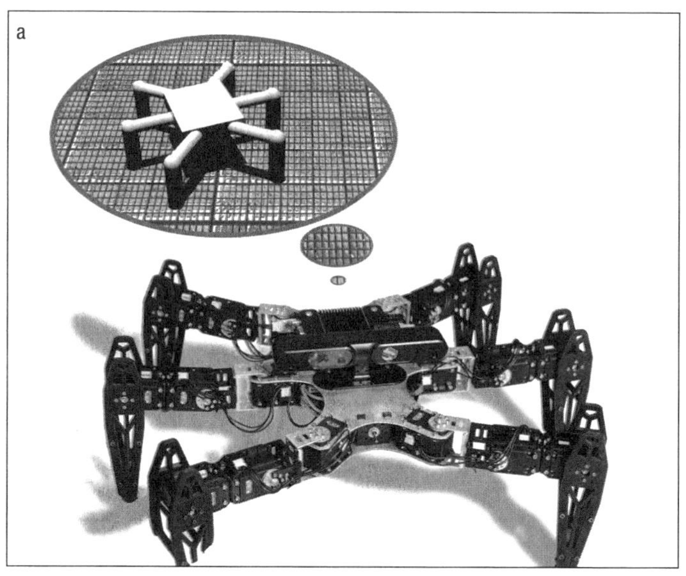

Figure 10.3 Using the Intelligent Trial and Error (IT&E) algorithm, robots, like animals, can quickly adapt to recover from damage. (a) An undamaged hexapod robot and an automatically generated map of the behavior-performance space. (b) Walking with a broken leg. To keep walking despite the damage, it uses a large map of the space of possible actions and their performance values. Specifically, the robot selects a behavior it thinks will perform well based on the previous (simulated) experience stored in the map. If that tested behavior does not work, it moves on to a different region of the map, which means an entirely different type of behavior. This new algorithm enables a damaged robot to get up and walk away in about a minute after trying only a handful of different behaviors. Images © Antoine Cully / UPMC-Université Pierre et Marie Curie (CC BY 4.0).

but we also help them, as when they are unable to complete a step leading to their goal (for example, pressing a button to call an elevator). This is, perhaps, not unlike the host–bacteria relationship that requires both bobtail squid and *Vibrio fischeri* to coordinate their activities to generate a mutually beneficial defense against predation. Rosenthal, Biswas, and Veloso (2010) purposely built the CoBot without arms, but it is able to communicate with people as it rolls around the multiple floors of its home building at Carnegie Mellon University to complete package deliveries. Because the CoBot cannot press buttons, when a delivery requires using the elevator, the robot must request help from a human to do so. With that assistance, the CoBot is able to complete the task, benefiting both the sender and recipient of the package.

Communication with the CoBot may be symbiotic, but it is fairly limited in its affective content. The field of affective computing has emerged along with the advent of affective sensing technologies, including algorithms for facial analysis (el Kaliouby and Robinson 2005) and sophisticated wristbands that detect and analyze electrodermal activity *in situ* (Gao et al. 2016). The possibility of robots capable of interpreting signals of human affect, such as facial expression, as well as making inferences from those expressions and grounding their meanings in the context of the nature of the world, has furthered the emergence of social robotics, especially in the work of Cynthia Breazeal at the Massachusetts Institute of Technology (MIT) (Breazeal 2003; Spaulding and Breazeal 2015). In the domain of social robots, *sociable* robots have the potential for symbiotic relationships with humans. Breazeal (2003) defines sociable robots as participative: they proactively engage people in a social manner to benefit the person, as well as themselves. An early example of a sociable robot in Breazeal's lab is Kismet, who was able to exchange turns during face-to-face interactions (Breazeal 2003). More recent sociable robots include Huggable, a robot teddy bear used in pediatric settings (Jeong et al. 2015) and sociable robots for early education.

Socially assistive robotics is a specialized field that combines assistive robotics, focused on aiding human users through interactions with robots, for example, as mobility assistants (see Chapter 9) and socially interactive robotics (Rabbitt, Kazdin, and Scassellatti 2015). The use of socially assistive robots (SARs) in mental health settings includes companions, such as the robotic seal Paro; therapeutic play partners for modeling social interaction, such as making eye contact (Scassellati, Admoni, and Mataric 2012); and coach / instructor for engaging patients in aspects of treatment, such as medication adherence (Fasola and Mataric 2013). In state-of-the-art use, SARs may be used along with a human therapist but may eventually be extended to the home (Rabbitt et al. 2015). Of particular interest for engaging children with autism spectrum disorders (ASDs) in social interaction and eliciting positive social responses is the use of an SAR as a partner in play.

Over more than a decade, Yale computer scientist and developmental psychologist, Brian Scassellati, has developed a set of SARs for promoting improved eye-to-eye gaze, facial expressions, and social engagement with autistic children (see Scassellati, Admoni, and Mataric 2012 for a review). Scassellati has found that SARs are more animate than a typical toy but less socially complex than a person, and so are effective in engaging a child with ASD in shared social activities. Once so engaged, the challenge is to guide the child to further participate in activities involving joint attention (directing shared interest toward objects by pointing or using eye contact), imitation, and turn-taking. Consider, for example, the SAR named Kaspar. This SAR has been used to initiate turn-taking games with autistic children. One person controls the robot's movement with a handheld remote, while the other mimics the robot's interactions. Because the child is able to see the effect of pressing buttons on Kaspar's motions, the remote becomes a means for control. In this context, the child engages in turn-taking through the act of passing the remote to another person (Scassellati et al. 2012).

Mutualism as a Foundation for Synthetic Biology and Organ-on-a-Chip Devices

Gut microbiome. The healthy human gut is an ecological community consisting of some hundred trillion single-celled microorganisms (bacteria, viruses, fungi, and in some cases, protozoa), making us holobionts (Charbonneau et al. 2016). The gut microbiome plays an essential role in human health, contributing to colonization resistance through four interrelated functions: direct inhibition, barrier maintenance, immune modulation, and metabolism (McKenney and Pamer 2015). From a developmental perspective, the functional biome of humans is an example of ecological succession: after initial colonization, communities of microbes undergo consecutive changes in composition and function until reaching a relatively stable climax community (Lozupone et al. 2012). During pregnancy, the mammalian fetus inhabits a largely sterile environment and is protected from infections by maternal immunity (McKenney and Pamer 2015). During early postnatal development, changes in gene expression in the neonatal small intestine and colon are believed to induce alterations in the immune system (Rakoff-Nahoum et al. 2015). Then, secretory antibodies in milk determine long-term intestinal microbiota composition (Rogier et al. 2014).

As Sonnenburg and Backhed (2016) have noted, gut microbes may be thought of as a "control center" for the modulation of human physiology. This can be illustrated by studies of childhood undernutrition in which healthy growth is defined from a microbial perspective (Blanton et al. 2016; Gordon et al. 2012; Subramanian et al. 2014, 2015). Subramanian et al. (2014, 2015) have conducted research in Bangladesh to examine the relationship between varying degrees of malnutrition and the maturity of the gut microbiome. Bangladeshi children with severe malnutrition, defined by weight for height measures, had significantly "younger" gut microbiota, and a similar, but less severe, immaturity was apparent in children with moderately acute malnutrition (Subramanian et al. 2014). But is malnutrition causally related to gut microbiome health? A causal role for gut microbiota in human malnutrition is supported by studies in which fecal samples from human donors are transplanted into germ free mice, thus transmitting the human donor's microbial community to the recipient animals (Blanton et al. 2016; Kau et al. 2015; Ridaura et al. 2013). Blanton et al. (2016) found, for example, that mice colonized with microbiota from healthy Malawian donor children gained significantly more weight and lean body mass than mice colonized with microbiota from undernourished donors, even though there was no difference in food consumption in the two groups of mice.

As another illustration of the potential role of gut microbes as a physiological control center, epidemiological studies have determined that maternal obesity during pregnancy increases the risk of later developmental disorders, such as ASD, in offspring (Gohir, Ratcliffe, and Sloboda 2015; Krakowiak et al. 2012). Maternal obesity, characterized by high macronutrients, may alter the gut microbiome of offspring in utero and disrupt fetal neurological development via the gut–brain axis (Cryan and Dinan 2012). A finding with profound implications for therapeutic interventions is that it may be possible to reverse these brain and behavioral deficits of offspring by manipulating their gut microbiome. In a study with mice, for example, Buffington et al. (2016) first demonstrated that maternal high-fat diet (MHFD)–induced obesity results not only in a reduction of specific bacterial species in the offspring gut microbiome, but also in their behavioral dysfunction. The MHFD-induced changes in offspring gut microbiota had specific effects, related to enhancing oxytocin levels, on the mesolimbic dopamine reward system in the ventral tegmental area (VTA). Buffington et al. (2016) discovered, however, that it was possible to correct the VTA synaptic dysfunction, as well as selectively reverse offspring social deficits, by means of oral treatment with a single commensal fecal bacterial species. Remarkably, then, fecal transfer effects on the microbiome transformed dysfunctional social behavior in these mice.

Microbiota-gut-brain axis. The bidirectional communication between the trillions of gut microorganisms and the nearly 100 billion neurons and billions more non-neural glial cells of the nervous system, called the gut-brain axis, consists of multiple direct and indirect endocrine, immune, and central and peripheral nervous system pathways (Collins, Surette, and Bercik 2012; Cryan and Dinan 2012, 2015). For example, the pathways linking the microbiome and the central nervous system include the vagus nerve, the circulatory system, and the immune system (Sampson and Mazmanian 2015). Gut microbiota appear to regulate host brain development and function via these pathways in health and disease (Fung, Olson, and Hsaio 2017; Vuong et al. 2017; Sampson and Mazmanian 2015; Sharon et al. 2016). The use of germ-free (GF) animals, such as mice, makes it possible to directly assess the influence of the microbiota on neurophysiology (Cryan and Dinan 2012). For example, compared to conventionally colonized controls, GF mice are characterized by disordered circuitry in stratum, hippocampus, amygdala, and hypothalamus, are significantly socially avoidant, and exhibit a defect in social cognition in which they do not recognize familiar versus unfamiliar mice (Collins et al. 2012; Sampson and Mazmanian 2015).

Synthetic biology and the gut. Because changes in the gut microbiota affect health, disease, and metabolism, a challenge for synthetic biology and other technologies is to develop devices that may become part of the microbiota consortia and nondestructively interrogate the gut (Arnold, Roach, and Azcarate-Peril (2016). A particular challenge for synthetic biology devices in the gut is to be able to survive, integrate into the complex gut microbiota, sense their environment, respond to the environment, provide something helpful, and report the state of the gut after excretion (Hays et al. 2015). One approach by Wyss faculty member Pam Silver and colleagues has been to engineer the cellular memory, the capability to convert a transient signal into a sustained response (Inniss and Silver 2013). Kotula et al. (2014) constructed engineered *Eschericia coli* bacteria with a synthetic memory system that survived in the mammalian gut and were able to sense and record antibiotic exposure during passage through the mouse gut. This work now lays a foundation for this and other synthetic genetic circuits to serve as diagnostic systems.

Human gut-on-a-chip. Another approach to the creation of a mutualistic relationship between the gut and a synthetic device for diagnostics and treatment is organ-on-chip technology (see Benam et al. 2015 for a review). The human gut-on-a-chip was developed to explore the complex set of interacting influences that comprise the gut microbiome. The signature characteristic of this biologically inspired device is that it provides an environment of physiologically relevant luminal flow as well as peristalsis-like mechanical deformations to promote formation of intestinal villi lined by epithelial cells (Kim et al. 2012; H. J. Kim et al. 2016). Like the human gut, the gut-on-a-chip is subject to disease, such as small intestine bacterial overgrowth and infection. Moreover, because it is possible to systematically manipulate the individual parameters from which gut microbiome function emerges, and because the microfluidic environment can sustain the living cells for up to two weeks, the gut-on-a-chip can be used to evaluate the use of probiotics on bacterial overgrowth and infection. That is exactly what H. J. Kim et al. (2016) did. They first demonstrated that experimentally halting peristaltic-like motions of the device, even while maintaining luminal flow, triggered bacterial overgrowth, similar to that seen in patients with ileus and inflammatory bowel disease. They then used the device to examine the protective effects of clinical probiotic and antibiotic therapies on bacterial overgrowth and inflammation. H. J. Kim et al. (2016) found that these therapies used in the gut-on-a-chip model system successfully protected the villi from injury.

Chip devices with integrated sensing. Early work that leveraged cardiomyocytes for controlled microscale actuation used lithography-based techniques to fabricate hybrid muscle-elastomer devices. As discussed in Chapter 4, these techniques require multiple hands-on steps, masks, and dedicated tools. For example, Feinberg et al. (2007) engineered 2D myocardial tissues by culturing rat ventricular cardiomyocytes on polydimethylsiloxane (PDMS) elastomer thin films. Tissues specifically engineered to be anisotropic exhibited contraction resulting in the bending of a PDMS thin film substrate. Subsequently, Grosberg et al. (2011) advanced this technology through batch fabrication of "heart on a chip" substrates. However, their approach still required a considerable amount of hands-on fabrication, including manually cutting the thin films, as well as using a stereoscopic microscope and image processing to make motility measurements. More recently,

Wyss faculty colleagues Jennifer Lewis and Kit Parker have developed a programmable multimaterial 3D printing technology capable of automated patterning and integration of materials for both tissue actuation and sensing. They call these integrated devices microphysiological systems (MPS) (Lind et al. 2016). During 3D printing of MPS, different types of specialized inks are deposited to form (1) multilayer cantilevers (a base layer, an embedded strain sensor, and tissue-guiding soft PDMS microfilaments), (2) electrical leads and contacts for sensor readout, and (3) wells with insulation covers holding the cardiomyocytes. The capability of MPS for integrated direct sensing, thus, greatly advances the technology beyond the need to use microscope-based kinematics to make measurements of tissue contractile stress.

Regenerative Medicine

Materials and devices that emulate biological growth. The term *regenerative medicine* may be used interchangeably with *tissue engineering*. Mao and Mooney (2015) propose three strategies for regenerative medicine: scaffold fabrication, vascularization and innervation, and altering the host environment (see Table 10.2). Attempts to fabricate materials capable of emulating tissue growth have presented numerous manufacturing challenges. Biological tissues transform in shape as they grow. This means that in applications that involve interactions with growing tissue, the architecture of any machine-printed device must be capable of dynamically transforming its shape in response to growth. If implanted, the device must also be capable of resorption by the body. Additive manufacturing, also called 3D printing, is a process in which three-dimensional objects are fabricated by the successive layering of materials under the control of a computer-designed printable model. The process of printing materials that additionally incorporate the capability for transformation *over time* (the fourth dimension), or materials capable of changing from one shape to another directly off the print bed, is called *4D printing* (Tibbits 2012).

One illustration of 4D printing inspired by biological growth is work by Wyss Core Faculty members, Jennifer Lewis and L. Ma-

Table 10.2 Strategies for regenerative medicine

Strategy	Description
Recapitulating organ and tissue structure via scaffold fabrication	Biodegradable synthetic scaffolds may be fabricated. Three-dimensional bioprinting (by ink jet or micro-extrusion) can create structures that combine high-resolution control over materials and cell placement within engineered constructs.
Integrating grafts with the host via vascularization and innervation	Strategies to promote graft vascularization include use of angiogenic factors and pre-vascularizing the graft before implanting. Promoting innervation by the host may be achieved via hydrogels patterned with channels and loaded with growth factors.
Altering the host environment to induce therapeutic responses	Infusion of cells (e.g., from human umbilical cord blood) may induce therapeutic responses, as in angiogenesis for aiding stroke recovery. Reducing inflammation due to implanted scaffolds may be achieved by changing the scaffold properties.

Source: A. Mao & D. Mooney (2015), Regenerative medicine: Current therapies and future directions, *Proceedings of the National Academy of Sciences, 112,* 14452–14459.

hadevan and colleagues, that has fabricated complex, plant-like, three-dimensional architectures that change shape when immersed in water (Gladman et al. 2016) (see Figure 10.4). Sped-up movies of changes in plant morphology as a function of hydration are always fun and surprising to watch because they reveal that sessile plants are, actually, always in motion (Awell, Kriedemann, and Turnbull 1999). The basis for morphological changes with hydration is due to differences in local swelling behavior arising from the directional orientation of stiff plant cell walls cellulose fibers (Gladman et al. 2016). By emulating the composition of plant cell walls with a hydrogel composite ink comprised of stiff cellulose fibrils within a soft acrylamide matrix, and controlling the local orientation of cellulose fibers along four-dimensional printing pathways, Gladman et al. (2016) have been able to programmably fabricate complex three-dimensional morphologies. In previous work, Mahadevan demonstrated that the shape of a long leaf (Liang and Mahadevan 2009) and the blooming of a lily (Liang and Mahadevan 2011) are governed by changes in the Gaussian curvature of elastic sheets, according to a mathematical model relating elastic modulus and thickness. Gladman et al. (2016) expand upon

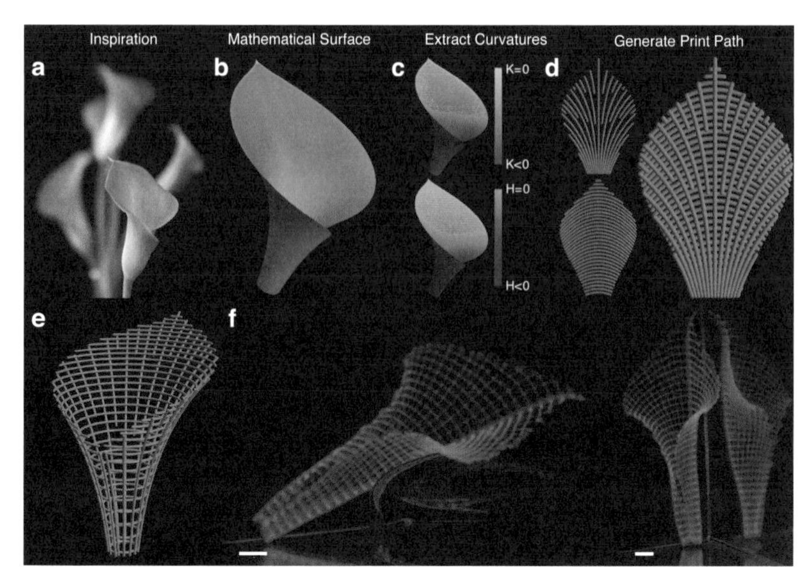

Figure 10.4 Predictive four-dimensional printing of biomimetic architectures. A native calla lily flower (a) inspires the mathematically generated model of the flower (b), with a well-defined curvature (c), that leads to the print path (d) obtained from the curvature model to create the geometry of the flower on swelling (see text and Supplementary Information). After swelling, the transformed calla lily (f) exhibits the same gradients of curvature as the predicted model (e), nozzle size = 410 μm (scale bars, 5 mm). A. Sydney Gladman, Elisabetta A. Matsumoto, Ralph G. Nuzzo, L. Mahadevan, and Jennifer A. Lewis (2016). Image reprinted with permission from Nature Publishing Group / Macmillan Publishers Ltd.

this model to guide the depositing of ink so that its cellulose fiber orientation provides precise control over the Gaussian curvature. Figure 10.4 illustrates how hydration transforms a printed lily morphology over time to emulate the growth of a biological lily, true 4D printing.

A somewhat different approach to 4D printing has been used to develop three-dimensional printed devices with architectures that change their morphology over time in response to biological growth. Morrison et al. (2015) report an early success with human infants in using a 4D printed, implanted, reusable dynamic splint to mitigate life-threatening tracheobroncho-malacia (TBM), a condition of excessive collapse of the airways during respiration. Unlike earlier airway stents and prosthetics, which typically fail because they are unable to accommodate airway growth, this airway splint device consists of a cylindrical architecture with an open section that allows gradual opening as the airway grows radially. Further, each splint is scaled to the airway anatomy dimensions of a patient to fabricate a truly personalized device (Morrison et al. 2015).

Bioprinting. Unlike 3D or 4D printing of biocompatible materials, such as hydrogels, bioprinting is a process that involves the precise layer-by-layer positioning of *biological materials, biochemical, and living cells,* with spatial control over functional components (Murphy and Atala 2014; Ozbolat 2015). One approach to bioprinting has been to emulate the fiber-spinning organs of spiders (Kang et al. 2011). These organs produce fibers by mixing fluids from a set of abdominal glands and spinning the resulting mixture. Each of the glands provides its own component, so that the spider is able to produce customized blends of silk with specific properties. Wyss faculty colleague Ali Khademhosseini and his coworkers have invented a way to emulate the flow of different components from spider glands in order to spin a synthetic material whose material composition and topography can be precisely controlled (Kang et al. 2011). Another of nature's solutions to fabrication is to spin an entire structure using one continuous fiber. The silkworm *Bombyx mori* constructs its cocoon by spinning a single kilometer-length fiber into a scaffold structure and then continues spinning the same individual fiber to cocoon itself within the confines of the scaffold, using a figure-8 pattern to accrete walls of increasing thickness (Zhao et al. 2005). At MIT, Neri Oxman and her colleagues use an extruder and robotic arm to emulate the process of cocooning for fabricating large-scale architectures (see, for example, Oxman 2012). The resulting structures are strong, lightweight, and aesthetically elegant.

A remarkably difficult challenge is to fabricate synthetic materials that are not only biologically compatible, but also consist of heterogeneous multicellular components. This requires a bioprinter that is not only capable of precisely positioning cells along a print path, but also requires being able to switch between different cell types and biomaterials. Different solutions have been proposed for this remarkably difficult challenge. Jennifer Lewis and colleagues use a microfluidic print head capable of seamlessly switching between two viscoelastic inks based on PDMS elastomers (Hardin et al. 2015). Further evolution of this work is the development of print heads that use a rotating impeller to actively mix multiple microscale materials for local control of 3D printing of viscoelastic inks (Ober, Foresti, and Lewis 2015).

An even greater challenge for bioprinting—what some consider to be in the realm of science fiction (Ozbolat 2015)—is the ability to create heterogeneous vascularized tissue constructs that mimic the geometries of natural tissue. Natural perfusable vascularized tissue provides nutrient, growth / signaling factor, and waste transport and so is a prerequisite for sustaining tissue constructs containing living cells. To create the precise geometries required for mimicking such tissue, Jennifer Lewis and Wyss colleagues (for example, Kolesky et al. 2016) custom-built a large-area 3D bioprinter with four independently controlled print heads. The printer heads release multiple inks (for example, an aqueous fugitive ink as well as cell-laden hydrogel inks) as its predefined sequence of movements lays down overlapping strands of the inks.

In another approach from the lab of Anthony Atala at the Wake Forest Institute for Regenerative Medicine, Kang et al. (2016) sequentially print cell-laden hydrogels with a synthetic polymer and temporary scaffolding. They use microchannels in a printed, cell-laden, hydrogel construct to promote transport of nutrients and oxygen for cell survival. Their human-scale constructs, such as a bioprinted skeletal muscle, are based upon a computer-generated, three-dimensional tissue model converted to a motion program that operates and guides the cell-laden, ink-dispensing printer nozzles. These may then be surgically implanted.

Regenerative Medicine and the Injured Nervous System

Cerebral organoids. *In vitro* modeling is a potent tool of twenty-first-century science for understanding disease and developing effective therapies. The complexity, cell heterogeneity, and gene expression of the 86 billion neurons and approximately equal number of glia of the human brain present a daunting challenge for *in vitro* models (Azevedo et al. 2009). Returning again to the earlier example of a "saturated reconstruction" of mouse neocortex at the nanoscale, which identifies every synaptic vesicle, is a reminder of the technical challenges of *in vitro* brain modeling (Kasthuri et al. 2015; see also Chapter 5). One major advance in studying *in vitro* growth of nervous systems has been the establishment of "cerebral organoid" cultures from pluripotent neural stem cells (Ader and Tanaka 2014; Benam et al. 2015; Lancaster et al. 2013; Tang-Schomer et al. 2014; Camp et al. 2015). The great potential of this approach follows from the similarity of gene expression between the fetal brain and organoid cells: more than 80 percent of genes expressed along the fetal cortex lineage are also expressed in organoids (Camp et al. 2015). This means that both organoid cells and their fetal counterparts use similar sets of genes in building structured cerebral tissue. It follows, then, that organoids may be used to investigate particular brain disease *in vitro*. For example, telencephalic organoids derived from individuals with autism spectrum disorder (ASD) were characterized by an overproduction of inhibitory neurons caused by increased FOXG1 gene expression (Mariani et al. 2015). This is consistent with the suggestion, discussed earlier, that excitation-inhibition imbalance is a developmental precursor of ASD.

Brain on a chip. A shortcoming of organoids for emulating organs is that they lack critical features of a fully functioning organ and may not be able to relate cell-level to organ-level behavior (Ingber 2016). An alternative *in vitro* technique for understanding the mechanical and chemical environments that guide wiring of the brain and synaptic plasticity is the use of microfluidic systems (Millet and Gillette 2012; Jain and Gillette 2015). State-of-the-art microdevices create microenvironments with dimensions as low as 10 nm around different subregions of networks of neurons—their synapses and pre- and postsynaptic processes. To illustrate, a device for studying synapses has three parallel primary channels, about 100 μm wide and 50 μm high, connected to even narrower, perpendicular, cross-communicating channels with inlet and outlet ports for independent control of fluids. Neurons are seeded into the two outer channels and, guided by the physical structure of the device, send axons and dendrites into the central channel, where synapse formation occurs (Jain and Gillette 2015).

Neuroscientists are now using microfluidic devices to study thin brain slice preparations that provide nutrient and oxygen delivery, metabolic waste removal, and access to both high-resolution microscopy and electrophysiological recording (see Huang, Williams, and Johnson 2012 for a review). One advanced technique integrates a segmented-flow microfluidic chamber with fast voltage-sensitive dye imaging and laser stimulation to map neural network activity in living brain slices (Ahar et al. 2013). Wireless optofluidic neural probes that combine ultrathin microfluidic drug delivery with cellular-scale inorganic LED (light-emitting diode) arrays are allowing study of neural circuits in awake, behaving animals (Jeong et al. 2015), and development of nanoelectronic networks with nanowire elements may soon make possible active monitoring and control of neuronal networks (Liu et al. 2013).

Blood-brain barrier-on-a-chip. Like the gut microbiome, the human nervous system is a community of cells, including neurons, glia, and endothelial cells. Nervous system function, including response to injury, emerges from the interactions of its cell members in the context of the complex web architecture of its circulatory system (see Chapters 5 and 6). The brain's circulatory system maintains a blood–brain barrier (BBB) whose function is to modulate local exchange of oxygen and nutrients as well as the local immune response in the brain (Abbott et al. 2010). A signature of the three-dimensional BBB on a chip is that it emulates the three-dimensional (cylindrical) architecture of the brain capillary structure, as well as fluid flow and extracellular matrix (ECM) mechanics (Herland et al. 2016). Moreover, the BBB organ-on-a-chip makes it possible to experimentally tease apart the independent contributions of the brain's microvascular endothelium, pericytes, and glia (astrocytes) to inflammatory stimuli, resulting in release of pro-inflammatory cytokines. To study the contribution of the different cell types to induced inflammatory cytokine release, Herland et al. (2016) first seeded human brain–derived microvascular endothelial cells, and either primary human pericytes or astrocytes, on the inner surface of the cylindrical collagen gel. This three-dimensional BBB organ-on-a-chip successfully formed a permeability barrier. They then added tumor necrosis factor-alpha (TNF-alpha) to this model system of cells in a three-dimensional cell–cell relationship and measured inflammatory cytokine release, implicated in Alzheimer's disease and traumatic brain injury. The BBB model system exhibited responses to the TNF-alpha inflammatory stimulus that more closely mimicked those observed in the living brain than the same cells co-cultured without a three-dimensional cell architecture.

Conclusions

To conclude, I propose five principles for building biologically inspired devices for emulating nature's assembly and repair process, based upon material in the chapters of this book.

1. **Self-assembly.** Emulate the process of self-assembly for building devices at all scales: from DNA origami, to organs on a chip, to smart composite manufacturing of microrobots and components for wearable robots with amorphous control.
2. **Consortia.** Build biological-synthetic hybrid devices from a consortium of interchangeable parts that are either biological or synthetic, and couple them in a social milieu.
3. **Decentralized control.** Build devices with components that individually have limited capabilities but collectively are able to self-assemble into complex architectures.
4. **Flexibility with stability.** Build devices from redundant sets of components (like Legos) so that they may be loosely assembled and stable for a particular function and then can be rapidly dissolved and reassembled for a new function.

5. **Emergent behavior.** Build devices from components that individually have specialized functions but collectively exhibit emergent behavioral dynamics.

What might we expect in the coming decades from biologically inspired devices for emulating assembly and repair that are built according to these principles? Some possibilities based on current work at the Wyss Institute include (1) DNA-based molecular robots, exemplified in the work of William Shih, Peng Yin, and Wesley Wong; (2) Homo chippiens, the integration of individual organ-on-a-chip microfluidic devices into complex systems that emulate the entire human body, as in work by Don Ingber; and (3) wearable robots with control systems seamlessly integrated with the nervous system, based upon smart composite manufacturing and amorphous computing in the work of Rob Wood and Radhika Nagpal. These are subjects for today's fiction and fantasy films, but they may be tomorrow's realities.

References

Index

References

Abbott, N. J., Patabendige, A. A. K., Dolman, D. E. M., Yusof, S. R., & Begley, D. J. (2010). Structure and function of the blood-brain barrier. *Neurobiology of Disease, 37*(1), 13–25. doi:10.1016/j.nbd.2009.07.030

Abelson, H., Beal, J., & Sussman, G. J. (2009). Amorphous computing. *Encyclopedia of complexity and applied systems science* (pp. 257–271).

Abrahams, B. S., & Geschwind, D. H. (2008). Advances in autism genetics: On the threshold of a new neurobiology. *Nat Rev Genet, 9*(5), 341–355. doi:10.1038/nrg2346

Abraira, V. E., & Ginty, D. D. (2013). The sensory neurons of touch. *Neuron, 79*(4), 618–639. doi:10.1016/j.neuron.2013.07.051

Ackman, J. B., Burbridge, T. J., & Crair, M. C. (2012). Retinal waves coordinate patterned activity throughout the developing visual system. *Nature, 490*(7419), 219–225. doi:10.1038/nature11529

Ackman, J. B., & Crair, M. C. (2014). Role of emergent neural activity in visual map development. *Curr Opin Neurobiol, 24*(1), 166–175. doi:10.1016/j.conb.2013.11.011

Ader, M., & Tanaka, E. M. (2014). Modeling human development in 3D culture. *Curr Opin Cell Biol, 31C*, 23–28. doi:10.1016/j.ceb.2014.06.013

Adolph, K. E., & Robinson, S. R. (2013). The road to walking. *The Oxford handbook of developmental psychology* (Vol. 1: *Body and Mind*, pp. 1–79). New York: Oxford University Press.

Ahn, A. N., & Full, R. J. (2002). A motor and a brake: Two leg extensor muscles acting at the same joint manage energy differently in a running insect. *J Exp Biol, 205*(Pt 3), 379–389.

Ahrar, S., Nguyen, T. V., Shi, Y., Ikrar, T., Xu, X., & Hui, E. E. (2013). Optical stimulation and imaging of functional brain circuitry in a segmented laminar flow chamber. *Lab Chip, 13*(4), 536–541. doi:10.1039/c2lc40689f

Ahrens, M. B., Li, J. M., Orger, M. B., Robson, D. N., Schier, A. F., Engert, F., & Portugues, R. (2012). Brain-wide neuronal dynamics during motor adaptation in zebrafish. *Nature, 485*(7399), 471–477. doi:10.1038/nature11057

Ahrens, M. B., Orger, M. B., Robson, D. N., Li, J. M., & Keller, P. J. (2013). Whole-brain functional imaging at cellular resolution using light-sheet microscopy. *Nat Methods, 10*(5), 413–420. doi:10.1038/nmeth.2434

Aizenberg, J., Weaver, J. C., Thanawala, M. S., Sundar, V. C., Morse, D. E., & Fratzl, P. (2005). Skeleton of *Euplectella* sp.: Structural hierarchy from the nanoscale to the macroscale. *Science, 309*(5732), 275–278. doi:10.1126/science.1112255

Ajiboye, A. B., Willett, F. R., Young, D. R., Memberg, W. D., Murphy, B. A., Miller, J. P., . . . Kirsch, R. F. (2017). Restoration of reaching and grasping movements through brain-controlled muscle stimulation in a person with tetraplegia: A proof-of-concept demonstration. *The Lancet, 389*(10081), 1821–1830. doi:10.1016/s0140-6736(17)30601-3

Akam, T., & Kullmann, D. M. (2010). Oscillations and filtering networks support flexible routing of information. *Neuron, 67*(2), 308–320. doi:10.1016/j.neuron.2010.06.019

Akam, T., & Kullmann, D. M. (2014). Oscillatory multiplexing of population codes for selective communication in the mammalian brain. *Nat Rev Neurosci, 15*(2), 111–122. doi:10.1038/nrn3668

Akay, T., Tourtellotte, W. G., Arber, S., & Jessell, T. M. (2014). Degradation of mouse locomotor pattern in the absence of proprioceptive sensory feedback. *Proc Natl Acad Sci USA, 111*(47), 16877–16882. doi:10.1073/pnas.1419045111

Alexander-Bloch, A., Giedd, J. N., & Bullmore, E. (2013). Imaging structural co-variance between human brain regions. *Nat Rev Neurosci, 14*(5), 322–336. doi:10.1038/nrn3465

Alivisatos, A. P., Chun, M., Church, G. M., Deisseroth, K., Donoghue, J. P., Greenspan, R. J., . . . Yuste, R. (2013). Neuroscience: The brain activity map. *Science, 339*(6125), 1284–1285. doi:10.1126/science.1236939

Alivisatos, A. P., Chun, M., Church, G. M., Greenspan, R. J., Roukes, M. L., & Yuste, R. (2012). The brain activity map project and the challenge of functional connectomics. *Neuron, 74*(6), 970–974. doi:10.1016/j.neuron.2012.06.006

Alsteens, D., Gaub, H. E., Newton, R., Pfreundschuh, M., Gerber, C., & Müller, D. J. (2017). Atomic force microscopy-based characterization and design of biointerfaces. *Nature Reviews Materials, 2*, 17008. doi:10.1038/natrevmats.2017.8

Alstermark, B., & Isa, T. (2012). Circuits for skilled reaching and grasping. *Annu Rev Neurosci, 35*, 559–578. doi:10.1146/annurev-neuro-062111-150527

Amador, A., Perl, Y. S., Mindlin, G. B., & Margoliash, D. (2013). Elemental gesture dynamics are encoded by song premotor cortical neurons. *Nature, 495*(7439), 59–64. doi:10.1038/nature11967

Amir, Y., Ben-Ishay, E., Levner, D., Ittah, S., Abu-Horowitz, A., & Bachelet, I. (2014). Universal computing by DNA origami robots in a living animal. *Nat Nanotechnol, 9*(5), 353–357. doi:10.1038/nnano.2014.58

Ampatzis, K., Song, J., Ausborn, J., & El Manira, A. (2014). Separate microcircuit modules of distinct v2a interneurons and motoneurons control the speed of locomotion. *Neuron, 83*(4), 934–943. doi:10.1016/j.neuron.2014.07.018

Amunts, K., Lepage, C., Borgeat, L., Mohlberg, H., Dickscheid, T., Rousseau, M. E., . . . Evans, A. C. (2013). BigBrain: an ultrahigh-resolution 3D human brain model. *Science, 340*(6139), 1472–1475. doi:10.1126/science.1235381

An, B., Miyashita, S., Tolley, M. T., Aukes, D. M., Meeker, l., Demaine, E., . . . Rus, D. (2014). *An end-to-end approach to making self-folded 3D surface shapes by uniform heating.* Paper presented at the IEEE International Conference on Robotics and Automation (ICRA), Hong Kong, China.

Andersen, R. A., & Cui, H. (2009). Intention, action planning, and decision making in parietal-frontal circuits. *Neuron, 63*(5), 568–583. doi:10.1016/j.neuron.2009.08.028

Andersen, R. A., Kellis, S., Klaes, C., & Aflalo, T. (2014). Toward more versatile and intuitive cortical brain-machine interfaces. *Curr Biol, 24*(18), R885–R897. doi:10.1016/j.cub.2014.07.068

Anderson, M. A., Burda, J. E., Ren, Y., Ao, Y., O'Shea, T. M., Kawaguchi, R., . . . Sofroniew, M. V. (2016). Astrocyte scar formation aids central nervous system axon regeneration. *Nature, 532*(7598), 195–200. doi:10.1038/nature17623

Angle, M. R., Cui, B., & Melosh, N. A. (2015). Nanotechnology and neurophysiology. *Curr Opin Neurobiol, 32*, 132–140. doi:10.1016/j.conb.2015.03.014

Araya, R., Vogels, T. P., & Yuste, R. (2014). Activity-dependent dendritic spine neck changes are correlated with synaptic strength. *Proc Natl Acad Sci USA, 111*(28), E2895–2904. doi:10.1073/pnas.1321869111

Arber, S. (2012). Motor circuits in action: Specification, connectivity, and function. *Neuron, 74*(6), 975–989. doi:10.1016/j.neuron.2012.05.011

Arnold, J. W., Roach, J., & Azcarate-Peril, M. A. (2016). Emerging technologies for gut microbiome research. *Trends Microbiol, 24*(11), 887–901. doi:10.1016/j.tim.2016.06.008

Artavanis-Tsakonas, S., Rand, M. D., & Lake, R. J. (1999). Notch signaling: Cell fate control and signal integration in development. *Science, 284*(5415), 770–776.

Arthur, W. (2006). D'Arcy Thompson and the theory of transformations. *Nature Reviews Genetics, 7*(5), 401. doi:10.1038/nrg1835

Ashmore, A. (2013). The man who illustrated the heavens. *Sky and Telescope, 126*(4), 72.

Aslin, R. N., Shukla, M., & Emberson, L. L. (2015). Hemodynamic correlates of cognition in human infants. *Annu Rev Psychol, 66*, 349–379. doi:10.1146/annurev-psych-010213-115108

Aukes, D. M., Goldberg, B., Cutkosky, M., & Wood, R. J. (2014). An analytical framework for developing inherently manufacturable pop-up laminate devices. *Smart Materials and Structures, 23*, 1–15.

Autumn, K., Dittmore, A., Santos, D., Spenko, M., & Cutkosky, M. (2006). Frictional adhesion: A new angle on gecko attachment. *J Exp Biol, 209*(Pt 18), 3569–3579. doi:209/18/3569 [pii] 10.1242/jeb.02486

Autumn, K., Sitti, M., Liang, Y. A., Peattie, A. M., Hansen, W. R., Sponberg, S., . . . Full, R. J. (2002). Evidence for van der Waals adhesion in gecko setae. *Proc Natl Acad Sci USA, 99*(19), 12252–12256. doi:10.1073/pnas.192252799

Ayoub, A., Oh, S. W., Xie, Y., Leng, J., Cotney, J., Dominguez, M. H., . . . Rakic, P. (2011). Transcriptional programs in transient embryonic zones of the cerebral cortex defined by high-resolution mRNA sequencing. *Proc Natl Acad Sci, 108,* 14950–14955.

Ayoub, A. E., & Rakic, P. (2015). Neuronal misplacement in schizophrenia. *Biol Psychiatry, 77*(11), 925–926. doi:10.1016/j.biopsych.2015.03.022

Azevedo, F. A., Carvalho, L. R., Grinberg, L. T., Farfel, J. M., Ferretti, R. E., Leite, R. E., . . . Herculano-Houzel, S. (2009). Equal numbers of neuronal and nonneuronal cells make the human brain an isometrically scaled-up primate brain. *J Comp Neurol, 513*(5), 532–541. doi:10.1002/cne.21974

Azim, E., & Alstermark, B. (2015). Skilled forelimb movements and internal copy motor circuits. *Curr Opin Neurobiol, 33,* 16–24. doi:10.1016/j.conb.2014.12.009

Azim, E., Jiang, J., Alstermark, B., & Jessell, T. M. (2014). Skilled reaching relies on a V2a propriospinal internal copy circuit. *Nature, 508*(7496), 357–363. doi:10.1038/nature13021

Azuma, R., Deeley, Q., Campbell, L. E., Daly, E. M., Giampietro, V., Brammer, M. J., . . . Murphy, D. G. M. (2015). An fMRI study of facial emotion processing in children and adolescents with 22q11.2 deletion syndrome. (Research) (Report). *Journal of Neurodevelopmental Disorders, 7,* 1.

Bachmann, L. C., Matis, A., Lindau, N. T., Felder, P., Gullo, M., & Schwab, M. E. (2013). Deep brain stimulation of the midbrain locomotor region improves paretic hindlimb function after spinal cord injury in rats. *Sci Transl Med, 5*(208), 208ra146. doi:10.1126/scitranslmed.3005972

Back, S. A., Luo, N. L., Borenstein, N. S., Levine, J. M., Volpe, J. J., & Kinney, H. C. (2001). Late oligodendrocyte progenitors coincide with the developmental window of vulnerability for human perinatal white matter injury. *J Neurosci, 21,* 1302–1312.

Back, S. A., & Miller, S. P. (2014). Brain injury in premature neonates: A primary cerebral dysmaturation disorder? *Ann Neurol, 75*(4), 469–486. doi:10.1002/ana.24132

Back, S. A., & Rivkees, S. A. (2004). Emerging concepts in periventricular white matter injury. *Seminars in Perinatology, 28*(6), 405–414.

Back, S. A., & Rosenberg, P. A. (2014). Pathophysiology of glia in perinatal white matter injury. *Glia, 62*(11), 1790–1815. doi:10.1002/glia.22658

Bae, B. I., Tietjen, I., Atabay, K. D., Evrony, G. D., Johnson, M. B., Asare, E., . . . Walsh, C. A. (2014). Evolutionarily dynamic alternative splicing of GPR56 regulates regional cerebral cortical patterning. *Science, 343*(6172), 764–768. doi:10.1126/science.1244392

Bagnall, M. W., & McLean, D. L. (2014). Modular organization of axial microcircuits in zebrafish. *Science, 343*(6167), 197–200. doi:10.1126/science.1245629

Baillet, S. (2017). Magnetoencephalography for brain electrophysiology and imaging. *Nat Neurosci, 20*(3), 327–339. doi:10.1038/nn.4504

Baisch, A. T., Ozcan, O., Goldberg, B., Ithier, D., & Wood, R. J. (2014). High speed locomotion for a quadrupedal microrobot. *Int J Rob Res, 33*(8), 1063–1082. doi:10.1177/0278364914521473

Balasubramaniam, R., & Turvey, M. T. (2004). Coordination modes in the multisegmental dynamics of hula hooping. *Biol Cybern, 90*(3), 176–190. doi:10.1007/s00422-003-0460-4

Ball, G., Boardman, J. P., Rueckert, D., Aljabar, P., Arichi, T., Merchant, N., . . . Counsell, S. J. (2012). The effect of preterm birth on thalamic and cortical development. *Cereb Cortex, 22,* 1016–1024.

Ball, M. P., Thakuria, J. V., Zaranek, A. W., . . . Church, G. M. (2012). A public resource facilitating clinical use of genomes. *Proc Natl Acad Sci, 109,* 11920–11927.

Bareyre, F. M., Kerschensteiner, M., Raineteau, O., Mettenleiter, T. C., Weinmann, O., & Schwab, M. E. (2004). The injured spinal cord spontaneously forms a new intraspinal circuit in adult rats. *Nat Neurosci, 7*(3), 269–277. doi:10.1038/nn1195

Bargmann, C. I. (2012). Beyond the connectome: How neuromodulators shape neural circuits. *Bioessays, 34*(6), 458–465. doi:10.1002/bies.201100185

Bargmann, C. I., & Marder, E. (2013). From the connectome to brain function. *Nat Methods, 10*(6), 483–490. doi:10.1038/nmeth.2451

Bargmann, C. I., Newsome, W. T., Anderson, D. G., Brown, E. H., Deisseroth, K., Donoghue, J. A., . . . Ugurbil, K. (2014). *Brain 2025: A scientific vision.* Bethesda, MD: National Institutes of Health.

Barth, F. G. (2004). Spider mechanoreceptors. *Curr Opin Neurobiol, 14*(4), 415–422. doi:10.1016/j.conb.2004.07.005

Barthelat, F., Yin, Z., & Buehler, M. J. (2016). Structure and mechanics of interfaces in biological materials. *Nat Rev Mater, 1*(4), 16007. doi:10.1038/natrevmats.2016.7

Bartlett, N. W., Tolley, M., Overvelde, J. T., Weaver, J. C., Mosadegh, B., Bertoldi, K., . . . Wood, R. J. (2015). A 3D-printed, functionally graded soft robot powered by combustion. *Science, 349*(6244), 161–165.

Bassett, D. S., & Sporns, O. (2017). Network neuroscience. *Nat Neurosci, 20*(3), 353–364. doi:10.1038/nn.4502

Bastian, A. J. (2006). Learning to predict the future: The cerebellum adapts feedforward movement control. *Curr Opin Neurobiol, 16*(6), 645–649. doi:10.1016/j.conb.2006.08.016

Bastian, A. J. (2011). Moving, sensing and learning with cerebellar damage. *Curr Opin Neurobiol, 21*(4), 596–601. doi:10.1016/j.conb.2011.06.007

Bastos, A. M., Vezoli, J., & Fries, P. (2015). Communication through coherence with inter-areal delays. *Curr Opin Neurobiol, 31*, 173–180. doi:10.1016/j.conb.2014.11.001

Bauer, U., & Federle, W. (2009). The insect-trapping rim of Nepenthes pitchers. *Plant Signaling and Behavior, 4*, 1019–1023.

Baumann, N., & Pham-Dinh, D. (2001). Biology of oligodendrocyte and myelin in the mammalian central nervous system. *Physiol Rev*, 81, 871–927.

Baumann, O., Borra, R. J., Bower, J. M., Cullen, K. E., Habas, C., Ivry, R. B., . . . Sokolov, A. A. (2015). Consensus paper: The role of the cerebellum in perceptual processes. *Cerebellum* (London), 197–220. doi:10.1007/s12311-014-0627-7

Bavelier, D., Levi, D. M., Li, R. W., Dan, Y., & Hensch, T. K. (2010). Removing brakes on adult brain plasticity: From molecular to behavioral interventions. *J Neurosci, 30*(45), 14964–14971. doi:10.1523/JNEUROSCI.4812-10.2010

Beek, P. J. (1989). Timing and phase locking in cascade juggling. *Ecological Psychology, 1*(1), 55–96.

Belmonte, J.-C. I., Callaway, E. M., Caddick, S. J., Churchland, P., Feng, G., Homanics, G. E., . . . Zhang, F. (2015). Brains, genes, and primates. *Neuron, 87*(3), 671. doi:10.1016/j.neuron.2015.07.021

Benam, K. H., Dauth, S., Hassell, B., Herland, A., Jain, A., Jang, K.-J., . . . Ingber, D. E. (2015). Engineered in vitro disease models. *Annual Review of Pathology: Mechanisms of Disease, 10*, 195–262. doi:10.1146/annurev-pathol-012414-040418

Benam, K. H., Villenave, R., Lucchesi, C., Varone, A., Hubeau, C., Lee, H. H., . . . Ingber, D. E. (2016). Small airway-on-a-chip enables analysis of human lung inflammation and drug responses in vitro. *Nat Methods, 13*(2), 151–157. doi:10.1038/nmeth.3697

Ben-Ishay, E., Abu-Horowitz, A., & Bachelet, I. (2013). Designing a bio-responsive robot from DNA origami. *J Vis Exp* (77), e50268. doi:10.3791/50268

Bensmaia, S. J., & Miller, L. E. (2014). Restoring sensorimotor function through intracortical interfaces: Progress and looming challenges. *Nat Rev Neurosci, 15*(5), 313–325. doi:10.1038/nrn3724

Berg, E. M., Hooper, S. L., Schmidt, J., & Buschges, A. (2015). A leg-local neural mechanism mediates the decision to search in stick insects. *Curr Biol, 25*(15), 2012–2017. doi:10.1016/j.cub.2015.06.017

Bergou, A. J., Ristroph, L., Guckenheimer, J., Cohen, I., & Wang, Z. J. (2010). Fruit flies modulate passive wing pitching to generate in-flight turns. *Phys Rev Lett, 104*(14). doi:10.1103/PhysRevLett.104.148101

Berni, J., Pulver, S. R., Griffith, L. C., & Bate, M. (2012). Autonomous circuitry for substrate exploration in freely moving Drosophila larvae. *Curr Biol, 22*(20), 1861–1870. doi:10.1016/j.cub.2012.07.048

Berthouze, L., & Goldfield, E. (2008). Assembly, tuning, and transfer of action systems in infants and robots. *Infant and Child Development, 17*, 25–42.

Beyeler, A., Metais, C., Combes, D., Simmers, J., & Le Ray, D. (2008). Metamorphosis-induced changes in the coupling of spinal thoraco-lumbar motor outputs during swimming in Xenopus laevis. *J Neurophysiol, 100*(3), 1372–1383. doi:10.1152/jn.00023.2008

Bhanpuri, N. H., Okamura, A. M., & Bastian, A. J. (2012). Active force perception depends on cerebellar function. *J Neurophysiol, 107*(6), 1612–1620. doi:10.1152/jn.00983.2011

Bhanpuri, N. H., Okamura, A. M., & Bastian, A. J. (2013). Predictive modeling by the cerebellum improves proprioception. *J Neurosci, 33*(36), 14301–14306. doi:10.1523/JNEUROSCI.0784-13.2013

Bhanpuri, N. H., Okamura, A. M., & Bastian, A. J. (2014). Predicting and correcting ataxia using a model of cerebellar function. *Brain, 137*(Pt 7), 1931–1944. doi:10.1093/brain/awu115

Bhatia, S. N., & Ingber, D. E. (2014). Microfluidic organs-on-chips. *Nat Biotechnol, 32*(8), 760–772. doi:10.1038/nbt.2989

Bialas, A. R., & Stevens, B. (2013). TGF-beta signaling regulates neuronal C1q expression and developmental synaptic refinement. *Nat Neurosci, 16*(12), 1773–1782. doi:10.1038/nn.3560

Bialek, W., Cavagna, A., Giardina, I., Mora, T., Pohl, O., Silvestri, E., . . . Walczak, A. M. (2014). Social interactions dominate speed control in poising natural flocks near criticality. *Proc Natl Acad Sci, 111*, 7212–7217.

Bicanski, A., Ryczko, D., Cabelguen, J. M., & Ijspeert, A. J. (2013a). From lamprey to salamander: An exploratory modeling study on the architecture of the spinal locomotor networks in the salamander. *Biol Cybern, 107*(5), 565–587. doi:10.1007/s00422-012-0538-y

Bicanski, A., Ryczko, D., Knuesel, J., Harischandra, N., Charrier, V., Ekeberg, O., . . . Ijspeert, A. J. (2013b). Decoding the mechanisms of gait generation in salamanders by combining neurobiology, modeling and robotics. *Biol Cybern, 107*(5), 545–564. doi:10.1007/s00422-012-0543-1

Biewener, A. A. (2016). Locomotion as an emergent property of muscle contractile dynamics. *J Exp Biol, 219*(Pt 2), 285–294. doi:10.1242/jeb.123935

Biewener, A. A., & Roberts, T. J. (2000). Muscle and tendon contributions to force, work, and elastic energy savings: A comparative perspective. *Exercise Sport Science Review, 28*, 99–107.

Biswal, B. B. (2012). Resting state fMRI: A personal history. *Neuroimage, 62*(2), 938–944. doi:10.1016/j.neuroimage.2012.01.090

Biswal, B. B., Mennes, M., Zuo, X. N., Gohel, S., Kelly, C., Smith, S. M., . . . Milham, M. P. (2010). Toward discovery science of human brain function. *Proc Natl Acad Sci USA, 107*(10), 4734–4739. doi:10.1073/pnas.0911855107

Bizzi, E., & Cheung, V. C. (2013). The neural origin of muscle synergies. *Front Comput Neurosci, 7*, 51. doi:10.3389/fncom.2013.00051

Bizzi, E., Cheung, V. C., d'Avella, A., Saltiel, P., & Tresch, M. (2008). Combining modules for movement. *Brain Res Rev, 57*(1), 125–133. doi:S0165-0173(07)00177-4 [pii] 10.1016/j.brainresrev.2007.08.004

Blanchard, G. B., & Adams, R. J. (2011). Measuring the multi-scale integration of mechanical forces during morphogenesis. *Curr Opin Genet Dev, 21*(5), 653–663. doi:10.1016/j.gde.2011.08.008

Blankenship, A. G., & Feller, M. B. (2009). Mechanisms underlying spontaneous patterned activity in developing neural circuits. *Nat Rev Neurosci, 11*(1), 18–29. doi:10.1038/nrn2759

Blanton, L. V., Barratt, M., Charbonneau, M., Ahmed, T., & Gordon, J. I. (2016). Childhood undernutrition, the gut microbiota, and microbiota-directed therapeutics. *Science, 352*(6293), 1533–1539. doi:10.1126/science

Bleyenheuft, Y., & Gordon, A. M. (2013). Precision grip control, sensory impairments and their interactions in children with hemiplegic cerebral palsy: A systematic review. *Res Dev Disabil, 34*(9), 3014–3028. doi:10.1016/j.ridd.2013.05.047

Blickhan, R., Seyfarth, A., Geyer, H., Grimmer, S., Wagner, H., & Gunther, M. (2007). Intelligence by mechanics. *Philos Trans A Math Phys Eng Sci, 365*(1850), 199–220. doi:10.1098/rsta.2006.1911

Bloch-Salisbury, E., Indic, P., Bednarek, F., & Paydarfar, D. (2009). Stabilizing immature breathing patterns of preterm infants using stochastic mechanosensory stimulation. *J Appl Physiol, 107*, 1017–1027.

Blumberg, M. S., Coleman, C. M., Sokoloff, G., Weiner, J. A., Fritzsch, B., & McMurray, B. (2015). Development of twitching in sleeping infant mice depends on sensory experience. *Curr Biol, 25*(12), 1672. doi:10.1016/j.cub.2015.05.050

Blumberg, M. S., Marques, H. G., & Iida, F. (2013). Twitching in sensorimotor development from sleeping rats to robots. *Curr Biol, 23*(12), R532–R537. doi:10.1016/j.cub.2013.04.075

Bobak, M., Chuan-Hsien, K., Yi-Chung, T., Yu-Suke, T., Tommaso, B.-B., Hossein, T., & Shuichi, T. (2010). Integrated elastomeric components for autonomous regulation of sequential and oscillatory flow switching in microfluidic devices. *Nat Phys, 6*(6), 433. doi:10.1038/nphys1637

Bolek, S., Wittlinger, M., & Wolf, H. (2012). Establishing food site vectors in desert ants. *J Exp Biol, 215*(Pt 4), 653. doi:10.1242/jeb.062406

Bongard, J. (2011). Morphological change in machines accelerates the evolution of robust behavior. *Proc Natl Acad Sci USA, 108*(4), 1234–1239. doi:10.1073/pnas.1015390108

Bongard, J., & Lipson, H. (2014). Evolved machines shed light on robustness and resilience. *Proc IEEE, 102*(5), 899–914.

Bongard, J., Zykov, V., & Lipson, H. (2006). Resilient machines through continuous self-modeling. *Science, 314*(5802), 1118–1121. doi:10.1126/science.1133687

Bonner, J. F., & Steward, O. (2015). Repair of spinal cord injury with neuronal relays: From fetal grafts to neural stem cells. *Brain Res, 1619*, 115–123. doi:10.1016/j.brainres.2015.01.006

Bonner, J. T. (2010). Brainless behavior: A myxomycete chooses a balanced diet. *Proc Natl Acad Sci USA, 107*(12), 5267–5268. doi:10.1073/pnas.1000861107

Borodinsky, L. N., Belgacem, Y. H., & Swapna, I. (2012). Electrical activity as a developmental regulator in the formation of spinal cord circuits. *Curr Opin Neurobiol, 22*(4), 624–630. doi:10.1016/j.conb.2012.02.004

Borton, D., Micera, S., Millan Jdel, R., & Courtine, G. (2013). Personalized neuroprosthetics. *Sci Transl Med, 5*(210), 1–12. doi:10.1126/scitranslmed.3005968

Bosch, M., & Hayashi, Y. (2012). Structural plasticity of dendritic spines. *Curr Opin Neurobiol, 22*(3), 383–388. doi:10.1016/j.conb.2011.09.002

Bouton, C. E., Shaikhouni, A., Annetta, N. V., Bockbrader, M. A., Friedenberg, D. A., Nielson, D. M., . . . Rezai, A. R. (2016). Restoring cortical control of functional movement in a human with quadriplegia. *Nature, 533*(7602), 247–250. doi:10.1038/nature17435

Boyan, G. S., & Reichert, H. (2011). Mechanisms for complexity in the brain: Generating the insect central complex. *Trends Neurosci, 34*(5), 247–257. doi:10.1016/j.tins.2011.02.002

Boyden, E. S. (2015). Optogenetics and the future of neuroscience. *Nat Neurosci, 18*(9), 1200–1201. doi:10.1038/nn.4094

Boyden, E. S., Zhang, F., Bamberg, E., Nagel, G., & Deisseroth, K. (2005). Millisecond-timescale, genetically targeted optical control of neural activity. *Nat Neurosci, 8*(9), 1263–1268. doi:10.1038/nn1525

Bradley, J. M., Paola, A., Joao, R. L. M., & Jeffrey, D. M. (2007). Neuronal subtype specification in the cerebral cortex. *Nat Rev Neurosci, 8*(6), 427. doi:10.1038/nrn2151

Brainard, M. S., & Doupe, A. J. (2013). Translating birdsong: Songbirds as a model for basic and applied medical research. *Annu Rev Neurosci, 36*, 489–517. doi:10.1146/annurev-neuro-060909-152826

Bramble, D. M., & Wake, D. B. (1985). Feeding mechanism of lower tetrapods. In M. Hildebrand, D. Bramble, K. Liem, & D. Wake (Eds.), *Functional vertebrate morphology* (pp. 230–261). Cambridge, MA: Harvard University Press.

Bramhall, Naomi F., Shi, F., Arnold, K., Hochedlinger, K., & Edge, Albert S. B. (2014). Lgr5-positive supporting cells generate new hair cells in the postnatal cochlea. *Stem Cell Reports, 2*(3). doi:10.1016/j.stemcr.2014.01.008

Brandman, O., & Meyer, T. (2008). Feedback loops shape cellular signals in space and time. *Science, 322*, 390–395.

Braun, H. A., Wissing, H., Schafer, K., & Hirsch, M. C. (1994). Oscillation and noise determine signal transduction in shark multimodal sensory cells. *Nature, 367*, 270–273.

Breakspear, M. (2017). Dynamic models of large-scale brain activity. *Nat Neurosci 20*, 340–352.

Breazeal, C. (2003). Toward sociable robots. *Robotics and Autonomous Systems, 42*(3), 167–175. doi:10.1016/S0921-8890(02)00373-1

Bressler, S. L., & Richter, C. G. (2014). Interareal oscillatory synchronization in top-down neocortical processing. *Curr Opin Neurobiol, 31C*, 62–66. doi:10.1016/j.conb.2014.08.010

Briffa, M., & Mowles, S. L. (2008). Hermit crabs. *Curr Biol, 18*(4), R144–R146. doi:10.1016/j.cub.2007.12.003

Briggs, D. E. (2015). The Cambrian explosion. *Curr Biol, 25*(19), R864–R868. doi:10.1016/j.cub.2015.04.047

Bright, M., Klose, J., & Nussbaumer, A. D. (2013). Giant tubeworms. *Curr Biol, 23*(6), R224–R225. doi:10.1016/j.cub.2013.01.039

Brindley, G. S., & Lewin, W. S. (1968). The sensations produced by electrical stimulation of the visual cortex. *J Physiol, 196*(2), 479.

Brooks, J. X., Carriot, J., & Cullen, K. E. (2015). Learning to expect the unexpected: Rapid updating in primate cerebellum during voluntary self-motion. *Nat Neurosci, 18*(9), 1310–1317. doi:10.1038/nn.4077

Bruderer, A. G., Danielson, D. K., Kandhadai, P., & Werker, J. F. (2015). Sensorimotor influences on speech perception in infancy. *Proc Natl Acad Sci USA, 112*(44), 13531–13536. doi:10.1073/pnas.1508631112

Brugués, J., & Needleman, D. (2014). Physical basis of spindle self-organization. *Proc Natl Acad Sci, 111*(52), 18496. doi:10.1073/pnas.1409404111

Bruno, A. M., Frost, W. N., & Humphries, M. D. (2015). Modular deconstruction reveals the dynamical and physical building blocks of a locomotion motor program. *Neuron, 86*(1), 304–318. doi:10.1016/j.neuron.2015.03.005

Buckner, R. L., Krienen, F. M., & Yeo, B. T. (2013). Opportunities and limitations of intrinsic functional connectivity MRI. *Nat Neurosci, 16*(7), 832–837. doi:10.1038/nn.3423

Buehler, M. J. (2010). Tu(r)ning weakness to strength. *Nano Today, 5*(5), 379–383. doi:10.1016/j.nantod.2010.08.001

Buehlmann, C., Hansson, B. S., & Knaden, M. (2012). Path integration controls nest-plume following in desert ants. *Curr Biol, 22*(7), 645–649. doi:10.1016/j.cub.2012.02.029

Buffington, S. A., Di Prisco, G. V., Auchtung, T. A., Ajami, N. J., Petrosino, J. F., & Costa-Mattioli, M. (2016). Microbial reconstitution reverses maternal diet-induced social and synaptic deficits in offspring. *Cell, 165*(7), 1762–1775. doi:10.1016/j.cell.2016.06.001

Bui, T. V., Akay, T., Loubani, O., Hnasko, T. S., Jessell, T. M., & Brownstone, R. M. (2013). Circuits for grasping: Spinal dI3 interneurons mediate cutaneous control of motor behavior. *Neuron, 78*(1), 191–204. doi:10.1016/j.neuron.2013.02.007

Bullmore, E., & Sporns, O. (2009). Complex brain networks: Graph theoretical analysis of structural and functional systems. *Nat Rev Neurosci, 10*(3), 186–198. doi:nrn2575 [pii] 10.1038/nrn2575

Bullmore, E., & Sporns, O. (2012). The economy of brain network organization. *Nat Rev Neurosci, 13*(5), 336–349. doi:10.1038/nrn3214

Bullmore, E. T., & Bassett, D. S. (2011). Brain graphs: Graphical models of the human brain connectome. *Annual Review of Clinical Psychology, 7*, 1–28.

Burke, A. C., Nelson, C. E., Morgan, B. A., & Tabin, C. (1995). Hox genes and the evolution of vertebrate axial morphology. *Development, 121*, 333–346.

Burykin, A., Costa, M. D., Citi, L., & Goldberger, A. L. (2014). Dynamical density delay maps: Simple, new method for visualising the behaviour of complex systems. *BMC Medical Informatics and Decision Making*. doi:10.1186/1472-6947-14-6

Buschges, A. (2012). Lessons for circuit function from large insects: Towards understanding the neural basis of motor flexibility. *Curr Opin Neurobiol, 22*(4), 602–608. doi:10.1016/j.conb.2012.02.003

Buschman, T. J., Denovellis, E. L., Diogo, C., Bullock, D., & Miller, E. K. (2012). Synchronous oscillatory neural ensembles for rules in the prefrontal cortex. *Neuron, 76*(4), 838–846. doi:10.1016/j.neuron.2012.09.029

Buser, J. R., Maire, J., Riddle, A., Gong, X., Nguyen, T., Nelson, K., . . . Back, S. A. (2012). Arrested preoligodendrocyte maturation contributes to myelination failure in premature infants. *Ann Neurol, 71*(1), 93–109. doi:10.1002/ana.22627

Bush, J. W. M., & Hu, D. L. (2006). Walking on water: Biolocomotion at the interface. *Ann Rev Fluid Mechanics, 38*, 339–369.

Buss, R. R., Sun, W., & Oppenheim, R. O. (2006). Adaptive roles of programmed cell death during nervous system development. *Annu Rev Neurosci, 29*, 1–35. doi:10.1146/

Buzsáki, G. (2006). *Rhythms of the brain*. New York: Oxford University Press.

Buzsaki, G., Anastassiou, C. A., & Koch, C. (2012). The origin of extracellular fields and currents—EEG, ECoG, LFP and spikes. *Nat Rev Neurosci, 13*(6), 407–420. doi:10.1038/nrn3241

Buzsaki, G., & Draguhn, A. (2004). Neuronal oscillations in cortical networks. *Science, 304*, 1926–1929.

Buzsaki, G., Logothetis, N., & Singer, W. (2013). Scaling brain size, keeping timing: evolutionary preservation of brain rhythms. *Neuron, 80*(3), 751–764. doi:10.1016/j.neuron.2013.10.002

Buzsáki, G., Peyrache, A., & Kubie, J. (2014). Emergence of cognition from action. *Cold Spring Harbor Symposia on Quantitative Biology, 79*, 41. doi:10.1101/sqb.2014.79.024679

Buzsaki, G., Stark, E., Berenyi, A., Khodagholy, D., Kipke, D. R., Yoon, E., & Wise, K. D. (2015). Tools for probing local circuits: high-density silicon probes combined with optogenetics. *Neuron, 86*(1), 92–105. doi:10.1016/j.neuron.2015.01.028

Buzsaki, G., & Wang, X.-J. (2012). Mechanisms of gamma oscillations. *Annu Rev Neurosci, 35*, 203–225.

Byrd, D., & Saltzman, E. (2003). The elastic phrase: Modeling the dynamics of boundary-adjacent lengthening. *J Phon, 31*(2), 149–180.

Bystron, I., Blakemore, C., & Rakic, P. (2008). Development of the human cerebral cortex: Boulder Committee revisited. *Nat Rev Neurosci, 9*(2), 110–122. doi:10.1038/nrn2252

Cai, D., Cohen, K. B., Luo, T., Lichtman, J., & Sanes, J. R. (2013). Improved tools for the brainbow toolbox. *Nat Methods*. doi:10.1038/nmeth.2450

Cai, L., Fong, A., Otoshi, C., Liang, Y., Burdick, J. W., Roy, R., & Edgerton, V. (2006). Implications of assist-as-needed robotic step training after a complete spinal cord injury on intrinsic strategies of motor learning. *J Neurosci, 26*(41), 10564–10568. doi:10.1523/JNEUROSCI.2266-06.2006

Callier, T., Schluter, E. W., Tabot, G. A., Miller, L. E., Tenore, F. V., & Bensmaia, S. J. (2015). Long-term stability of sensitivity to intracortical microstimulation of somatosensory cortex. *J Neural Eng, 12*(5), 056010. doi:10.1088/1741-2560/12/5/056010

Camazine, S., Deneubourg, J.-L., & al. (Eds.). (2001). *Self-organization in biological systems*. Princeton, NJ: Princeton University Press.

Camp, J. G., Badsha, F., Florio, M., Kanton, S., Gerber, T., Wilsch-Brauninger, M., . . . Treutlein, B. (2015). Human cerebral organoids recapitulate gene expression programs of fetal neocortex development. *Proc Natl Acad Sci USA, 112*(51), 15672–15677.

Campas, O. (2016). A toolbox to explore the mechanics of living embryonic tissues. *Semin Cell Dev Biol, 55*, 119–130. doi:10.1016/j.semcdb.2016.03.011

Campas, O., Mammoto, T., Hasso, S., Sperling, R. A., O'Connell, D., Bischof, A. G., . . . Ingber, D. E. (2014). Quantifying cell-generated mechanical forces within living embryonic tissues. *Nat Methods, 11*(2), 183–189. doi:10.1038/nmeth.2761

Cannon, T. D. (2015). How schizophrenia develops: Cognitive and brain mechanisms underlying onset of psychosis. *Trends Cogn Sci, 19*(12), 744–756. doi:10.1016/j.tics.2015.09.009

Cannon, W. B. (1929). *A laboratory course in physiology* (7th ed.). Cambridge, MA: Harvard University Press.

Card, G., & Dickinson, M. H. (2008a). Performance trade-offs in the flight initiation of Drosophila. *J Exp Biol, 211*(Pt 3), 341–353. doi:10.1242/jeb.012682

Card, G., & Dickinson, M. H. (2008b). Visually mediated motor planning in the escape response of Drosophila. *Curr Biol, 18*(17), 1300–1307. doi:10.1016/j.cub.2008.07.094

Card, G. M. (2012). Escape behaviors in insects. *Curr Opin Neurobiol, 22*(2), 180–186. doi:10.1016/jconb.2011.12.009

Carello, C., Grosofsky, A., Reichel, F., Solomon, H., & Turvey, M. T. (1989). Visually perceiving what is reachable. *Ecol Psychol, 1*, 27–54.

Carello, C., & Turvey, M. (2015). Dynamic (effortful) touch. *Scholarpedia, 10*(4), 8242. doi:10.4249/scholarpedia.8242

Carlson, J. M., & Doyle, J. (2002). Complexity and robustness. *Proc Natl Acad Sci USA, 99 Suppl 1*, 2538–2545. doi:10.1073/pnas.012582499

Caron, J. B., Morris, S. C., & Cameron, C. B. (2013). Tubicolous enteropneusts from the Cambrian period. *Nature, 495*(7442), 503–506. doi:10.1038/nature12017

Caron, J. B., Scheltema, A., Schander, C., & Rudkin, D. (2006). A soft-bodied mollusc with radula from the Middle Cambrian Burgess Shale. *Nature, 442*(7099), 159–163. doi:10.1038/nature04894

Carrillo-Reid, L., Yang, W., Kang Miller, J. E., Peterka, D. S., & Yuste, R. (2017). Imaging and optically manipulating neuronal ensembles. *Annu Rev Biophys.* doi:10.1146/annurev-biophys-070816-033647

Carroll, S. B. (2008). Evo-devo and an expanding evolutionary synthesis: A genetic theory of morphological evolution. *Cell, 134*(1), 25–36. doi:10.1016/j.cell.2008.06.030

Carroll, S. B., Grenier, J. K., & Weatherbee, S. D. (2005). *From DNA to diversity* (2nd ed.). Malden, MA: Blackwell.

Carter, A. R., Shulman, G. L., & Corbetta, M. (2012). Why use a connectivity-based approach to study stroke and recovery of function? *Neuroimage, 62*(4), 2271–2280. doi:10.1016/j.neuroimage.2012.02.070

Carus-Cadavieco, M., Gorbati, M., Li, Y., Bender, F., Van Der, V., S., Kosse, C., . . . Korotkova, T. (2017). Gamma oscillations organize top-down signalling to hypothalamus and enable food seeking. *Nature, 542*(7640). doi:10.1038/nature21066

Castro, C. E., Kilchherr, F., Kim, D. N., Shiao, E. L., Wauer, T., Wortmann, P., . . . Dietz, H. (2011). A primer to scaffolded DNA origami. *Nat Methods, 8*(3), 221–229. doi:10.1038/nmeth.1570

Catania, K. C. (2012). Tactile sensing in specialized predators—from behavior to the brain. *Curr Opin Neurobiol, 22*(2), 251–258. doi:10.1016/j.conb.2011.11.014

Catania, K. C., & Henry, E. C. (2006). Touching on somatosensory specializations in mammals. *Curr Opin Neurobiol, 16*, 467–473.

Catania, K. C., Leitch, D. B., & Gauthier, D. (2011). A star in the brainstem reveals the first step of cortical magnification. *PLoS ONE, 6*, 1–9.

Cerda, E., & Mahadevan, L. (2003). Geometry and physics of wrinkling. *Phys Rev Lett, 90*(7), 074302.

Cerminara, N. L., Lang, E. J., Sillitoe, R. V., & Apps, R. (2015). Redefining the cerebellar cortex as an assembly of non-uniform Purkinje cell microcircuits. *Nat Rev Neurosci, 16*(2), 79–93. doi:10.1038/nrn3886

Chakravarthy, V. S., Joseph, D., & Bapi, R. S. (2010). What do the basal ganglia do? A modeling perspective. *Biol Cybern, 103*(3), 237–253. doi:10.1007/s00422-010-0401-y

Chan, V., Asada, H. H., & Bashir, R. (2014). Utilization and control of bioactuators across multiple length scales. *Lab Chip, 14*(4), 653–670. doi:10.1039/c3lc50989c

Chan, V., Park, K., Collens, M. B., Kong, H., Saif, T. A., & Bashir, R. (2012). Development of miniaturized walking biological machines. *Sci Rep, 2,* 857. doi:10.1038/srep00857

Chao, D. L., Ma, L., & Shen, K. (2009). Transient cell-cell interactions in neural circuit formation. *Nat Rev Neurosci, 10,* 262–271.

Chapple, W. (2012). Kinematics of walking in the hermit crab, *Pagurus pollicarus. Arthropod Structure and Development, 41*(2), 119–131. doi:10.1016/j.asd.2011.11.004

Charbonneau, M. R., Blanton, L. V., DiGiulio, D. B., Relman, D. A., Lebrilla, C. B., Mills, D. A., & Gordon, J. I. (2016). A microbial perspective of human developmental biology. *Nature, 535*(7610), 48–55. doi:10.1038/nature18845

Che, J., & Dorgan, K. M. (2010). It's tough to be small: Dependence of burrowing kinematics on body size. *J Exp Biol, 213*(Pt 8), 1241–1250. doi:10.1242/jeb.038661

Chhetri, R. K., Amat, F., Wan, Y., Hockendorf, B., Lemon, W. C., & Keller, P. J. (2015). Whole-animal functional and developmental imaging with isotropic spatial resolution. *Nat Methods, 12*(12), 1171–1178. doi:10.1038/nmeth.3632

Chialvo, D. R. (2010). Emergent complex neural dynamics. *Nat Physics, 6*(10), 744–750. doi:10.1038/nphys1803

Chirarattananon, P., Chen, Y., Helbling, E. F., Ma, K. Y., Cheng, R., & Wood, R. J. (2017). Dynamics and flight control of a flapping-wing robotic insect in the presence of wind gusts. *Interface Focus, 7*(1), 20160080. doi:10.1098/rsfs.2016.0080

Chittka, L., & Niven, J. (2009). Are bigger brains better? *Curr Biol, 19*(21), R995–R1008. doi:10.1016/j.cub.2009.08.023Nat.

Choe, M. S., Ortiz-Mantilla, S., Makris, N., Gregas, M., Bacic, J., Haehn, D., . . . Grant, P. E. (2013). Regional infant brain development: An MRI-based morphometric analysis in 3 to 13 month olds. *Cereb Cortex, 23*(9), 2100–2117. doi:10.1093/cercor/bhs197

Choi, H. J., & Mark, L. S. (2004). Scaling affordances for human reach actions. *Hum Mov Sci, 23*(6), 785–806. doi:10.1016/j.humov.2004.08.004

Chouard, T., & Venema, A. (2015). Nature insight: Machine intelligence. *Nature, 521*(7553), 435.

Chrastil, E. R., Sherrill, K. R., Hasselmo, M. E., & Stern, C. E. (2016). Which way and how far? Tracking of translation and rotation information for human path integration. *Hum Brain Mapp, 37*(10), 3636–3655. doi:10.1002/hbm.23265

Christensen, D., Van Naarden Braun, K., Doernberg, N. S., Maenner, M. J., Arneson, C. L., Durkin, M. S., . . . Yeargin-Allsopp, M. (2014). Prevalence of cerebral palsy, co-occurring autism spectrum disorders, and motor functioning—Autism and Developmental Disabilities Monitoring Network, USA, 2008. *Dev Med Child Neurol, 56*(1), 59–65. doi:10.1111/dmcn.12268

Chu, D. M., Ma, J., Prince, A. L., Antony, K. M., Seferovic, M. D., & Aagaard, K. M. (2017). Maturation of the infant microbiome community structure and function across multiple body sites and in relation to mode of delivery. *Nat Med, 23*(3), 314–326. doi:10.1038/nm.4272

Chung, W. S., & Barres, B. A. (2012). The role of glial cells in synapse elimination. *Curr Opin Neurobiol, 22*(3), 438–445. doi:10.1016/j.conb.2011.10.003

Chung, W. S., Welsh, C. A., Barres, B. A., & Stevens, B. (2015). Do glia drive synaptic and cognitive impairment in disease? *Nat Neurosci, 18*(11), 1539–1545. doi:10.1038/nn.4142

Churchland, M. M., Afshar, A., & Shenoy, K. V. (2006). A central source of movement variability. *Neuron, 52*(6), 1085–1096. doi:10.1016/j.neuron.2006.10.034

Churchland, M. M., & Cunningham, J. P. (2014). A dynamical basis set for generating reaches. *Cold Spring Harb Symp Quant Biol, 79,* 67–80. doi:10.1101/sqb.2014.79.024703

Churchland, M. M., Cunningham, J. P., Kaufman, M. T., Foster, J. D., Nuyujukian, P., Ryu, S. I., & Shenoy, K. V. (2012). Neural population dynamics during reaching. *Nature, 487*(7405), 51–56. doi:10.1038/nature11129

Churchland, M. M., Cunningham, J. P., Kaufman, M. T., Ryu, S. I., & Shenoy, K. V. (2010). Cortical preparatory activity: Representation of movement or first cog in a dynamical machine? *Neuron, 68*(3), 387–400. doi:10.1016/j.neuron.2010.09.015

Cisek, P. (2012). Making decisions through a distributed consensus. *Curr Opin Neurobiol, 22*, 1–10.

Cisek, P., & Kalaska, J. F. (2010). Neural mechanisms for interacting with a world full of action choices. *Annu Rev Neurosci, 33*, 269–298. doi:10.1146/annurev.neuro.051508.135409

Cisek, P., & Pastor-Bernier, A. (2014). On the challenges and mechanisms of embodied decisions. *Philos Trans R Soc Lond B Biol Sci, 369*(1655). doi:10.1098/rstb.2013.0479

Clark, A. (1997). *Being there: Putting brain, body, and world together again*. Cambridge, MA: MIT Press.

Clark, A. (2003). *Natural-born cyborgs: Minds, technologies, and the future of human intelligence*. New York: Oxford University Press.

Clark, A. (2006). Language, embodiment, and the cognitive niche. *Trends Cog Sci, 10*, 370–374.

Clark, A. (2008). *Supersizing the mind: Embodiment, action, and cognitive extension*. New York: Oxford University Press.

Clarke, L. E., & Barres, B. A. (2013). Emerging roles of astrocytes in neural circuit development. *Nat Rev Neurosci, 14*(5), 311–321. doi:10.1038/nrn3484

Clause, A., Kim, G., Sonntag, M., Weisz, C. J., Vetter, D. E., Rubsamen, R., & Kandler, K. (2014). The precise temporal pattern of prehearing spontaneous activity is necessary for tonotopic map refinement. *Neuron, 82*(4), 822–835. doi:10.1016/j.neuron.2014.04.001

Clawson, T. S., Ferrari, S., Fuller, S. B., & Wood, R. J. (2016). *Spiking neural network (SNN) control of a flapping insect-scale robot*. Paper presented at the 2016 IEEE 55th Conference on Decision and Control, Las Vegas, NV.

Clouchoux, C., & Limperopoulos, C. (2012). Novel applications of quantitative MRI for the fetal brain. *Pediatr Radiol, 42 Suppl 1*, S24–S32. doi:10.1007/s00247-011-2178-0

Clouchoux, C., Riviere, D., Mangin, J. F., Operto, G., Regis, J., & Coulon, O. (2010). Model-driven parameterization of the cortical surface for localization and inter-subject matching. *Neuroimage, 50*(2), 552–566. doi:10.1016/j.neuroimage.2009.12.048

Coen, P., Clemens, J., Weinstein, A. J., Pacheco, D. A., Deng, Y., & Murthy, M. (2014). Dynamic sensory cues shape song structure in Drosophila. *Nature, 507*(7491). doi:10.1038/nature13131

Cohen, A. E., & Mahadevan, L. (2003). Kinks, rings, and rackets in filamentous structures. *Proc Natl Acad Sci, 100*, 12141–12146.

Collin, G., & van den Heuvel, M. P. (2013). The ontogeny of the human connectome: Development and dynamic changes of brain connectivity across the life span. *Neuroscientist, 19*(6), 616–628. doi:10.1177/1073858413503712

Collinger, J. L., Foldes, S., Bruns, T. M., Wodlinger, B., Gaunt, R., & Weber, D. J. (2013). Neuroprosthetic technology for individuals with spinal cord injury. *J Spinal Cord Med, 36*(4), 258–272. doi:10.1179/2045772313Y.0000000128

Collinger, J. L., Kryger, M. A., Barbara, R., Betler, T., Bowsher, K., Brown, E. H. P., . . . Boninger, M. L. (2014). Collaborative approach in the development of high-performance brain–computer interfaces for a neuroprosthetic arm: Translation from animal models to human control. *Clin Transl Sci, 7*(1), 52–59. doi:10.1111/cts.12086

Collinger, J. L., Wodlinger, B., Downey, J. E., Wang, W., Tyler-Kabara, E. C., Weber, D. J., . . . Schwartz, A. B. (2012). High-performance neuroprosthetic control by an individual with tetraplegia. *The Lancet, 381*(9866), 557–564. doi:10.1016/S0140-6736(12)61816-9

Collins, J. J., Imhoff, T. T., & Grigg, P. (1996). Noise-enhanced information transmission in rat SA1 cutaneous mechanoreceptors via aperiodic stochastic resonance. *J Neurophysiol, 76*, 642–645.

Collins, J. J., & Stewart, I. (1993). Coupled nonlinear oscillators and the symmetries of animal gaits. *Journal of Nonlinear Science, 3*(3), 349–392.

Collins, S. M., Surette, M., & Bercik, P. (2012). The interplay between the intestinal microbiota and the brain. *Nat Rev Microbiol, 10*(11), 735–742. doi:10.1038/nrmicro2876

Colonnese, M., Kaminska, A., Minlabaev, M., Milh, M., Bloem, B., Lescure, S., . . . Khazipov, R. (2010). A conserved switch in sensory processing prepares developing neocortex for vision. *Neuron, 67*, 480–498.

Combes, D., Merrywest, S. D., Simmers, J., & Sillar, K. T. (2004). Developmental segregation of spinal networks driving axial- and hindlimb-based locomotion in metamorphosing Xenopus laevis. *J Physiol, 559*(Pt 1), 17–24. doi:10.1113/jphysiol.2004.069542

Combes, S. A., & Daniel, T. L. (2003). Flexural stiffness in insect wings. I. Scaling and the influence of wing venation. *J Exp Biol, 206*(Pt 17), 2979.

Combes, S. A., Rundle, D. E., Iwasaki, J. M., & Crall, J. D. (2012). Linking biomechanics and ecology through predator-prey interactions: flight performance of dragonflies and their prey. *J Exp Biol, 215*(Pt 6), 903–913. doi:10.1242/jeb.059394

Concha, A., Mellado, P., Morera-Brenes, B., Sampaio Costa, C., Mahadevan, L., & Monge-Najera, J. (2015). Oscillation of the velvet worm slime jet by passive hydrodynamic instability. *Nat Commun, 6*, 6292. doi:10.1038/ncomms7292

Conway Morris, S. (1998). *The crucible of creation: The Burgess Shale and the rise of animals*. New York: Oxford University Press.

Coombs, S. (2014). *Lateral line system*. New York: Springer.

Coombs, S., & Bleckmann, H. (2006). Lateral line research: Recent advances and new opportunities. *J Acoust Soc Amer, 120*(5), 3056–3056. doi:10.1121/1.4787292

Coombs, S., Görner, P., and Münz, H. (Eds.) (1989). *Mechanosensory lateral line: Neurobiology and evolution*. New York: Springer.

Cooper, K. L., Sears, K. E., Uygur, A., Maier, J., Baczkowski, K. S., Brosnahan, M., . . . Tabin, C. J. (2014). Patterning and post-patterning modes of evolutionary digit loss in mammals. *Nature, 511*(7507), 41–45. doi:10.1038/nature13496

Corbett, D., Jeffers, M., Nguemeni, C., Gomez-Smith, M., & Livingston-Thomas, J. (2015). Lost in translation: Rethinking approaches to stroke recovery. *Prog Brain Res, 218*, 413–434. doi:10.1016/bs.pbr.2014.12.002

Corbett, E. A., Ethier, C., Oby, E. M., Kording, K., Perreault, E. J., & Miller, L. E. (2013). Advanced user interfaces for upper limb functional electrical stimulation. In D. Farina, W. Jensen, & M. Akay (Eds.), *Introduction to neural engineering for motor rehabilitation* (pp. 377–399). New York: Wiley.

Corbetta, M. (2010). Functional connectivity and neurological recovery. *Dev Psychobiol, 54*, 239–253.

Cordo, P., Inglis, J. T., Verschueren, S., Collins, J. J., Merfeld, D. M., Rosenblum, S., . . . Moss, F. (1996). Noise in human muscle spindles. *Nature, 383*(6603), 769–770.

Correll, N., Onal, C., Liang, H., Schoenfeld, E., & Rus, D. (2014). Soft autonomous materials using active elasticity and embedded distributred computation. *Experimental Robotics, Springer Tracts in Advanced Robotics, 79*, 227–240.

Courtine, G., Gerasimenko, Y., van den Brand, R., Yew, A., Musienko, P., Zhong, H., . . . Edgerton, V. R. (2009). Transformation of nonfunctional spinal circuits into functional states after the loss of brain input. *Nat Neurosci, 12*(10), 1333–1342. doi:10.1038/nn.2401

Courtine, G., Micera, S., DiGiovanna, J., & del R Millán, J. (2013). Brain–machine interface: Closer to therapeutic reality? *The Lancet, 381*(9866), 515–517. doi:10.1016/s0140-6736(12)62164-3

Courtine, G., Song, B., Roy, R. R., Zhong, H., Herrmann, J. E., Ao, Y., . . . Sofroniew, M. V. (2008). Recovery of supraspinal control of stepping via indirect propriospinal relay connections after spinal cord injury. *Nat Med, 14*(1), 69–74. doi:10.1038/nm1682

Couzin-Fuchs, E., Kiemel, T., Gal, O., Ayali, A., & Holmes, P. (2015). Intersegmental coupling and recovery from perturbations in freely running cockroaches. *J Exp Biol, 218*(Pt 2), 285–297. doi:10.1242/jeb.112805

Cowan, N. J., Ankarali, M. M., Dyhr, J. P., Madhav, M. S., Roth, E., Sefati, S., . . . Daniel, T. L. (2014). Feedback control as a framework for understanding tradeoffs in biology. *Integrative and Comparative Biology, 54*(2), 223–237. doi:10.1093/icb/icu050

Coyne, J., Boussy, I., Prout, T., Bryant, S., Jones, J., & Moore, J. (1982). Long-distance migration of "Drosophila." *American Naturalist, 119*(4), 589–595.

Coyne, J. A. (2005). Switching on evolution: How does evo-devo explain the huge diversity of life on Earth? *Nature, 435*, 1029–1030.

Craddock, R. C., Jbabdi, S., Yan, C. G., Vogelstein, J. T., Castellanos, F. X., Di Martino, A., . . . Milham, M. P. (2013). Imaging human connectomes at the macroscale. *Nat Methods, 10*(6), 524–539. doi:10.1038/nmeth.2482

Cramer, S. C., Sur, M., & Dobkin, B. H. (2011). Harnessing neuroplasticity for clinical applications. *Brain, 134*, 1591–1609.

Cranford, S. W., Tarakanova, A., Pugno, N. M., & Buehler, M. J. (2012). Nonlinear material behaviour of spider silk yields robust webs. *Nature, 482*(7383), 72–76. doi:10.1038/nature10739

Crapse, T. B., & Sommer, M. A. (2008). Corollary discharge circuits in the primate brain. *Curr Opin Neurobiol, 18*(6), 552–557. doi:10.1016/j.conb.2008.09.017

Cregg, J. M., Depaul, M. A., Filous, A. R., Lang, B. T., Tran, A., & Silver, J. (2014). Functional regeneration beyond the glial scar. *Experimental Neurology, 253*, 197–207. doi:10.1016/j.expneurol.2013.12.024

Crespi, A., Karakasiliotis, K., Guignard, A., & Ijspeert, A. J. (2013). Salamandra Robotica II: An amphibious robot to study salamander-like swimming and walking gaits. *Robotics, IEEE Transactions on, 29*(2), 308–320. doi:10.1109/TRO.2012.2234311

Crompton, A. W., German, R. Z., & Thexton, A. J. (2008). Development of the movement of the epiglottis in infant and juvenile pigs. *Zoology (Jena), 111*(5), 339–349. doi:10.1016/j.zool.2007.10.002

Crompton, A. W., & Musinsky, C. (2011). How dogs lap: Ingestion and intraoral transport in *Canis familiaris. Biol Lett, 7*(6), 882–884. doi:10.1098/rsbl.2011.0336

Crutchfield, J. P. (2012). Between order and chaos. *Nat Phys, 8*, 17–24.

Cryan, J. F., & Dinan, T. G. (2012). Mind-altering microorganisms: The impact of the gut microbiota on brain and behaviour. *Nat Rev Neurosci, 13*(10), 701–712. doi:10.1038/nrn3346

Cryan, J. F., & Dinan, T. G. (2015). Gut microbiota: Microbiota and neuroimmune signalling—Metchnikoff to microglia. *Nat Rev Gastroenterol Hepatol, 12*(9), 494–496. doi:10.1038/nrgastro.2015.127

Cully, A., Clune, J., Tarapore, D., & Mouret, J. B. (2015). Robots that can adapt like animals. *Nature, 521*(7553), 503–507. doi:10.1038/nature14422

Cunningham, J. P., & Yu, B. M. (2014). Dimensionality reduction for large-scale neural recordings. *Nat Neurosci, 17*(11), 1500–1509. doi:10.1038/nn.3776

Currie, C. R., Scott, J. A., Summerbell, R. C., & Malloch, D. (1999). Fungus-growing ants use antibiotic-producing bacteria to control garden parasites. *Nature, 398*, 701–704.

Cvetkovic, C., Raman, R., Chan, V., Williams, B. J., Tolish, M., Bajaj, P., . . . Bashir, R. (2014). Three-dimensionally printed biological machines powered by skeletal muscle. *Proc Natl Acad Sci USA, 111*(28), 10125–10130. doi:10.1073/pnas.1401577111

Dacke, M., Baird, E., Byrne, M., Scholtz, C. H., & Warrant, E. J. (2013). Dung beetles use the Milky Way for orientation. *Curr Biol, 23*(4), 298–300. doi:10.1016/j.cub.2012.12.034

Daeschler, E. B., Shubin, N. H., & Jenkins, F. A., Jr. (2006). A Devonian tetrapod-like fish and the evolution of the tetrapod body plan. *Nature, 440*(7085), 757–763. doi:nature04639 [pii] 10.1038/nature04639

Dai, X., Zhou, W., Gao, T., Liu, J., & Lieber, C. M. (2016). Three-dimensional mapping and regulation of action potential propagation in nanoelectronics-innervated tissues. *Nat Nanotechnol, 11*(9), 776–782. doi:10.1038/nnano.2016.96

Dale, L., Smith, J. C., & Slack, J. M. (1985). Mesoderm induction in Xenopus laevis: A quantitative study using a cell lineage label and tissue-specific antibodies. *Journal of Embryology and Experimental Morphology, 89*, 289.

Daley, M. A., Felix, G., & Biewener, A. A. (2007). Running stability is enhanced by a proximo-distal gradient in joint neuromechanical control. *J Exp Biol, 210*(Pt 3), 383–394. doi:10.1242/jeb.02668

Damen, W. G., Saridaki, T., & Averof, M. (2002). Diverse adaptations of an ancestral gill: A common evolutionary origin for wings, breathing organs, and spinnerets. *Curr Biol, 12*(19), 1711–1716. doi:S0960982202011260 [pii]

Dammann, O., & Leviton, A. (2014). Intermittent or sustained systemic inflammation and the preterm brain. *Pediatr Res, 75*(3), 376–380. doi:10.1038/pr.2013.238

Daniels, J. T. (2016). Visionary stem-cell therapies: Stem-cell engineering has allowed successful cornea transplantations in rabbits and the regeneration of transparent lens tissue in children, demonstrating the therapeutic potential of this approach. (Biomedicine) (Report). *Nature, 531*(7594), 309.

Dantu, K., Berman, S., Kate, B., & Nagpal, R. (2012). *A comparison of deterministic and stochastic approaches for allocating spatially dependent tasks in micro-aerial vehicle collectives.* Paper presented at the IEEE / RSJ International Conference on Intelligent Robots and Systems, Vilamoura, Portugal, October 2012, pp. 793–800.

Davidson, E. H., & Erwin, D. H. (2006). Gene regulatory networks and the evolution of animal body plans. *Science, 311*(5762), 796–800. doi:10.1126/science.1113832

Davis, M. F., Figueroa Velez, D. X., Guevarra, R. P., Yang, M. C., Habeeb, M., Carathedathu, M. C., & Gandhi, S. P. (2015). Inhibitory neuron transplantation into adult visual cortex creates a new critical period that rescues impaired vision. *Neuron, 86*(4), 1055–1066. doi:10.1016/j.neuron.2015.03.062

Dean, J. M., McClendon, E., Hansen, K., Azimi-Zonooz, A., Chen, K., Riddle, A., . . . Back, S. A. (2013). Prenatal cerebral ischemia disrupts MRI-defined cortical microstructure through disturbances in neuronal arborization. *Sci Transl Med, 5*(168), 1–11. doi:10.1126/scitranslmed.3004669

Deck, M., Lokmane, L., Chauvet, S., Mailhes, C., Keita, M., Niquille, M., . . . Garel, S. (2013). Pathfinding of corticothalamic axons relies on a rendezvous with thalamic projections. *Neuron, 77*(3), 472–484. doi:10.1016/j.neuron.2012.11.031

Deco, G., Jirsa, V., & McIntosh, A. (2011). Emerging concepts for the dynamical organization of resting-state activity in the brain. *Nat Rev Neurosci, 12*, 43–56.

Deco, G., Jirsa, V., & McIntosh, A. (2013). Resting brains never rest: Computational insights into potential cognitive architectures. *Trends Neurosci, 36*(5), 268–274. doi:10.1016/j.tins.2013.03.001

Deco, G., Jirsa, V., McIntosh, A., Sporns, O., & Kotter, R. (2009). Key role of coupling, delay, and noise in resting state brain fluctuations. *Proc Natl Acad Sci, 106*, 10302–10307.

Deco, G., & Kringelbach, M. L. (2014). Great expectations: Using whole-brain computational connectomics for understanding neuropsychiatric disorders. *Neuron, 84*(5), 892–905. doi:10.1016/j.neuron.2014.08.034

Deco, G., & Kringelbach, M. L. (2016). Metastability and coherence: Extending the communication through coherence hypothesis using a whole-brain computational perspective. *Trends Neurosci, 39*(3), 125–135. doi:10.1016/j.tins.2016.01.001

Dehmelt, L., & Bastiaens, P. I. (2010). Spatial organization of intracellular communication: Insights from imaging. *Nat Rev Mol Cell Biol, 11*(6), 440–452. doi:10.1038/nrm2903

Deisseroth, K. (2010). Controlling the brain with light. *Scientific American*, 49–55.

Deisseroth, K. (2014). Circuit dynamics of adaptive and maladaptive behaviour. *Nature, 505*(7483), 309–317. doi:10.1038/nature12982

Deisseroth, K. (2015). Optogenetics: 10 years of microbial opsins in neuroscience. *Nat Neurosci, 18*(9), 1213 1225.

Deisseroth, K., Etkin, A., & Malenka, R. C. (2015). Optogenetics and the circuit dynamics of psychiatric disease. *JAMA, 313*(20), 2019–2020. doi:10.1001/jama.2015.2544

de Jeu, M., & De Zeeuw, C. I. (2012). Video-oculography in mice. *J Vis Exp* (65), e3971. doi:10.3791/3971

Dekkers, M. P., & Barde, Y. A. (2013). Developmental biology: Programmed cell death in neuronal development. *Science, 340*(6128), 39–41. doi:10.1126/science.1236152

Del Negro, C. A., Wilson, C. G., Butera, R. J., Rigatto, H., & Smith, J. C. (2002). Periodicity, mixed-mode oscillations, and quasiperiodicity in a rhythm-generating neural network. *Biophys J, 82*, 206–214.

Deluca, C., Golzar, A., Santandrea, E., Lo Gerfo, E., Eštočinová, J., Moretto, G., . . . Chelazzi, L. (2014). The cerebellum and visual perceptual learning: Evidence from a motion extrapolation task. *Cortex, 58*, 52–71. doi:10.1016/j.cortex.2014.04.017

Demaine, E. D., & O'Rourke, J. (2007). A survey of folding and unfolding in computational geometry. In J. E. Goodman, J. Pach, & E. Welzl (Eds.), *Combinatorial and Computational Geometry* (pp. 167–211). Cambridge, UK: Cambridge University Press.

De Robertis, E. M. (2006). Spemann's organizer and self-regulation in amphibian embryos. *Nat Rev Mol Cell Biol, 7*, 296–302.

De Robertis, E. M. (2009). Spemann's organizer and the self-regulation of embryonic fields. *Mech Dev, 126*, 925–941.

Deschenes, M., Moore, J., & Kleinfeld, D. (2012). Sniffing and whisking in rodents. *Curr Opin Neurobiol, 22*, 243–250.

Devor, A., Bandettini, P. A., Boas, D. A., Bower, J. M., Buxton, R. B., Cohen, L. B., . . . Yodh, A. G. (2013). The challenge of connecting the dots in the B.R.A.I.N. *Neuron, 80*(2), 270–274. doi:10.1016/j.neuron.2013.09.008

de Vries, J. I. P., Visser, G. H. A., & Prechtl, H. F. R. (1982). The emergence of fetal behaviour. I. Qualitative aspects. *Early Human Development, 7*(4), 301–322. doi:10.1016/0378-3782(82)90033-0

De Zeeuw, C. I., Hoebeek, F. E., Bosman, L. W., Schonewille, M., Witter, L., & Koekkoek, S. K. (2011). Spatiotemporal firing patterns in the cerebellum. *Nat Rev Neurosci, 12*(6), 327–344. doi:10.1038/nrn3011

De Zeeuw, C. I., & Ten Brinke, M. M. (2015). Motor learning and the cerebellum. *Cold Spring Harb Perspect Biol, 7*(9), a021683. doi:10.1101/cshperspect.a021683

Dhindsa, R. S., & Goldstein, D. B. (2016). Schizophrenia: From genetics to physiology at last. *Nature, 530*(7589), 162–163. doi:10.1038/nature16874

Diamond, M. E., von Heimendahl, M., Knutsen, P. M., Kleinfeld, D., & Ahissar, E. (2008). "Where" and "what" in the whisker sensorimotor system. *Nat Rev Neurosci, 9*(8), 601–612. doi:10.1038/nrn2411

Diaz Quiroz, J. F., & Echeverri, K. (2013). Spinal cord regeneration: Where fish, frogs and salamanders lead the way, can we follow? *Biochem J, 451*(3), 353–364. doi:10.1042/BJ20121807

Dickinson, M., & Moss, C. F. (2012). Neuroethology. *Curr Opin Neurobiol, 22*, 177–179.

Dickinson, M. H. (1999). Wing rotation and the aerodynamic basis of insect flight. *Science, 284*(5422), 1954–1960. doi:10.1126/science.284.5422.1954

Dickinson, M. H. (2014). Death Valley, Drosophila, and the Devonian toolkit. *Annu Rev Entomol, 59*, 51–72. doi:10.1146/annurev-ento-011613-162041

Dickinson, M. H., Farley, C. T., Full, R. J., Koehl, M. A. R., Kram, R., & Lehman, S. (2000). How animals move: An integrative view. *Science, 288*(5463), 100–106. doi:10.1126/science.288.5463.100

Di Fiore, J. M., Martin, R. J., & Gauda, E. B. (2013). Apnea of prematurity—perfect storm. *Respir Physiol Neurobiol, 189*(2), 213–222. doi:10.1016/j.resp.2013.05.026

Dinan, T. G., & Cryan, J. F. (2017). Gut-brain axis in 2016: Brain-gut-microbiota axis—mood, metabolism and behaviour. *Nat Rev Gastroenterol Hepatol, 14*(2), 69–70. doi:10.1038/nrgastro.2016.200

Ding, Y., Sharpe, S. S., Wiesenfeld, K., & Goldman, D. I. (2013). Emergence of the advancing neuromechanical phase in a resistive force dominated medium. *Proc Natl Acad Sci USA, 110*(25), 10123–10128. doi:10.1073/pnas.1302844110

Dinstein, I., Pierce, K., Eyler, L., Solso, S., Malach, R., Behrmann, M., & Courchesne, E. (2011). Disrupted neural synchronization in toddlers with autism. *Neuron, 70*(6), 1218–1225. doi:10.1016/j.neuron.2011.04.018

Do, K. Q., Cuenod, M., & Hensch, T. K. (2015). Targeting oxidative stress and aberrant critical period plasticity in the developmental trajectory to schizophrenia. *Schizophr Bull, 41*(4), 835–846. doi:10.1093/schbul/sbv065

Dobkin, B. H., & Duncan, P. W. (2012). Should body weight-supported treadmill training and robotic-assistive steppers for locomotor training trot back to the starting gate? *Neurorehabil Neural Repair, 26*(4), 308–317. doi:10.1177/1545968312439687

Dominici, N., Ivanenko, Y. P., Cappellini, G., d'Avella, A., Mondi, V., Cicchese, M., . . . Lacquaniti, F. (2011). Locomotor primitives in newborn babies and their development. *Science, 334*(6058), 997–999. doi:10.1126/science.1210617

Dominici, N., Keller, U., Vallery, H., Friedli, L., van den Brand, R., Starkey, M. L., . . . Courtine, G. (2012). Versatile robotic interface to evaluate, enable and train locomotion and balance after neuromotor disorders. *Nat Med, 18*(7), 1142–1147. doi:10.1038/nm.2845

Dong, X., Shen, K., & Bulow, H. E. (2015). Intrinsic and extrinsic mechanisms of dendritic morphogenesis. *Annu Rev Physiol, 77*, 271–300. doi:10.1146/annurev-physiol-021014-071746

Donnelly, J. L., Clark, C. M., Leifer, A. M., Pirri, J. K., Haburcak, M., Francis, M. M., . . . Alkema, M. J. (2013). Monoaminergic orchestration of motor programs in a complex C. elegans behavior. *PLoS Biol, 11*(4), e1001529. doi:10.1371/journal.pbio.1001529

Dorgan, K. M. (2015). The biomechanics of burrowing and boring. *J Exp Biol, 218*(Pt 2), 176–183. doi:10.1242/jeb.086983

Dorgan, K. M., Jumars, P. A., Johnson, B., Boudreau, B. P., & Landis, E. (2005). Burrow extension by crack propagation. *Nature, 433*, 475.

Douglas, R. J., & Martin, K. A. (2004). Neuronal circuits of the neocortex. *Annu Rev Neurosci, 27*, 419–451. doi:10.1146/annurev.neuro.27.070203.144152

Douglas, R. J., & Martin, K. A. (2012). Behavioral architecture of the cortical sheet. *Curr Biol, 22*(24), R1033–R1038. doi:10.1016/j.cub.2012.11.017

Douglas, S. M., Bachelet, I., & Church, G. M. (2012). A logic-gated nanorobot for targeted transport of molecular payloads. *Science, 335*(6070), 831–834. doi:10.1126/science.1214081

Douglas, S. M., Dietz, H., Liedl, T., Hogberg, B., Graf, F., & Shih, W. M. (2009). Self-assembly of DNA into nanoscale three-dimensional shapes. *Nature, 459*(7245), 414–418. doi:10.1038/nature08016

Downey, J. E., Weiss, J. M., Muelling, K., Venkatraman, A., Valois, J. S., Hebert, M., . . . Collinger, J. L. (2016). Blending of brain-machine interface and vision-guided autonomous robotics improves neuroprosthetic arm performance during grasping. *J Neuroeng Rehabil, 13*, 28. doi:10.1186/s12984-016-0134-9

Downs, J., Daeschler, E., Jenkins, F., & Shubin, N. (2008). The cranial endoskeleton of *Tiktaalik rosae. Nature, 455*, 925–929.

Dreier, T., Wolff, P. H., Cross, E. E., & Cochran, W. D. (1979). Patterns of breath intervals during non-nutritive sucking in full-term and "at risk" preterm infants with normal neurological examinations. *Early Hum Develop, 3*, 187–199.

Drew, P. J., Duyn, J. H., Golanov, E., & Kleinfeld, D. (2008). Finding coherence in spontaneous oscillations. *Nature Neuroscience, 11*(9), 991. doi:10.1038/nn0908-991

Driggers, R. W., Ho, C. Y., Korhonen, E. M., Kuivanen, S., Jaaskelainen, A. J., Smura, T., . . . Vapalahti, O. (2016). Zika virus infection with prolonged maternal viremia and fetal brain abnormalities. *N Engl J Med, 374*(22), 2142–2151. doi:10.1056/NEJMoa1601824

Dubilier, N., Bergin, C., & Lott, C. (2008). Symbiotic diversity in marine animals: The art of harnessing chemosynthesis. *Nat Rev Microbiol, 6*(10), 725–740. doi:10.1038/nrmicro1992

Dubois, J., Hertz-Pannier, L., Dehaene-Lambertz, G., Cointepas, Y., & Le Bihan, D. (2006). Assessment of the early organization and maturation of infants' cerebral white matter fiber bundles: A feasibility study using quantitative diffusion tensor imaging and tractography. *Neuroimage, 30*(4), 1121–1132. doi:10.1016/j.neuroimage.2005.11.022

Dudley, R., & Yanoviak, S. P. (2011). Animal aloft: The origins of aerial behavior and flight. *Integr Comp Biol, 51*(6), 926–936. doi:10.1093/icb/icr002

Dugas, R. (1958). *Mechanics in the seventeenth century, from the scholastic antecedents to classical thought.* New York: Central Book Co.

Dunlop, J. W. C., & Fratzl, P. (2010). Biological composites. *Ann Rev Mater Res, 40*(1), 1–24. doi:10.1146/annurev-matsci-070909-104421

Dunlop, J. W. C., Weinkamer, R., & Fratzl, P. (2011). Artful interfaces within biological materials. *Materials Today, 14*(3), 70–78. doi:10.1016/s1369-7021(11)70056-6

Dupre, C., & Yuste, R. (2017). Non-overlapping neural networks in *Hydra vulgaris. Curr Biol, 27*(8), 1085–1097. doi:10.1016/j.cub.2017.02.049

Duque, A., Krsnik, Z., Kostović, I., & Rakic, P. (2016). Secondary expansion of the transient subplate zone in the developing cerebrum of human and nonhuman primates. *Proc Natl Acad Sci USA, 113*(35), 9892–9897. doi:10.1073/pnas.1610078113

Dyke, G., de Kat, R., Palmer, C., van der Kindere, J., Naish, D., & Ganapathisubramani, B. (2013). Aerodynamic performance of the feathered dinosaur Microraptor and the evolution of feathered flight. *Nat Commun, 4,* 2489. doi:10.1038/ncomms3489

Eberhard, W. G., & Wcislo, W. T. (2011). Grade changes in brain-body allometry: Morphological and behavioural correlates of brain size in miniature spiders, insects and other vertebrates. *Adv Insect Physiol, 40,* 155–213.

Ebert, D. H., & Greenberg, M. E. (2013). Activity-dependent neuronal signalling and autism spectrum disorder. *Nature, 493*(7432), 327–337. doi:10.1038/nature11860

Edelman, G. M., & Gally, J. A. (2001). Degeneracy and complexity in biological systems. *Proc Natl Acad Sci USA, 98*(24), 13763–13768. doi:10.1073/pnas.231499798

Edgar, J., Khan, S., Blaskey, L., Chow, V., Rey, M., Gaetz, W., . . . Roberts, T. (2015). Neuromagnetic oscillations predict evoked-response latency delays and core language deficits in autism spectrum disorders. *J Autism Dev Disord, 45*(2), 395–405. doi:10.1007/s10803-013-1904-x

Ehrlich, P. J., & Lanyon, L. E. (2002). Mechanical strain and bone cell function: A review. *Osteoporosis International, 13,* 688–700.

Ehrsson, H. H. (2012). The concept of body ownership and its relation to multisensory integration. In B. E. Stein (Ed.), *New handbook of multisensory processing* (pp. 775–792). Cambridge, MA: MIT Press.

Ehrsson, H. H., Rosén, B., Stockselius, A., Ragnö, C., Köhler, P., & Lundborg, G. (2008). Upper limb amputees can be induced to experience a rubber hand as their own. *Brain, 131*(12), 3443–3452. doi:10.1093/brain/awn297

Eiben, A. E., & Smith, J. (2015). From evolutionary computation to the evolution of things. *Nature, 521*(7553), 476–482. doi:10.1038/nature14544

Einspieler, C., & Prechtl, H. F. (2005). Prechtl's assessment of general movements: A diagnostic tool for the functional assessment of the young nervous system. *Ment Retard Dev Disabil Res Rev, 11*(1), 61–67. doi:10.1002/mrdd.20051

Eisner, T., & Aneshansley, D. (2000). Defense by foot adhesion in a beetle (Hemisphaerota cynea). *Proc Natl Acad Sci, 97,* 6568–6573.

Eklof-Ljunggren, E., Haupt, S., Ausborn, J., Dehnisch, I., Uhlen, P., Higashijima, S., & El Manira, A. (2012). Origin of excitation underlying locomotion in the spinal circuit of zebrafish. *Proc Natl Acad Sci USA, 109*(14), 5511–5516. doi:10.1073/pnas.1115377109

El Kaliouby, R., & Robinson, P. (2005). Generalization of a vision-based computational model of mind-reading. *Lecture Notes Comput Sci, 3784,* 582–589.

Emery, B. (2010). Regulation of oligodendrocyte differentiation and myelination. *Science, 330*(6005), 779–782. doi:10.1126/science.1190927

Engel, A. K., & Fries, P. (2010). Beta-band oscillations—signalling the status quo? *Curr Opin Neurobiol, 20*(2), 156–165. doi:10.1016/j.conb.2010.02.015

Engel, A. K., König, P., Kreiter, A. K., & Singer, W. (1991). Interhemispheric synchronization of oscillatory neuronal responses in cat visual cortex. *Science, 252*(5009), 1177–1179.

Engel, M. S. (2015). Insect evolution. *Curr Biol, 25*(19), R868–R872. doi:10.1016/j.cub.2015.07.059

Engel, P. (2009). *10-fold origami: Fabulous paperfolds you can make in just 10 steps!* North Clarendon, VT: Tuttle.

Engert, F. (2014). The big data problem: Turning maps into knowledge. *Neuron, 83*(6), 1246–1248. doi:10.1016/j.neuron.2014.09.008

Erny, D., Hrabe de Angelis, A. L., Jaitin, D., Wieghofer, P., Staszewski, O., David, E., . . . Prinz, M. (2015). Host microbiota constantly control maturation and function of microglia in the CNS. *Nat Neurosci, 18*(7), 965–977. doi:10.1038/nn.4030

Erwin, D. H., & Davidson, E. (2009). The evolution of hierarchical gene regulatory networks. *Nat Genetics, 10,* 141–148.

Espinosa, J. S., & Stryker, M. P. (2012). Development and plasticity of the primary visual cortex. *Neuron, 75*(2), 230–249. doi:10.1016/j.neuron.2012.06.009

Esposito, M. S., Capelli, P., & Arber, S. (2014). Brainstem nucleus MdV mediates skilled forelimb motor tasks. *Nature, 508*(7496), 351–356. doi:10.1038/nature13023

Estes, M. L., & McAllister, A. K. (2016). Maternal immune activation: Implications for neuropsychiatrc disorders. *Science, 353*(6301), 772–777.

Fajen, B. R., Riley, M., & Turvey, M. T. (2008). Information, affordances, and the control of action in sport. *Int J Sports Psychol, 40,* 79–107.

Fajen, B. R., & Warren, W. H. (2003). Behavioral dynamics of steering, obstacle avoidance, and route selection. *J Exp Psychol: Hum Percep Perform, 29*(2), 343–362. doi:10.1037/0096-1523.29.2.343

Fan, J. M., Nuyujukian, P., Kao, J. C., Chestek, C. A., Ryu, S. I., & Shenoy, K. V. (2014). Intention estimation in brain–machine interfaces. *J Neural Eng, 11*(1), 016004. doi:10.1088/1741-2560/11/1/016004

Farah, M. J. (2015). An ethics toolbox for neurotechnology. *Neuron, 86*(1), 34–37. doi:10.1016/j.neuron.2015.03.038

Fasola, J., & Mataric, M. (2013). A socially assistive robot exercise coach for the elderly. *Journal of Human-Robot Interaction, 2*(2), 32. doi:10.5898/JHRI.2.2.Fasola

Fasoli, S. E., Fragala-Pinkham, M., Hughes, R., Hogan, N., Krebs, H. I., & Stein, J. (2008). Upper limb robotic therapy for children with hemiplegia. *Am J Phys Med Rehabil, 87*(11), 929–936. doi:10.1097/PHM.0b013e31818a6aa4

Fee, M. S. (2014). The role of efference copy in striatal learning. *Curr Opin Neurobiol, 25,* 194–200. doi:10.1016/j.conb.2014.01.012

Feinberg, A. W. (2015). Biological soft robotics. *Annu Rev Biomed Eng, 17,* 243–265. doi:10.1146/annurev-bioeng-071114-040632

Feinberg, A. W., Feigel, A., Shevkoplyas, S., Sheehy, S., Whitesides, G. M., & Parker, K. K. (2007). Muscular thin films for building actuators and powering devices. *Science, 317,* 1366–1370.

Feldman, D. E. (2009). Synaptic mechanisms for plasticity in neocortex. *Annu Rev Neurosci, 32,* 33–55.

Feldman, J. L., Del Negro, C. A., & Gray, P. A. (2013). Understanding the rhythm of breathing: So near, yet so far. *Annu Rev Physiol, 75,* 423–452. doi:10.1146/annurev-physiol-040510-130049

Felton, S., Tolley, M., Demaine, E., Rus, D., & Wood, R. (2014). Applied origami: A method for building self-folding machines. *Science, 345*(6197), 644–646. doi:10.1126/science.1252610

Felton, S. M., Tolley, M. T., Shin, B., Onal, C. D., Demaine, E. D., Rus, D., & Wood, R. J. (2013). Self-folding with shape memory composites. *Soft Matter, 9*(32), 7688. doi:10.1039/c3sm51003d

Fenno, L., Yizhar, O., & Deisseroth, K. (2011). The development and application of optogenetics. *Annu Rev Neurosci, 34,* 389–412. doi:10.1146/annurev-neuro-061010-113817

Ferenczi, E. A., Zalocusky, K. A., Liston, C., Grosenick, L., Warden, M. R., Amatya, D., . . . Deisseroth, K. (2016). Prefrontal cortical regulation of brainwide circuit dynamics and reward-related behavior. *Science, 351*(6268), aac9698. doi:10.1126/science.aac9698

Ferrell, J. E., Jr. (2012). Bistability, bifurcations, and Waddington's epigenetic landscape. *Curr Biol, 22*(11), R458–R466. doi:10.1016/j.cub.2012.03.045

Fets, L., Kay, R., & Velazquez, F. (2010). Dictyostelium. *Curr Biol, 20,* R1008–R1010.
Fetters, L., Chen, Y. P., Jonsdottir, J., & Tronick, E. Z. (2004). Kicking coordination captures differences between full-term and premature infants with white matter disorder. *Human Movement Science, 22*(6), 729–748.
Feynman, R. P. (2011). There's plenty of room at the bottom. *Resonance, 16,* 890–905.
Fidelin, K., & Wyart, C. (2014). Inhibition and motor control in the developing zebrafish spinal cord. *Curr Opin Neurobiol, 26C,* 103–109. doi:10.1016/j.conb.2013.12.016
Fields, R. D., Araque, A., Johansen-Berg, H., Lim, S. S., Lynch, G., Nave, K. A., . . . Wake, H. (2013). Glial biology in learning and cognition. *Neuroscientist.* doi:10.1177/1073858413504465
Finio, B. M., & Wood, R. J. (2010). Distributed power and control actuation in the thoracic mechanics of a robotic insect. *Bioinspir Biomim, 5*(4), 045006. doi:10.1088/1748-3182/5/4/045006
Fisher, M., Loewy, R., Hardy, K., Schlosser, D., & Vinogradov, S. (2013). Cognitive interventions targeting brain plasticity in the prodromal and early phases of schizophrenia. *Annu Rev Clin Psychol, 9,* 435–463. doi:10.1146/annurev-clinpsy-032511-143134
Flash, T., & Hochner, B. (2005). Motor primitives in vertebrates and invertebrates. *Curr Opin Neurobiol, 15*(6), 660–666.
Fleiss, B., & Gressens, P. (2011). Tertiary mechanisms of brain damage: A new hope for treatment of cerebral palsy? *Lancet Neurol, 11*(6), 556–566. doi:10.1016/S1474-4422(12)70058-3
Flemming, H. C., Wingender, J., Szewzyk, U., Steinberg, P., Rice, S. A., & Kjelleberg, S. (2016). Biofilms: An emergent form of bacterial life. *Nat Rev Microbiol, 14*(9), 563–575. doi:10.1038/nrmicro.2016.94
Flesher, S. N., Collinger, J. L., Foldes, S. T., Weiss, J. M., Downey, J. E., Tyler-Kabara, E. C., . . . Gaunt, R. (2016). Intracortical microstimulation of human somatosensort cortex. *Sci Transl Med, 8,* 1–10.
Floreano, D., Ijspeert, A. J., & Schaal, S. (2014). Robotics and neuroscience. *Curr Biol, 24*(18), R910-R920. doi:10.1016/j.cub.2014.07.058
Floreano, D., & Keller, L. (2010). Evolution of adaptive behaviour in robots by means of Darwinian selection. *PLoS Biol, 8*(1), e1000292. doi:10.1371/journal.pbio.1000292
Floreano, D., Pericet-Camara, R., Viollet, S., Ruffier, F., Bruckner, A., Leitel, R., . . . Franceschini, N. (2013). Miniature curved artificial compound eyes. *Proc Natl Acad Sci USA, 110*(23), 9267–9272. doi:10.1073/pnas.1219068110
Floreano, D., & Wood, R. J. (2015). Science, technology and the future of small autonomous drones. *Nature, 521*(7553), 460–466. doi:10.1038/nature14542
Fluet, G., Merians, A. S., Qiu, Q., Davidow, A., & Adamovich, S. (2014). Comparing integrated training of the hand and arm with isolated training of the same effectors in persons with stroke using haptically rendered virtual environments, a randomized clinical trial. *J Neuroeng Rehabil, 11,* 1–11.
Fonseca, S. T., Holt, K. G., Fetters, L., & Saltzman, E. (2004). Dynamic resources used in ambulation by children with spastic hemiplegic cerebral palsy: Relationship to kinematics, energetics, and asymmetries. *Phys Ther, 84*(4), 344–354; discussion 355–358.
Fonseca, S. T., Holt, K. G., Saltzman, E., & Fetters, L. (2001). A dynamical model of locomotion in spastic hemiplegic cerebral palsy: Influence of walking speed. *Clin Biomech (Bristol, Avon), 16*(9), 793–805. doi:S0268003301000675
Forger, D. B., & Paydarfar, D. (2004). Starting, stopping, and resetting biological oscillators: In search of optimum perturbations. *J Theoret Biol, 230*(4), 521–532.
Fornito, A., & Bullmore, E. T. (2015). Connectomics: A new paradigm for understanding brain disease. *Eur Neuropsychopharmacol, 25*(5), 733–748. doi:10.1016/j.euroneuro.2014.02.011
Fornito, A., Zalesky, A., & Breakspear, M. (2015). The connectomics of brain disorders. *Nat Rev Neurosci, 16*(3), 159–172. doi:10.1038/nrn3901
Forterre, Y., Skotheim, J., Dumais, J., & Mahadevan, L. (2005). How the Venus flytrap snaps. *Nature, 433*(7024), 417–421. doi:10.1038/nature03072

Foster, P. C., Mlot, N. J., Lin, A., & Hu, D. L. (2014). Fire ants actively control spacing and orientation within self-assemblages. *J Exp Biol, 217*(Pt 12), 2089–2100. doi:10.1242/jeb.093021

Fox, M. D., & Raichle, M. E. (2007). Spontaneous fluctuations in brain activity observed with functional magnetic resonance imaging. *Nat Rev Neurosci, 8*(9), 700–711.

Franceschini, N. (2014). Small brains, smart machines: From fly vision to robot vision and back again. *Proc IEEE, 102*(5), 751–781. doi:10.1109/JPROC.2014.2312916

Franchak, J. M., Celano, E. C., & Adolph, K. E. (2012). Perception of passage through openings depends on the size of the body in motion. *Exp Brain Res, 223*(2), 301–310. doi:10.1007/s00221-012-3261-y

Franco, S. J., & Muller, U. (2013). Shaping our minds: Stem and progenitor cell diversity in the mammalian neocortex. *Neuron, 77*(1), 19–34. doi:10.1016/j.neuron.2012.12.022

Frankel, F., and Whitesides, G. M. (2007). *On the surface of things: Images of the extraordinary in science.* Cambridge, MA: Harvard University Press.

Fransson, P. (2005). Spontaneous low-frequency BOLD signal fluctuations: An fMRI investigation of the resting-state default mode of brain function hypothesis. *Human Brain Mapping, 26*(1), 15–29. doi:10.1002/hbm.20113

Fransson, P., Aden, U., Blennow, M., & Lagercrantz, H. (2011). The functional architecture of the infant brain as revealed by resting-state fMRI. *Cereb Cortex, 21*(1), 145–154. doi:10.1093/cercor/bhq071

Franze, K. (2013). The mechanical control of nervous system development. *Development, 140*(15), 3069–3077. doi:10.1242/dev.079145

Fratzl, P., & Barth, F. G. (2009). Biomaterial systems for mechanosensing and actuation. *Nature, 462*(7272), 442–448. doi:10.1038/nature08603

Friedli, L., Rosenzweig, E. S., Barraud, Q., Schubert, M., Dominici, N., Awai, L., . . . Courtine, G. (2015). Pronounced species divergence in corticospinal tract reorganization and functional recovery after lateralized spinal cord injury favors primates. *Sci Transl Med, 7*(302), 1–12.

Fries, P. (2005). A mechanism for cognitive dynamics: Neuronal communication through neuronal coherence. *Trends Cogn Sci, 9*(10), 474–480. doi:10.1016/j.tics.2005.08.011

Fries, P. (2009). Neuronal gamma-band synchronization as a fundamental process in cortical computation. *Annu Rev Neurosci, 32,* 209–224. doi:10.1146/annurev.neuro.051508.135603

Fries, P. (2015). Rhythms for cognition: Communication through coherence. *Neuron, 88*(1), 220–235. doi:10.1016/j.neuron.2015.09.034

Fries, P., Nikolić, D., & Singer, W. (2007). The gamma cycle. *Trends in Neurosciences, 30*(7), 309–316. doi:10.1016/j.tins.2007.05.005

Friesen, L. M., Shannon, R. V., Baskent, D., & Wang, X. (2001). Speech recognition in noise as a function of the number of spectral channels: Comparison of acoustic hearing and cochlear implants. *J Acoust Soc Amer, 110*(2), 1150. doi:10.1121/1.1381538

Frisch, K. V., Wenner, A. M., & Johnson, D. L. (1967). Honeybees: Do they use direction and distance information provided by their dancers? *Science, 158*(3804), 1072–1077.

Frith, U., & Happé, F. (2005). Autism spectrum disorder. *Curr Biol, 15*(19), R786–R790. doi:10.1016/j.cub.2005.09.033

Frohnhofer, H. G., & Nusslein-Volhard, C. (1986). Organization of anterior pattern in the Drosophila embryo by the maternal gene bicoid. *Nature*(6093), 120–125.

Frye, M. A., Tarsitano, M., & Dickinson, M. H. (2003). Odor localization requires visual feedback during free flight in Drosophila melanogaster. *J Exp Biol, 206*(Pt 5), 843.

Fu, T. M., Hong, G., Zhou, T., Schuhmann, T. G., Viveros, R. D., & Lieber, C. M. (2016). Stable long-term chronic brain mapping at the single-neuron level. *Nat Methods, 13*(10), 875–882. doi:10.1038/nmeth.3969

Fuchs, A., Goldner, B., Nolte, I., & Schilling, N. (2014). Ground reaction force adaptations to tripedal locomotion in dogs. *Veterinary Journal, 201*(3), 307–315. doi:10.1016/j.tvjl.2014.05.012

Fujioka, M., Okano, H., & Edge, A. S. (2015). Manipulating cell fate in the cochlea: A feasible therapy for hearing loss. *Trends Neurosci, 38*(3), 139–144. doi:10.1016/j.tins.2014.12.004

Full, R. J., & Koditschek, D. E. (1999). Templates and anchors: Neuromechanical hypotheses of legged locomotion on land. *J Exper Biol, 202*(Pt 23), 3325–3332.
Fuller, S. B., Karpelson, M., Censi, A., Ma, K. Y., & Wood, R. J. (2014). Controlling free flight of a robotic fly using an onboard vision sensor inspired by insect ocelli. *J R Soc Interface, 11*(97), 20140281. doi:10.1098/rsif.2014.0281
Fung, T. C., Olson, C. A., & Hsiao, E. Y. (2017). Interactions between the microbiota, immune and nervous systems in health and disease. *Nat Neurosci, 20*(2), 145–155. doi:10.1038/nn.4476
Gaige, T. A., Benner, T., Wang, R., Wedeen, V. J., & Gilbert, R. J. (2007). Three dimensional myoarchitecture of the human tongue determined in vivo by diffusion tensor imaging with tractography. *Journal of Magnetic Resonance Imaging, 26*(3), 654–661. doi:10.1002/jmri.21022
Galantucci, B., Fowler, C., & Turvey, M. (2006). The motor theory of speech perception reviewed (vol 13, pg 361, 2006). *Psychonomic Bulletin & Review, 13*(4), 361–377.
Gallo, V., & Deneen, B. (2014). Glial development: The crossroads of regeneration and repair in the CNS. *Neuron, 83*(2), 283–308. doi:10.1016/j.neuron.2014.06.010
Gao, P., & Ganguli, S. (2015). On simplicity and complexity in the brave new world of large-scale neuroscience. *Curr Opin Neurobiol, 32*, 148–155. doi:10.1016/j.conb.2015.04.003
Gao, P., Sultan, K. T., Zhang, X.-J., & Shi, S.-H. (2013). Lineage-dependent circuit assembly in the neocortex. *Development* (Cambridge, England), *140*(13), 2645. doi:10.1242/dev.087668
Gao, W., Emaminejad, S., Nyein, H. Y., Challa, S., Chen, K., Peck, A., . . . Javey, A. (2016). Fully integrated wearable sensor arrays for multiplexed in situ perspiration analysis. *Nature, 529*(7587), 509–514. doi:10.1038/nature16521
Gao, Z., van Beugen, B. J., & De Zeeuw, C. I. (2012). Distributed synergistic plasticity and cerebellar learning. *Nat Rev Neurosci, 13*(9), 619–635. doi:10.1038/nrn3312
Garate, V. R., Parri, A., Yan, T., Munih, M., Lova, R. M., Vitiello, N., & Ronsse, R. (2016). Walking assistance using artificial primitives. *IEEE Robotics & Automation Magazine, 23*(1), 83–95. doi:10.1109/MRA.2015.2510778
Gardner, D. L., Mark, L. S., Ward, J. A., & Edkins, H. (2001). How do task characteristics affect the transitions between seated and standing reaches? *Ecol Psychol, 13*, 245–274.
Garel, S., & Lopez-Bendito, G. (2014). Inputs from the thalamocortical system on axon pathfinding mechanisms. *Curr Opin Neurobiol, 27*, 143–150. doi:10.1016/j.conb.2014.03.013
Gart, S., Socha, J. J., Vlachos, P., & Jung, S. (2015). Dogs lap using acceleration-driven open pumping. *Proc Natl Acad Sci, 112*(52), 15798. doi:10.1073/pnas.1514842112
Gates, B. D., Xu, Q., Love, J. C., Wolfe, D. B., & Whitesides, G. M. (2004). Unconventional nanofabrication. *Ann Rev Mater Res, 34*(1), 339–372. doi:10.1146/annurev.matsci.34.052803.091100
Gauda, E. B., & Martin, R. J. (2012). Control of breathing. In C. A. Gleason & S. U. Devaskar (Eds.), *Avery's diseases of the newborn* (Chapter 43, pp. 584–597). New York: Elsevier.
Gavelis, G. S., Hayakawa, S., White, R. A., III, Gojobori, T., Suttle, C. A., Keeling, P. J., & Leander, B. S. (2015). Eye-like ocelloids are built from different endosymbiotically acquired components. *Nature, 523*(7559), 204–207. doi:10.1038/nature14593
Gee, H. (2013). Tubular worms from the Burgess Shale. *Nature, 495*(7442), 458–459.
George, P. M., & Steinberg, G. K. (2015). Novel stroke therapeutics: Unraveling stroke pathophysiology and its impact on clinical treatments. *Neuron, 87*(2), 297–309. doi:10.1016/j.neuron.2015.05.041
Gérard, K., & Mathieu, F. (2012). The contribution of bone to whole-organism physiology. *Nature, 481*(7381), 314. doi:10.1038/nature10763
Gerits, A., Farivar, R., Rosen, B. R., Wald, L. L., Boyden, E. S., & Vanduffel, W. (2012). Optogenetically induced behavioral and functional network changes in primates. *Curr Biol, 22*(18), 1722–1726. doi:10.1016/j.cub.2012.07.023
Geschwind, D. H., & Levitt, P. (2007). Autism spectrum disorders: Developmental disconnection syndromes. *Curr Opin Neurobiol, 17*(1), 103–111. doi:10.1016/j.conb.2007.01.009

Geschwind, D. H., & Rakic, P. (2013). Cortical evolution: Judge the brain by its cover. *Neuron, 80*(3), 633–647. doi:10.1016/j.neuron.2013.10.045

Gesell, A. (1946). *The child from five to ten* (2nd ed.). New York: Harper and Brothers.

Gibson, J. (1966). *The senses considered as perceptual systems.* Boston: Houghton Mifflin.

Gibson, J. J. (1986). *The ecological approach to visual perception.* Hillsdale, NJ: Lawrence Erlbaum. Originally published in 1979.

Gibson, M. C., Patel, A. B., Nagpal, R., & Perrimon, N. (2006). The emergence of geometric order in proliferating metazoan epithelia. *Nature, 442*(7106), 1038–1041. doi:10.1038/nature 05014

Gilbert, R. J., Napadow, V. J., Gaige, T. A., & Wedeen, V. J. (2007). Anatomical basis of lingual hydrostatic deformation. *J Exper Biol, 210*(Pt 23), 4069–4082.

Gilbert, S. F. (2001). Ecological developmental biology: Developmental biology meets the real world. *Dev Biol, 233*(1), 1–12. doi:10.1006/dbio.2001.0210 S0012-1606(01)90210-6 [pii]

Gillespie, P. G., & Muller, U. (2009). Mechanotransduction by hair cells: Models, molecules, and mechanisms. *Cell, 139*(1), 33–44. doi:10.1016/j.cell.2009.09.010

Gilmour, D., Rembold, M., & Leptin, M. (2017). From morphogen to morphogenesis and back. *Nature, 541*(7637), 311–320. doi:10.1038/nature21348

Giszter, S. F. (2015). Motor primitives—new data and future questions. *Curr Opin Neurobiol, 33,* 156–165. doi:10.1016/j.conb.2015.04.004

Gjorgjieva, J., Drion, G., & Marder, E. (2016). Computational implications of biophysical diversity and multiple timescales in neurons and synapses for circuit performance. *Curr Opin Neurobiol, 37,* 44–52. doi:10.1016/j.conb.2015.12.008

Gladman, A. S., Matsumoto, E. A., Nuzzo, R. G., Mahadevan, L., & Lewis, J. A. (2016). Biomimetic 4D printing. *Nat Mater, 15,* 413–418. doi:10.1038/nmat4544

Glass, L. (1988). *From clocks to chaos: The rhythms of life.* Princeton, NJ: Princeton University Press.

Glasser, M. F., Coalson, T. S., Robinson, E. C., Hacker, C. D., Harwell, J., Yacoub, E., . . . Van Essen, D. C. (2016). A multi-modal parcellation of human cerebral cortex. *Nature, 536*(7615), 171–178. doi:10.1038/nature18933

Goaillard, J. M., Taylor, A. L., Schulz, D. J., & Marder, E. (2009). Functional consequences of animal-to-animal variation in circuit parameters. *Nat Neurosci, 12*(11), 1424–1430. doi:10.1038/nn.2404

Goehring, L., Mahadevan, L., & Morris, S. W. (2009). Nonequilibrium scale selection mechanism for columnar jointing. *Proc Natl Acad Sci, 106,* 387–392.

Goehring, N. W., & Grill, S. W. (2013). Cell polarity: Mechanochemical patterning. *Trends Cell Biol, 23*(2), 72–80. doi:10.1016/j.tcb.2012.10.009

Gohir, W., Ratcliffe, E. M., & Sloboda, D. M. (2015). Of the bugs that shape us: Maternal obesity, the gut microbiome, and long-term disease risk. *Pediatr Res, 77*(1-2), 196–204. doi:10.1038/pr.2014.169

Goldberger, A. L., Amaral, L. A., Hausdorff, J. M., Ivanov, P., Peng, C. K., & Stanley, H. E. (2002). Fractal dynamics in physiology: Alterations with disease and aging. *Proc Natl Acad Sci USA, 99 Suppl 1,* 2466–2472.

Goldberger, A. L., & West, B. J. (1987). Applications of nonlinear dynamics to clinical cardiology. *Ann N Y Acad Sci, 504,* 195.

Goldfarb, M., Lawson, B., & Shultz, A. (2013). Realizing the promise of robotic leg prostheses. *Sci Transl Med,* 5(210), 1–4.

Goldfield, E. C. (1989). Transition from rocking to crawling: Postural constraints on infant movement. *Dev Psychol, 25*(6), 913–919.

Goldfield, E. C. (1995). *Emergent forms: Origins and early development of human action and perception.* New York: Oxford University Press.

Goldfield, E. C. (2007). A dynamical systems approach to infant oral feeding and dysphagia: From model system to therapeutic medical device. *EcolPsychol, 19*(1), 21–48.

Goldfield, E. C. (2016). *Developmental foundations of technology for pediatric neurorehabilitation.* Paper presented at the NIH workshop Can Technology Make a Difference in Pediatric Rehabilitation?, Bethesda, MD.

Goldfield, E. C., Kay, B. A., & Warren, W. H., Jr. (1993). Infant bouncing: The assembly and tuning of action systems. *Child Devel, 64*(4), 1128–1142.

Goldfield, E. C., & Michel, G. F. (1986a). The ontogeny of infant bimanual reaching during the first year. *Infant Behavior and Development, 9*(1), 81–89. doi:10.1016/0163-6383(86)90040-8

Goldfield, E. C., & Michel, G. F. (1986b). Spatiotemporal linkage in infant interlimb coordination. *Dev Psychobiol, 19*(3), 259–264. doi:10.1002/dev.420190311

Goldfield, E. C., Park, Y.-L., Chen, B.-R., Hsu, W.-H., Young, D., Wehner, M., . . . Wood, R. J. (2012). Bio-inspired design of soft robotic assistive devices: The interface of physics, biology, and behavior. *Ecological Psychol, 24*(4), 300–327. doi:10.1080/10407413.2012.726179

Goldfield, E. C., Perez, J., & Engstler, K. (2017). Neonatal feeding behavior as a complex dynamical system. *Seminars in Speech and Language, 38*(2), 77.

Goldfield, E. C., Richardson, M. J., Lee, K. G., & Margetts, S. (2006). Coordination of sucking, swallowing, and breathing and oxygen saturation during early infant breast-feeding and bottle-feeding. *Pediatr Res, 60*(4), 450–455. doi:01.pdr.0000238378.24238.9d [pii] 10.1203/01.pdr.0000238378.24238.9d

Goldfield, E. C., Schmidt, R. C., & Fitzpatrick, P. (1999). Coordination dynamics of abdomen and chest during infant breathing: A comparison of full-term and preterm infants at 38 weeks postconceptional age. *Ecological Psychol, 11*(3), 209.

Goldfield, E., & Wolff, P. H. (2002). Motor development in infancy. In A. Slater and M. Lewis (Eds.), *Introduction to infant development.* Oxford: Oxford University Press.

Goldstein, J. (1999). Emergence as a construct: History and issues. *Emergence: Complexity and Organization, 1*(1), 49.

Goldstein, L., Byrd, D., & Saltzman, E. (2006). The role of vocal tract gestural action units in understanding the evolution of phonology. In M. Arbib (Ed.), *Action to language via the mirror neuron system* (pp. 215–249). New York: Cambridge University Press.

Golubitsky, M., Stewart, I., Buono, P.-L., & Collins, J. J. (1998). A modular network for legged locomotion. *Physica D, 115,* 56–72.

Golubitsky, M., Stewart, I., Buono, P.-L., & Collins, J. J. (1999). Symmetry in locomotor central pattern generators and animal gaits. *Nature, 401,* 693–695.

Gonzalez-Fernandez, M. (2014). Development of upper limb prostheses: Current progress and areas for growth. *Arch Phys Med Rehabil, 95*(6), 1013–1014. doi:10.1016/j.apmr.2013.11.021

Gordon, J. I., Dewey, K. G., Mills, D. A., & Medzhitov, R. M. (2012). The human gut microbiota and undernutrition. *Sci Transl Med, 4*(137), 137ps112. doi:10.1126/scitranslmed.3004347

Gottlieb, G. (1998). Normally occurring environmental and behavioral influences on gene activity: From central dogma to probabilistic epigenesis. *Psychol Rev, 105*(4), 792.

Gottlieb, G. (2007). Probabilistic epigenesis. *Dev Sci, 10*(1), 1–11. doi:DESC556 [pii] 10.1111/j.1467-7687.2007.00556.x

Goulding, M. (2009). Circuits controlling vertebrate locomotion: Moving in a new direction. *Nat Rev Neurosci, 10*(7), 507–518. doi:10.1038/nrn2608

Goulding, M. (2012). Motor neurons that multitask. *Neuron, 76*(4), 669–670. doi:10.1016/j.neuron.2012.11.011

Graeber, M. B. (2010). Changing face of microglia. *Science, 330*(6005), 783–788. doi:10.1126/science.1190929

Grant, R. A., Mitchinson, B., Fox, C. W., & Prescott, T. J. (2009). Active touch sensing in the rat: Anticipatory and regulatory control of whisker movements during surface exploration. *J Neurophysiol, 101*(2), 862–874. doi:10.1152/jn.90783.2008

Grant, S. G. (2012). Synaptopathies: Diseases of the synaptome. *Curr Opin Neurobiol, 22*(3), 522–529. doi:10.1016/j.conb.2012.02.002

Graule, M. A., Chirarattananon, P., Fuller, S. B., Jafferis, N. T., Ma, K. Y., Spenko, M., . . . Wood, R. J. (2016). Perching and takeoff of a robotic insect on overhangs using switchable electrostatic adhesion. *Science, 352*(6288), 978–982.

Graybiel, A. M. (2008). Habits, rituals, and the evaluative brain. *Annu Rev Neurosci, 31,* 359–387. doi:10.1146/annurev.neuro.29.051605.112851

Graziano, M. S., & Aflalo, T. N. (2007). Rethinking cortical organization: Moving away from discrete areas arranged in hierarchies. *Neuroscientist, 13*(2), 138–147. doi:10.1177/1073858406295918

Graziano, M. S., Aflalo, T. N., & Cooke, D. F. (2005). Arm movements evoked by electrical stimulation in the motor cortex of monkeys. *J Neurophysiol, 94*(6), 4209–4223. doi:10.1152/jn.01303.2004

Green, M. F., Horan, W. P., & Lee, J. (2015). Social cognition in schizophrenia. *Nat Rev Neurosci, 16*(10), 620–631. doi:10.1038/nrn4005

Grefkes, C., & Fink, G. R. (2014). Connectivity-based approaches in stroke and recovery of function. *Lancet Neurol, 13*(2), 206–216. doi:10.1016/s1474-4422(13)70264-3

Gremillion, G., Humbert, J. S., & Krapp, H. G. (2014). Bio-inspired modeling and implementation of the ocelli visual system of flying insects. *Biol Cybern, 108*(6), 735–746. doi:10.1007/s00422-014-0610-x

Grice, E. A., & Segre, J. A. (2012). The human microbiome: Our second genome. *Annu Rev Genomics Hum Genet, 13,* 151–170. doi:10.1146/annurev-genom-090711-163814

Grienberger, C., & Konnerth, A. (2012). Imaging calcium in neurons. *Neuron, 73*(5), 862–885. doi:10.1016/j.neuron.2012.02.011

Griffith, L. C. (2012). Identifying behavioral circuits in Drosophila melanogaster: Moving targets in a flying insect. *Curr Opin Neurobiol, 22*(4), 609–614. doi:10.1016/j.conb.2012.01.002

Grill, S., & Hyman, A. A. (2005). Spindle positioning by cortical pulling forces. *Dev Cell, 8*(4), 461–465. doi:10.1016/j.devcel.2005.03.014

Grillner, S., Hellgren, J., Menard, A., Saitoh, K., & Wikstrom, M. A. (2005). Mechanisms for selection of basic motor programs—roles for the striatum and pallidum. *Trends Neurosci, 28*(7), 364–370. doi:S0166-2236(05)00129-3 [pii] 10.1016/j.tins.2005.05.004

Grillner, S., Ip, N., Koch, C., Koroshetz, W., Okano, H., Polachek, M., . . . Sejnowski, T. J. (2016). Worldwide initiatives to advance brain research. *Nat Neurosci, 19*(9), 1118–1122. doi:10.1038/nn.4371

Grillner, S., Wallen, P., Saitoh, K., Kozlov, A., & Robertson, B. (2008). Neural bases of goal-directed locomotion in vertebrates—an overview. *Brain Res Rev, 57*(1), 2–12. doi:10.1016/j.brainresrev.2007.06.027

Grimaldi, D. A., & Engel, M. S. (2005). *Evolution of the insects.* New York: Cambridge University Press.

Grimes, D. T., Boswell, C. W., Morante, N. F., Henkelman, R. M., Burdine, R. D., & Ciruna, B. (2016). Zebrafish models of idiopathic scoliosis link cerebrospinal fluid flow defects to spinal curvature. *Science,* 352(6291), 1341–1344.

Grinthal, A., Noorduin, W. L., & Aizenberg, J. (2016). A constructive chemical conversation. *Amer Scientist, 104,* 228–235.

Grosberg, A., Alford, P. W., McCain, M. L., & Parker, K. K. (2011). Ensembles of engineered cardiac tissues for physiological and pharmacological study: Heart on a chip. *Lab Chip, 11*(24), 4165–4173. doi:10.1039/c1lc20557a

Grosenick, L., Marshel, J. H., & Deisseroth, K. (2015). Closed-loop and activity-guided optogenetic control. *Neuron, 86*(1), 106–139. doi:10.1016/j.neuron.2015.03.034

Grover, D., Katsuki, T., & Greenspan, R. J. (2016). Flyception: Imaging brain activity in freely walking fruit flies. *Nat Methods, 13*(7), 569–572. doi:10.1038/nmeth.3866

Grzybowski, B. A., & Huck, W. T. (2016). The nanotechnology of life-inspired systems. *Nat Nanotechnol, 11*(7), 585–592. doi:10.1038/nnano.2016.116

Guillot, C., & Lecuit, T. (2013). Mechanics of epithelial tissue homeostasis and morphogenesis. *Science, 340,* 1185–1189.

Gurney, J. (1992). *Dinotopia: A land apart from time.* Nashville, TN: Turner.

Gutfreund, Y., Matzner, H., Flash, T., & Hochner, B. (2006). Patterns of motor activity in the isolated nerve cord of the octopus arm. *Biol Bull, 211*(3), 212.

Guttmacher, A. E., Maddox, Y. T., & Spong, C. Y. (2014). The Human Placenta Project: Placental structure, development, and function in real time. *Placenta, 35*(5), 303–304. doi:10.1016/j.placenta.2014.02.012

Habas, P. A., Scott, J. A., Roosta, A., Rajagopalan, V., Kim, K., Rousseau, F., . . . Studholme, C. (2012). Early folding patterns and asymmetries of the normal human brain detected from in utero MRI. *Cereb Cortex, 22*(1), 13–25. doi:10.1093/cercor/bhr053

Hafed, Z. M., Stingl, K., Bartz-Schmidt, K. U., Gekeler, F., & Zrenner, E. (2016). Oculomotor behavior of blind patients seeing with a subretinal visual implant. *Vision Res, 118,* 119–131. doi:10.1016/j.visres.2015.04.006

Hagglund, M., Dougherty, K. J., Borgius, L., Itohara, S., Iwasato, T., & Kiehn, O. (2013). Optogenetic dissection reveals multiple rhythmogenic modules underlying locomotion. *Proc Natl Acad Sci USA, 110*(28), 11589–11594. doi:10.1073/pnas.1304365110

Hagmann, P., Grant, P. E., & Fair, D. A. (2012). MR connectomics: A conceptual framework for studying the developing brain. *Front Syst Neurosci, 6,* 43. doi:10.3389/fnsys.2012.00043

Haith, A. M., & Krakauer, J. W. (2013). Model-based and model-free mechanisms of human motor learning. *Adv Exp Med Biol, 782,* 1–21. doi:10.1007/978-1-4614-5465-6_1

Haiyi, L., & Mahadevan, L. (2009). The shape of a long leaf. *Proc Natl Acad Sci, 106*(52), 22049. doi:10.1073/pnas.0911954106

Haken, H. (1983). *Synergetics: An introduction* (3rd rev. and enl. ed.). Berlin: Springer.

Haken, H., Kelso, J. A., & Bunz, H. (1985). A theoretical model of phase transitions in human hand movements. *Biol Cybern, 51*(5), 347–356.

Halder, G., Dupont, S., & Piccolo, S. (2012). Transduction of mechanical and cytoskeletal cues by YAP and TAZ. *Nat Rev Mol Cell Biol, 13*(9), 591–600. doi:10.1038/nrm3416

Hall, B. K. (2006). *Fins into limbs: Evolution, development and transformation.* Chicago: University of Chicago Press.

Hamm, J. P., Peterka, D. S., Gogos, J. A., & Yuste, R. (2017). Altered cortical ensembles in mouse models of schizophrenia. *Neuron, 94*(1), 153–167 e158. doi:10.1016/j.neuron.2017.03.019

Hansell, M. H. (2005). *Animal architecture.* New York: Oxford University Press.

Haotian, L., Hong, O., Jie, Z., Shan, H., Zhenzhen, L., Shuyi, C., . . . Yizhi, L. (2016). Lens regeneration using endogenous stem cells with gain of visual function. *Nature, 531*(7594), 323. doi:10.1038/nature17181

Harada, S., & Rodan, G. A. (2003). Control of osteoblast function and regulation of bone mass. *Nature, 423,* 349–355.

Hardin, J. O., Ober, T. J., Valentine, A. D., & Lewis, J. A. (2015). Microfluidic printheads for multimaterial 3D printing of viscoelastic inks. *Adv Mater, 27*(21), 3279–3284. doi:10.1002/adma.201500222

Hardwick, R. M., Rottschy, C., Miall, R. C., & Eickhoff, S. B. (2013). A quantitative meta-analysis and review of motor learning in the human brain. *Neuroimage, 67,* 283–297. doi:10.1016/j.neuroimage.2012.11.020

Hargrove, L. J., Simon, A. M., Young, A. J., Lipschutz, R. D., Finucane, S. B., Smith, D. G., & Kuiken, T. A. (2013). Robotic leg control with EMG decoding in an amputee with nerve transfers. *N Engl J Med, 369*(13), 1237–1242. doi:10.1056/NEJMoa1300126

Harkema, S., Gerasimenko, Y., Hodes, J., Burdick, J., Angeli, C., Chen, Y., . . . Edgerton, V. R. (2011). Effect of epidural stimulation of the lumbosacral spinal cord on voluntary movement, standing, and assisted stepping after motor complete paraplegia: A case study. *The Lancet, 377*(9781), 1938–1947. doi:10.1016/s0140-6736(11)60547-3

Harvey, C. D., Coen, P., & Tank, D. W. (2012). Choice-specific sequences in parietal cortex during a virtual-navigation decision task. *Nature, 484*(7392), 62–68. doi:10.1038/nature10918

Harvey, C. D., & Svoboda, K. (2007). Locally dynamic synaptic learning rules in pyramidal neuron dendrites. *Nature, 450*(7173), 1195–1200. doi:10.1038/nature06416

Hashimoto, K., Ichikawa, R., Kitamura, K., Watanabe, M., & Kano, M. (2009). Translocation of a "winner" climbing fiber to the Purkinje cell dendrite and subsequent elimination of "losers" from the soma in developing cerebellum. *Neuron, 63*(1), 106–118. doi:10.1016/j.neuron.2009.06.008

Hashimoto, K., & Kano, M. (2013). Synapse elimination in the developing cerebellum. *Cell Mol Life Sci, 70*(24), 4667–4680. doi:10.1007/s00018-013-1405-2

Hatten, M. E. (2002). New directions in neuronal migration. *Science, 297*(5587), 1660–1663. doi:10.1126/science.1074572

Hauser, H., Sumioka, H., Fuchslin, R. M., & Pfeifer, R. (2013). Introduction to the special issue on morphological computation. *Artif Life, 19,* 1–8.

Hawkes, E., An, B., Benbernou, N. M., Tanaka, H., Kim, S., Demaine, E. D., . . . Wood, R. J. (2010). Programmable matter by folding. *Proc Natl Acad Sci USA, 107*(28), 12441–12445. doi:10.1073/pnas.0914069107

Hawkes, E. W., Eason, E. V., Christensen, D. L., & Cutkosky, M. R. (2015). Human climbing with efficiently scaled gecko-inspired dry adhesives. *J R Soc Interface, 12*(102), 20140675. doi:10.1098/rsif.2014.0675

Hawrylycz, M., Anastassiou, C., Arkhipov, A., Berg, J., Buice, M., Cain, N., . . . MindScope. (2016). Inferring cortical function in the mouse visual system through large-scale systems neuroscience. *Proc Natl Acad Sci USA, 113*(27), 7337–7344. doi:10.1073/pnas.1512901113

Hayashi, R., Ishikawa, Y., Sasamoto, Y., Katori, R., Nomura, N., Ichikawa, T., . . . Nishida, K. (2016). Co-ordinated ocular development from human iPS cells and recovery of corneal function. *Nature, 531*(7594), 376–380. doi:10.1038/nature17000

Hays, S. G., Patrick, W. G., Ziesack, M., Oxman, N., & Silver, P. A. (2015). Better together: Engineering and application of microbial symbioses. *Curr Opin Biotechnol, 36,* 40–49. doi:10.1016/j.copbio.2015.08.008

Hazlett, B. A. (1981). Daily movements of the hermit crab *Clibanarius vittatus. Bulletin of Marine Science, 31*(1), 177–183.

Heath-Heckman, E. A., Peyer, S. M., Whistler, C. A., Apicella, M. A., Goldman, W. E., & McFall-Ngai, M. J. (2013). Bacterial bioluminescence regulates expression of a host cryptochrome gene in the squid-Vibrio symbiosis. *MBio, 4*(2). doi:10.1128/mBio.00167-13

Heepe, L., & Gorb, S. N. (2014). Biologically inspired mushroom-shaped adhesive microstructures. *Ann Rev Mater Res, 44*(1), 173–203. doi:10.1146/annurev-matsci-062910-100458

Heinze, S., & Homberg, U. (2007). Maplike representation of celestial e-vector orientations in the brain of an insect. *Science, 315,* 995–997. doi:10.1126/science.1135531

Hensch, T. K. (2005). Critical period plasticity in local cortical circuits. *Nat Rev Neurosci, 6*(11), 877–888. doi:10.1038/nrn1787

Hensch, T. K. (2014). Bistable parvalbumin circuits pivotal for brain plasticity. *Cell, 156*(1–2), 17–19. doi:10.1016/j.cell.2013.12.034

Herland, A., van der Meer, A. D., FitzGerald, E. A., Park, T. E., Sleeboom, J. J., & Ingber, D. E. (2016). Distinct contributions of astrocytes and pericytes to neuroinflammation identified in a 3D human blood-brain barrier on a chip. *PLoS ONE, 11*(3), e0150360. doi:10.1371/journal.pone.0150360

Hermann, D. M., & Chopp, M. (2012). Promoting brain remodelling and plasticity for stroke recovery: Therapeutic promise and potential pitfalls of clinical translation. *Lancet Neurol, 11*(4), 369–380. doi:10.1016/s1474-4422(12)70039-x

Herron, M., Hackett, J., Aylward, F., & Michod, R. (2009). Triassic origin and early radiation of multicellular volvocine algae. *Proc Natl Acad Sci, 106,* 3254–3258.

Higham, T. E., & Biewener, A. A. (2011). Functional and architectural complexity within and between muscles: Tegional variation and intermuscular force transmission. *Philos Trans R Soc Lond B Biol Sci, 366*(1570), 1477–1487. doi:10.1098/rstb.2010.0359

Hiiemae, K. M., & Crompton, A. W. (1985). Mastication, food transport and swallowing. In M. Hildebrand, D. Bramble, K. Liem, & D. Wake (Eds.), *Functional vertebrate morphology* (pp. 262–290). Cambridge, MA: Harvard University Press.

Hilgetag, C., & Barbas, H. (2006). Role of mechanical factors in the morphology of the primate cerebral cortex. *PLOS Comp Biol, 2*, 146–159.

Hill, D. N., Curtis, J. C., Moore, J. D., & Kleinfeld, D. (2011). Primary motor cortex reports efferent control of vibrissa motion on multiple timescales. *Neuron, 72*(2), 344–356. doi:10.1016/j.neuron.2011.09.020

Hochberg, L. R., Bacher, D., Jarosiewicz, B., Masse, N., Simeral, J. D., Vogel, J., . . . Donoghue, J. (2012). Reach and grasp by people with tetraplegia using a neurally controlled robot arm. *Nature, 485*, 372–377. doi:10.1016/j.neuron.2012.06.006, 10.1021/nn304724q, 10.1038/483397a, 10.3171/2011.10.JNS102122, 10.1038/nature11076.

Hochner, B. (2008). Octopuses. *Curr Biol, 18*(19), R897–R898. doi:10.1016/j.cub.2008.07.057

Hochner, B. (2012). An embodied view of octopus neurobiology. *Curr Biol, 22*(20), R887–R892. doi:10.1016/j.cub.2012.09.001

Hoerder-Suabedissen, A., & Molnar, Z. (2013). Molecular diversity of early-born subplate neurons. *Cereb Cortex, 23*(6), 1473–1483. doi:10.1093/cercor/bhs137

Hoerder-Suabedissen, A., & Molnar, Z. (2015). Development, evolution and pathology of neocortical subplate neurons. *Nat Rev Neurosci, 16*(3), 133–146. doi:10.1038/nrn3915

Hoffman, K. L., & Wood, R. J. (2011). Myriapod-like ambulation of a segmented microrobot. *Auton Robots, 31*(1), 103–114. doi:10.1007/s10514-011-9233-4

Hogan, N., & Sternad, D. (2012). Dynamic primitives of motor behavior. *Biol Cybern, 106*(11–12), 727–739. doi:10.1007/s00422-012-0527-1

Hogan, N., & Sternad, D. (2013). Dynamic primitives in the control of locomotion. *Front Comput Neurosci, 7*, 71. doi:10.3389/fncom.2013.00071

Hogy, S. M., Worley, D. R., Jarvis, S. L., Hill, A. E., Reiser, R. F., & Haussler, K. K. (2013). Kinematic and kinetic analysis of dogs during trotting after amputation of a pelvic limb. *American Journal of Veterinary Research, 74*(9), 1164. doi:10.2460/ajvr.74.9.1164

Hölldobler, B., & Wilson, E. O. (2009). *The superorganism: The beauty, elegance, and strangeness of insect societies.* New York: Norton.

Holmes, P., Full, R. J., Koditschek, D., & Guckenheimer, J. (2006). The dynamics of legged locomotion: Models, analyses, and challenges. *SIAM Review, 48*(2), 207–304.

Holst, E. V. (1973). *The behavioural physiology of animals and man: The collected papers of Erich von Holst.* Coral Gables, FL: University of Miami Press.

Holt, K. G., Fonseca, S. T., & LaFiandra, M. E. (2000). The dynamics of gait in children with spastic hemiplegic cerebral palsy: Theoretical and clinical implications. *Human Movement Science, 19*(3), 375–405.

Holt, K. G., Obusek, J. P., & Fonseca, S. T. (1996). Constraints on disordered locomotion: A dynamical systems perspective on spastic cerebral palsy. *Hum Movement Sci, 15*(2), 177–202.

Holt, K. G., Saltzman, E., Ho, C. L., Kubo, M., & Ulrich, B. D. (2006). Discovery of the pendulum and spring dynamics in the early stages of walking. *J Mot Behav, 38*(3), 206–218.

Holtmaat, A., Randall, J., & Cane, M. (2013). Optical imaging of structural and functional synaptic plasticity in vivo. *Eur J Pharmacol, 719*(1–3), 128–136. doi:10.1016/j.ejphar.2013.07.020

Holtmaat, A., & Svoboda, K. (2009). Experience-dependent structural synaptic palsticity in the mammalian brain. *Nat Rev Neurosci, 10*, 647–658.

Holzapfel, B. M., Reichert, J. C., Schantz, J. T., Gbureck, U., Rackwitz, L., Noth, U., . . . Hutmacher, D. W. (2013). How smart do biomaterials need to be? A translational science and clinical point of view. *Adv Drug Deliv Rev, 65*(4), 581–603. doi:10.1016/j.addr.2012.07.009

Hooper, S. L. (2012). Body size and the neural control of movement. *Curr Biol, 22*(9), R318–R322. doi:10.1016/j.cub.2012.02.048

Hooper, S. L., Guschlbauer, C., Blümel, M., Rosenbaum, P., Gruhn, M., Akay, T., & Büschges, A. (2009). Neural control of unloaded leg posture and of leg swing in stick insect, cockroach, and mouse differs from that in larger animals. *J Neurosci, 29*(13), 4109. doi:10.1523/JNEURO SCI.5510-08.2009

Howard, J., Grill, S. W., & Bois, J. S. (2011). Turing's next steps: The mechanochemical basis of morphogenesis. *Nat Rev Mol Cell Biol, 12*, 392–398.

Hsiao, E. Y., McBride, S. W., Hsien, S., Sharon, G., Hyde, E. R., McCue, T., . . . Mazmanian, S. K. (2013). Microbiota modulate behavioral and physiological abnormalities associated with neurodevelopmental disorders. *Cell, 155*(7), 1451–1463. doi:10.1016/j.cell.2013.11.024

Hsu, W. H., Miranda, D., Young, D., Cakert, K., Qureshi, M., & Goldfield, E. (2014). Developmental changes in coordination of infant arm and leg movements and the emergence of function. *Journal of Motor Learning and Development, 2*(4), 69–79. doi:10.1123/jmld.2013-0033

Hu, D. L., & Bush, J. W. M. (2010). The hydrodynamics of water-walking arthropods. *J Fluid Mech, 644*, 5–33. doi:10.1017/S0022112009992205

Hu, D. L., Chan, B., & Bush, J. W. M. (2003). The hydrodynamics of water strider locomotion. *Nature, 424*(6949), 663. doi:10.1038/nature01793

Hu, H., Gan, J., & Jonas, P. (2014). Fast-spiking, parvalbumin$^+$ GABAergic interneurons: From cellular design to microcircuit function. *Science, 345*(6196), 1255263. doi:10.1126/science .1255263

Huang, B.-L., & Mackem, S. (2014). Use it or lose it. *Nature, 511*, 34–35.

Huang, H., & Vasung, L. (2014). Gaining insight of fetal brain development with diffusion MRI and histology. *Int J Dev Neurosci, 32*, 11–22. doi:10.1016/j.ijdevneu.2013.06.005

Huang, Y., Williams, J. C., & Johnson, S. M. (2012). Brain slice on a chip: Opportunities and challenges of applying microfluidic technology to intact tissues. *Lab Chip, 12*(12), 2103–2117. doi:10.1039/c2lc21142d

Huber, D., Gutnisky, D. A., Peron, S., O'Connor, D. H., Wiegert, J. S., Tian, L., . . . Svoboda, K. (2012). Multiple dynamic representations in the motor cortex during sensorimotor learning. *Nature, 484*(7395), 473–478. doi:10.1038/nature11039

Hudspeth, A. J. (1989). How the ear's works work. *Nature, 341*, 397–404.

Hudspeth, A. J. (2014). Integrating the active process of hair cells with cochlear function. *Nat Rev Neurosci, 15*(9), 600–614. doi:10.1038/nrn3786

Huh, D., Matthews, B. D., Mammoto, A., Montoya-Zavala, M., Hsin, H. Y., & Ingber, D. E. (2010). Reconstituting organ-level lung functions on a chip. *Science, 328*(5986), 1662–1668. doi:10.1126/science.1188302

Humayun, M. S., de Juan, E., & Dagnelie, G. (2016). The bionic eye: A quarter century of retinal prosthesis research and development. *Ophthalmology, 123*(10), S89–S97. doi:10.1016/j.ophtha.2016.06.044

Huys, R., Perdikis, D., & Jirsa, V. K. (2014). Functional architectures and structured flows on manifolds: A dynamical framework for motor behavior. *Psychol Rev, 121*(3), 302–336. doi:10.1037/a0037014

Hwang, E. J. (2013). The basal ganglia, the ideal machinery for the cost-benefit analysis of action plans. *Front Neural Circuits, 7*, 121. doi:10.3389/fncir.2013.00121

Hwang, E. J., Bailey, P. M., & Andersen, R. A. (2013). Volitional control of neural activity relies on the natural motor repertoire. *Curr Biol, 23*(5), 353–361. doi:10.1016/j.cub.2013.01.027

Ijspeert, A. J. (2008). Central pattern generators for locomotion in animals and robots: A review. *Neural Networks, 21*, 642–653.

Ijspeert, A. J. (2014). Biorobotics: Using robots to emulate and investigate agile locomotion. *Science, 346*(6206), 196–203. doi:10.1126/science.1254486

Ijspeert, A. J., Crespi, A., Ryczko, D., & Cabelguen, J. M. (2007). From swimming to walking with a salamander robot driven by a spinal cord model. *Science, 315*(5817), 1416–1420.

Ijspeert, A., Nakanishi, J., Hoffman, H., Pastor, P., & Schaal, S. (2013). Dynamical movement primitives: Learning attractor models for motor behaviors. *Neural Computation, 25,* 328–373.
Ilievski, F., Mazzeo, A. D., Shepherd, R. F., Chen, X., & Whitesides, G. M. (2011). Soft robots for chemists. *Angewandte Chemie, 50,* 1890–1895.
Im, K., Paldino, M. J., Poduri, A., Sporns, O., & Grant, P. E. (2014). Altered white matter connectivity and network organization in polymicrogyria revealed by individual gyral topology-based analysis. *Neuroimage, 86,* 182–193. doi:10.1016/j.neuroimage.2013.08.011
Imamura, F., Ayoub, A., Rakic, P., & Greer, C. (2011). Timing of neurogenesis is a determinant of olfactory circuitry. *Nat Neurosci, 14,* 331–337.
Ingber, D. (1998). In search of cellular control: Signal transduction in context. *Journal of Cellular Biochemistry, 72,* 232–237.
Ingber, D. E. (2006). Cellular mechanotransduction: Putting all the pieces together again. *FASEB J, 20*(7), 811–827. doi:10.1096/fj.05-5424rev
Ingber, D. E. (2016). Reverse engineering human pathophysiology with organs-on-chips. *Cell, 164*(6), 1105–1109. doi:10.1016/j.cell.2016.02.049
Ingber, D. E., & Jamieson, J. D. (1985). *Cells as tensegrity structures: Architectural regulation of histodifferentiation by physical forces transduced over basement membrane.* London: Academic Press.
Inniss, M. C., & Silver, P. A. (2013). Building synthetic memory. *Curr Biol, 23*(17), R812–R816. doi:10.1016/j.cub.2013.06.047
Innocenti, G. M., & Price, D. J. (2005). Exuberance in the development of cortical networks. *Nat Rev Neurosci, 6,* 955–965.
Isaacson, J. S., & Scanziani, M. (2011). How inhibition shapes cortical activity. *Neuron, 72*(2), 231–243. doi:10.1016/j.neuron.2011.09.027
Iverson, J. M. (2010). Multimodality in infancy: Vocal-motor and speech-gesture coordinations in typical and atypical development. *Enfance, 2010*(3), 257. doi:10.4074/S0013754510003046
Iwasaki, T., & Chen, A. (2014). Biological clockwork underlying adaptive rhythmic movements. *Proc Natl Acad Sci, 111,* 978–983.
Izquierdo, E. J., & Beer, R. D. (2016). The whole worm: Brain-body-environment models of C. elegans. *Curr Opin Neurobiol, 40,* 23–30. doi:10.1016/j.conb.2016.06.005
Jacob, F. (1977). Evolution and tinkering. *Science, 196*(4295), 1161–1166.
Jacobson, M., & Rao, M. S. (2005). *Developmental neurobiology* (4th ed.). New York: Kluwer Academic / Plenum.
Jain, A., & Gillette, M. U. (2015). Development of microfluidic devices for the manipulation of neuronal synapses. In E. Biffi (Ed.), *Microfluidic and compartmentalized platforms for neurobiological research. Neuromethods* (Vol. 103, pp. 127–137). New York: Springer Science and Business Media. doi:10.1007/978-1-4939-2510-0_7
Janssen, P., & Scherberger, H. (2015). Visual guidance in control of grasping. *Annu Rev Neurosci, 38,* 69–86. doi:10.1146/annurev-neuro-071714-034028
Jarvis, S. L., Worley, D. R., Hogy, S. M., Hill, A. E., Haussler, K. K., & Reiser, R. F., II. (2013). Kinematic and kinetic analysis of dogs during trotting after amputation of a thoracic limb. *American Journal of Veterinary Research, 74*(9), 1155–1163. doi:10.2460/ajvr.74.9.1155
Jean, A. (2001). Brain stem control of swallowing: Neuronal network and cellular mechanisms. *Physiol Rev, 81*(2), 929–969.
Jenkinson, N., & Brown, P. (2011). New insights into the relationship between dopamine, beta oscillations and motor function. *Trends Neurosci, 34*(12), 611–618. doi:10.1016/j.tins.2011.09.003
Jennings, J. H., & Stuber, G. D. (2014). Tools for resolving functional activity and connectivity within intact neural circuits. *Curr Biol, 24*(1), R41–R50. doi:10.1016/j.cub.2013.11.042
Jeong, J. W., McCall, J. G., Shin, G., Zhang, Y., Al-Hasani, R., Kim, M., . . . Rogers, J. A. (2015). Wireless optofluidic systems for programmable in vivo pharmacology and optogenetics. *Cell, 162*(3), 662–674. doi:10.1016/j.cell.2015.06.058
Ji, N., & Flavell, S. W. (2017). Hydra: Imaging nerve nets in action. *Curr Biol, 27*(8), R294–R295. doi:10.1016/j.cub.2017.03.040

Jin, X., Tecuapetla, F., & Costa, R. M. (2014). Basal ganglia subcircuits distinctively encode the parsing and concatenation of action sequences. *Nat Neurosci, 17*(3), 423–430. doi:10.1038/nn.3632

Johansen-Berg, H., & Rushworth, M. F. (2009). Using diffusion imaging to study human connectional anatomy. *Annu Rev Neurosci, 32*, 75–94. doi:10.1146/annurev.neuro.051508.135735

Jones, E. G., & Rakic, P. (2010). Radial columns in cortical architecture: t is the composition that counts. *Cereb Cortex, 20*(10), 2261–2264. doi:10.1093/cercor/bhq127

Jones, G. (2015). Sensory biology: Acoustic reflectors attract bats to roost in pitcher plants. *Curr Biol, 25*(14), R609–R610. doi:10.1016/j.cub.2015.06.004

Jordan, M. I., & Mitchell, T. M. (2015). Machine learning: Trends, perspectives, and prospects. *Science, 349*(6245), 255–260.

Jorgenson, L. A., Newsome, W. T., Anderson, D. J., Bargmann, C. I., Brown, E. N., Deisseroth, K., . . . Wingfield, J. C. (2015). The BRAIN Initiative: Developing technology to catalyse neuroscience discovery. *Philos Trans R Soc Lond B Biol Sci, 370*(1668). doi:10.1098/rstb.2014.0164

Kadoya, K., Lu, P., Nguyen, K., Lee-Kubli, C., Kumamaru, H., Yao, L., . . . Tuszynski, M. H. (2016). Spinal cord reconstitution with homologous neural grafts enables robust corticospinal regeneration. *Nat Med, 22*(5), 479–487. doi:10.1038/nm.4066

Kahn, A. (2000). *Kind of blue*. New York: Da Capo Press.

Kalaska, J. F. (2009). From intention to action: Motor cortex and the control of reaching movements. *Adv Exp Med Biol, 629*, 139–178. doi:10.1007/978-0-387-77064-2_8

Kandel, E. R., Markram, H., Matthews, P. M., Yuste, R., & Koch, C. (2013). Neuroscience thinks big (and collaboratively). *Nat Rev Neurosci, 14*(9), 659–664. doi:10.1038/nrn3578

Kang, E., Jeong, G. S., Choi, Y. Y., Lee, K. H., Khademhosseini, A., & Lee, S. H. (2011). Digitally tunable physicochemical coding of material composition and topography in continuous microfibres. *Nat Mater, 10*(11), 877–883. doi:10.1038/nmat3108

Kang, H. W., Lee, S. J., Ko, I. K., Kengla, C., Yoo, J. J., & Atala, A. (2016). A 3D bioprinting system to produce human-scale tissue constructs with structural integrity. *Nat Biotechnol, 34*(3), 312–319. doi:10.1038/nbt.3413

Kannan, S., Dai, H., Raghavendra, S., Navatah, B., Jyoti, A., Janisse, J., . . . Kannan, R. (2012). Dendrimer-based postnatal therapy for neuroinflammation and cerebral palsy in a rabbit model. *Sci Transl Med, 4*, 1–11.

Kano, M., & Hashimoto, K. (2009). Synapse elimination in the central nervous system. *Curr Opin Neurobiol, 19*, 154–161.

Kano, M., & Watanabe, M. (2013). Cerebellar circuits. In J. Rubenstein and P. Rakic (Eds.), *Neural circuit development and function in the brain: Comprehensive developmental neuroscience* (Vol. 3, pp. 75–93). Amsterdam: Academic Press. doi:10.1016/b978-0-12-397267-5.00028-5

Kantz, H., & Schreiber, T. (1997). *Nonlinear time series analysis*. New York: Cambridge University Press.

Kao, J. C., Nuyujukian, P., Ryu, S. I., Churchland, M. M., Cunningham, J. P., & Shenoy, K. V. (2015). Single-trial dynamics of motor cortex and their applications to brain-machine interfaces. *Nat Commun, 6*, 7759. doi:10.1038/ncomms8759

Kapur, S., Friedman, J., Zatsiorsky, V. M., & Latash, M. L. (2010). Finger interaction in a three-dimensional pressing task. *Exp Brain Res, 203*(1), 101–118. doi:10.1007/s00221-010-2213-7

Kargo, W. J., & Giszter, S. F. (2008). Individual premotor drive pulses, not time-varying synergies, are the units of adjustment for limb trajectories constructed in spinal cord. *J Neurosci, 28*(10), 2409–2425. doi:10.1523/JNEUROSCI.3229-07.2008

Karsenti, E. (2004). Spindle saga. *Nature, 432*, 563–564.

Karsenty, G. (2003). The complexities of skeletal biology. *Nature, 423*, 316–318.

Karsenty, G., & Oury, F. (2012). Biology without walls: The novel endocrinology of bone. *Annu Rev of Physiol, 74*, 87–105.

Kastanenka, K. V., & Landmesser, L. T. (2010). In vivo activation of channelrhodopsin-2 reveals that normal patterns of spontaneous activity are required for motoneuron guidance and maintenance of guidance molecules. *J Neurosci, 30*(31), 10575–10585. doi:10.1523/JNEUROSCI.2773-10.2010

Kasthuri, N., Hayworth, K. J., Berger, D. R., Schalek, R. L., Conchello, J. A., Knowles-Barley, S., . . . Lichtman, J. W. (2015). Saturated reconstruction of a volume of neocortex. *Cell, 162*(3), 648–661. doi:10.1016/j.cell.2015.06.054

Kato, S., Kaplan, H. S., Schrodel, T., Skora, S., Lindsay, T. H., Yemini, E., . . . Zimmer, M. (2015). Global brain dynamics embed the motor command sequence of Caenorhabditis elegans. *Cell, 163*(3), 656–669. doi:10.1016/j.cell.2015.09.034

Kato, S., Xu, Y., Cho, C. E., Abbott, L. F., & Bargmann, C. I. (2014). Temporal responses of *C. elegans* chemosensory neurons are preserved in behavioral dynamics. *Neuron, 81*(3), 616–628. doi:10.1016/j.neuron.2013.11.020

Katsuki, T., & Greenspan, R. J. (2013). Jellyfish nervous systems. *Curr Biol, 23*(14), R592–R594. doi:10.1016/j.cub.2013.03.057

Katz, L. C., & Shatz, C. J. (1996). Synaptic activity and the construction of cortical circuits. *Science, 274*(5290), 1133–1138.

Kau, A. L., Planer, J. D., Liu, J., Rao, S., Yatsunenko, T., Trehan, I., . . . Gordon, J. I. (2015). Functional characterization of IgA-targeted bacterial taxa from undernourished Malawian children that produce diet-dependent enteropathy. *Sci Transl Med, 7*(276), 276ra224. doi:10.1126/scitranslmed.aaa4877

Kaufman, M. T., Churchland, M. M., Ryu, S. I., & Shenoy, K. V. (2014). Cortical activity in the null space: Permitting preparation without movement. *Nat Neurosci, 17*(3), 440–448. doi:10.1038/nn.3643

Kay, B. A., Kelso, J. A. S., Saltzman, E. L., & Schöner, G. (1987). Space-time behavior of single and bimanual rhythmical movements: Data and limit cycle model. *J Exp Psychol: Hum Percep Perform, 13*(2), 178–192. doi:10.1037/0096-1523.13.2.178

Keedwell, P. A., Andrew, C., Williams, S. C. R., Brammer, M. J., & Phillips, M. L. (2005). A double dissociation of ventromedial prefrontal cortical responses to sad and happy stimuli in depressed and healthy individuals. *Biol Psychiatry, 58*(6), 495–503. doi:10.1016/j.biopsych.2005.04.035

Keller, P. J. (2013). Imaging morphogenesis: Technological advances and biological insights. *Science, 340*, 1184–1194.

Keller, P. J., & Ahrens, M. B. (2015). Visualizing whole-brain activity and development at the single-cell level using light-sheet microscopy. *Neuron, 85*(3), 462–483. doi:10.1016/j.neuron.2014.12.039

Kelley, M. W. (2006). Regulation of cell fate in the sensory epithelia of the inner ear. *Nat Rev Neurosci, 7*(11), 837–849. doi:10.1038/nrn1987

Kelso, J. A. S. (1995). Extending the basic picture: Breaking away. In *Dynamic patterns: The self-organization of brain and behavior* (pp. 97–135). Cambridge, MA: MIT Press.

Kelso, J. A. S., & Engstrom, D. A. (2006). *The complementary nature.* Cambridge, MA: MIT Press.

Kelso, J. A. S., Holt, K. G., Rubin, P., & Kugler, P. N. (1981). Patterns of human interlimb coordination emerge from the properties of non-linear, limit cycle oscillatory processes: Theory and data. *J Mot Behav, 13*(4), 226–261. doi:10.1080/00222895.1981.10735251

Kelso, J. A. S., Tuller, B., Vatikiotis-Bateson, E., & Fowler, C. A. (1984). Functionally specific articulatory cooperation following jaw perturbations during speech: Evidence for coordinative structures. *J Exp Psychol: Hum Percep Perform, 10*(6), 812–832. doi:10.1037/0096-1523.10.6.812

Kelty-Stephen, D. G., & Dixon, J. A. (2013). Temporal correlations in postural sway moderate effects of stochastic resonance on postural stability. *Hum Mov Sci, 32*(1), 91–105. doi:10.1016/j.humov.2012.08.006

Kelty-Stephen, D. G., Palatinus, K., Saltzman, E., & Dixon, J. A. (2013). A tutorial on multifractality, cascades, and interactivity for empirical time series in ecological science. *Ecolog Psychol, 25*(1), 1–62. doi:10.1080/10407413.2013.753804

Kempf, A., & Schwab, M. E. (2013). Nogo-A represses anatomical and synaptic plasticity in the central nervous system. *Physiology (Bethesda), 28*(3), 151–163. doi:10.1152/physiol.00052.2012

Khalil, A. S., & Collins, J. J. (2010). Synthetic biology: applications come of age. *Nat Rev Genet, 11*(5), 367–379. doi:10.1038/nrg2775

Khazipov, R., Colonnese, M., & Minlebaev, M. (2013). Neonatal cortical rhythms. In J. Rubenstein & P. Rakic (Eds.), *Neural circuit development and function in the brain: Comprehensive developmental neuroscience* (Vol. 3, pp. 131–153). Amsterdam: Academic Press. doi:10.1016/b978-0-12-397267-5.00141-2

Khundrakpam, B. S., Reid, A., Brauer, J., Carbonell, F., Lewis, J., Ameis, S., . . . Brain Development Cooperative Group. (2013). Developmental changes in organization of structural brain networks. *Cereb Cortex, 23*(9), 2072–2085. doi:10.1093/cercor/bhs187

Khwaja, O., & Volpe, J. J. (2008). Pathogenesis of cerebral white matter injury of prematurity. *Arch Dis Child Fetal Neonatal Ed, 93*(2), F153–F161. doi:93/2/F153 [pii] 10.1136/adc.2006.108837

Kiecker, C., & Lumsden, A. (2012). The role oforganizers in patterning the nervous system. *Annu Rev Neurosci, 35*, 347–367.

Kiehn, O. (2011). Development and functional organization of spinal locomotor circuits. *Curr Opin Neurobiol, 21*(1), 100–109. doi:10.1016/j.conb.2010.09.004

Kiehn, O. (2016). Decoding the organization of spinal circuits that control locomotion. *Nat Rev Neurosci, 17*(4), 224–238. doi:10.1038/nrn.2016.9

Kier, W. M. (2012). The diversity of hydrostatic skeletons. *J Exper Biol 215*, 1247–1257.

Kier, W. M., & Smith, K. K. (1985). Tongues, tentacles and trunks: The biomechanics of movement in muscular-hydrostats. *Zoological Journal of the Linnean Society, 83*(4), 307–324. doi:10.1111/j.1096-3642.1985.tb01178.x

Kim, C. K., Adhikari, A., & Deisseroth, K. (2017). Integration of optogenetics with complementary methodologies in systems neuroscience. *Nat Rev Neurosci, 18*(4), 222–235. doi:10.1038/nrn.2017.15

Kim, H. J., Huh, D., Hamilton, G., & Ingber, D. E. (2012). Human gut-on-a-chip inhabited by microbial flora that experiences intestinal peristalsis-like motions and flow. *Lab Chip, 12*(12), 2165–2174. doi:10.1039/c2lc40074j

Kim, H. J., Li, H., Collins, J. J., & Ingber, D. E. (2016). Contributions of microbiome and mechanical deformation to intestinal bacterial overgrowth and inflammation in a human gut-on-a-chip. *Proc Natl Acad Sci USA, 113*(1), E7–E15. doi:10.1073/pnas.1522193112

Kim, J. S., Greene, M. J., Zlateski, A., Lee, K., Richardson, M., Turaga, S. C., . . . EyeWirers. (2014). Space-time wiring specificity supports direction selectivity in the retina. *Nature, 509*(7500), 331–336. doi:10.1038/nature13240

Kim, S., Laschi, C., & Trimmer, B. (2013). Soft robotics: A bioinspired evolution in robotics. *Trends Biotechnol, 31*(5), 287–294. doi:10.1016/j.tibtech.2013.03.002

Kim, S. Y., Chung, K., & Deisseroth, K. (2013). Light microscopy mapping of connections in the intact brain. *Trends Cogn Sci, 17*(12), 596–599. doi:10.1016/j.tics.2013.10.005

Kim, T. I., McCall, J. G., Jung, Y. H., Huang, X., Siuda, E. R., Li, Y., . . . Bruchas, M. R. (2013). Injectable, cellular-scale optoelectronics with applications for wireless optogenetics. *Science, 340*(6129), 211–216. doi:10.1126/science.1232437

King, H. M., Shubin, N. H., Coates, M. I., & Hale, M. E. (2011). Behavioral evidence for the evolution of walking and bounding before terrestriality in sarcopterygian fishes. *Proc Natl Acad Sci, 108*(52), 21146–21151.

Kinkhabwala, A., Riley, M., Koyama, M., Monen, J., Satou, C., Kimura, Y., . . . Fetcho, J. (2011). A structural and functional ground plan for neurons in the hindbrain of zebrafish. *Proc Natl Acad Sci, 108*, 1164–1169.

Kirk, D. L. (2005). A twelve-step program for evolving multicellularity and a division of labor. *Bioessays, 27*, 299–310.

Kirkby, L. A., Sack, G. S., Firl, A., & Feller, M. B. (2013). A role for correlated spontaneous activity in the assembly of neural circuits. *Neuron, 80*(5), 1129–1144. doi:10.1016/j.neuron.2013.10.030

Kirschner, M., & Gerhart, J. (2005). *The plausibility of life*. New Haven, CT: Yale University Press.

Klamroth-Marganska, V., Blanco, J., Campen, K., Curt, A., Dietz, V., Ettlin, T., . . . Riener, R. (2014). Three-dimensional, task-specific robot therapy of the arm after stroke: A multicentre, parallel-group randomised trial. *Lancet Neurol, 13*(2), 159–166. doi:10.1016/s1474-4422(13)70305-3

Kleinfeld, D., & Deschenes, M. (2011). Neuronal basis for object location in the vibrissa scanning sensorimotor system. *Neuron, 72*(3), 455–468. doi:10.1016/j.neuron.2011.10.009

Kleinfeld, D., Deschenes, M., Wang, F., & Moore, J. D. (2014). More than a rhythm of life: Breathing as a binder of orofacial sensation. *Nat Neurosci, 17*(5), 647–651. doi:10.1038/nn.3693

Knierim, J. J., & Zhang, K. (2012). Attractor dynamics of spatially correlated neural activity in the limbic system. *Annu Rev Neurosci, 35*, 267. doi:10.1146/annurev-neuro-062111-150351

Knoll, A. H. (2003). *Life on a young planet: The first three billion years of evolution on earth*. Princeton, NJ: Princeton University Press.

Knoll, A. H., & Carroll, S. B. (1999). Early animal evolution: emerging views from comparative biology and geology. *Science, 284*(5423), 2129–2137. doi:7621 [pii]

Knopfel, T. (2012). Genetically encoded optical indicators for the analysis of neuronal circuits. *Nat Rev Neurosci, 13*(10), 687–700. doi:10.1038/nrn3293

Knowlton, N. (2008). Coral reefs. *Curr Biol, 18*, R18–R21.

Knuesel, I., Chicha, L., Britschgi, M., Schobel, S. A., Bodmer, M., Hellings, J., . . . Prinssen, E. (2014). Maternal immune activation and abnormal brain development across CNS disorders. *Nat Rev Neurol, 10*, 643–660.

Koch, H., Garcia, A. J., III, & Ramirez, J. M. (2011). Network reconfiguration and neuronal plasticity in rhythm-generating networks. *Integr Comp Biol, 51*(6), 856–868. doi:10.1093/icb/icr099

Koehl, M. A. (2004). Biomechanics of microscopic appendages: Functional shifts caused by changes in speed. *J Biomech, 37*(6), 789–795. doi:10.1016/j.jbiomech.2003.06.001

Koehl, M. A., Silk, W. K., Liang, H., & Mahadevan, L. (2008). How kelp produce blade shapes suited to different flow regimes: A new wrinkle. *Integr Comp Biol, 48*(6), 834–851. doi:10.1093/icb/icn069

Koenig, J. E., Spor, A., Scalfone, N., Fricker, A. D., Stombaugh, J., Knight, R., . . . Ley, R. E. (2011). Succession of microbial consortia in the developing infant gut microbiome. *Proc Natl Acad Sci, 108*, 4578–4585.

Koh, J.-S., Yang, E., Jung, G., Jung, S.-P., Son, J., Lee, S.-I., . . . Cho, K.-J. (2015). Jumping on water: Surfacetension–dominated jumping of water striders and robotic insects. *Science, 349*, 517–521.

Kolasinski, J., Takahashi, E., Stevens, A. A., Benner, T., Fischl, B., Zollei, L., & Grant, P. E. (2013). Radial and tangential neuronal migration pathways in the human fetal brain: Anatomically distinct patterns of diffusion MRI coherence. *Neuroimage, 79*, 412–422. doi:10.1016/j.neuroimage.2013.04.125

Kolesky, D. B., Homan, K. A., Skylar-Scott, M. A., & Lewis, J. A. (2016). Three-dimensional bioprinting of thick vascularized tissues. *Proc Natl Acad Sci USA, 113*(12), 3179–3184. doi:10.1073/pnas.1521342113

Kollmannsberger, P., Bidan, C. M., Dunlop, J. W. C., & Fratzl, P. (2011). The physics of tissue patterning and extracellular matrix organisation: How cells join forces. *Soft Matter, 7*(20), 9549. doi:10.1039/c1sm05588g

Kopell, N. J., Gritton, H. J., Whittington, M. A., & Kramer, M. A. (2014). Beyond the connectome: The dynome. *Neuron, 83*(6), 1319–1328. doi:10.1016/j.neuron.2014.08.016

Kory, J. M., Jeong, S., & Breazeal, C. (2013). *Robotic learning companions for early language development*. Paper presented at the ICMI 2013, Sydney, Australia.

Koser, D. E., Thompson, A. J., Foster, S. K., Dwivedy, A., Pillai, E. K., Sheridan, G. K., . . . Franze, K. (2016). Mechanosensing is critical for axon growth in the developing brain. *Nat Neurosci, 19*(1592–1598). doi:10.1038/nn.4394

Kostovic, I., & Jovanov-Milosevic, N. (2006). The development of cerebral connections during the first 20–45 weeks' gestation. *Semin Fetal Neonatal Med, 11*(6), 415–422. doi:10.1016/j.siny.2006.07.001

Kostovic, I., & Vasung, L. (2009). Insights from in vitro fetal magnetic resonance imaging of cerebral development. *Semin Perinatol*, 33(4), 220–233. doi:10.1053/j.semperi.2009.04.003

Kotter, R. (2004). Online retrieval, processing, and visualization of primate connectivity data from the CoCoMac Database. (Author abstract) (Report). *Neuroinformatics*, 2(2), 127.

Kotula, J. W., Kerns, S. J., Shaket, l. A., Siraj, L., Collins, J. J., Way, J. C., & Silver, P. A. (2014). Programmable bacteria detect and record an environmental signal in the mammalian gut. *Proc Natl Acad Sci, 111*, 4838–4843.

Kovac, M. (2016). Learning from nature how to land aerial robots. *Science, 352*(6288), 895–896.

Krakauer, J. W. (2006). Motor learning: Its relevance to stroke recovery and neurorehabilitation. *Curr Opin Neurol, 19*, 84–90.

Krakauer, J. W., Ghazanfar, A. A., Gomez-Marin, A., Maciver, M. A., & Poeppel, D. (2017). Neuroscience needs behavior: Correcting a reductionist bias. *Neuron*, 93(3), 480–490. doi:10.1016/j.neuron.2016.12.041

Krakowiak, P., Walker, C. K., Bremer, A. A., Baker, A. S., Ozonoff, S., Hansen, R. L., & Hertz-Picciotto, I. (2012). Maternal metabolic conditions and risk for autism and other neurodevelopmental disorders. *Pediatrics, 129*(5), e1121. doi:10.1542/peds.2011-2583

Krapp, H. G. (2009). Ocelli. *Curr Biol, 19*(11), R435–R437. doi:10.1016/j.cub.2009.03.034

Krebs, H. I., Hogan, N., Durfee, W. K., & Herr, H. (2006). Rehabilitation robotics, orthotics, and prosthetics. In M. E. Selzer, S. Clarke, L. Cohen, P. Duncan, & F. Gage (Eds.), *Textbook of neural repair and rehabilitation* (Vol. 2, pp. 165–181). Cambridge, UK: Cambridge University Press.

Krebs, H. I., Volpe, B., & Hogan, N. (2009). A working model of stroke recovery from rehabilitation robotics practitioners. *J Neuroeng Rehabil, 6*, 6. doi:10.1186/1743-0003-6-6

Krieg, M., Helenius, J., Heisenberg, C.-P., & Muller, D. J. (2008). A bond for a lifetime: Employing membrane nanotubes from living cells to determine receptor-ligand kinetics. *Angewandte Chemie* (International ed. in English), *47*(50), 9775. doi:10.1002/anie.200803552

Krigger, K. W. (2006). Cerebral palsy: An overview. *American Family Physician, 73*(1), 91.

Kronenberg, H. M. (2003). Developmental regulation of the growth plate. *Nature, 423*, 332–336.

Kuban, K. C., Allred, E. N., O'Shea, T. M., Paneth, N., Pagano, M., Dammann, O., . . . Keller, C. E. (2009). Cranial ultrasound lesions in the NICU predict cerebral palsy at age 2 years in children born at extremely low gestational age. *J Child Neurol, 24*, 63–72.

Kubow, T. M., & Full, R. J. (1999). The role of the mechanical system in control: A hypothesis of selfstabilization in hexapedal runners. *Philos Trans R Soc B: Biological Sciences, 354*(1385), 849–861. doi:10.1098/rstb.1999.0437

Kugler, P. N., & Turvey, M. T. (1987). *Information, natural laws, and the self-assembly of rhythmic movement*. Hillsdale, NJ: Lawrence Erlbaum.

Kuhlman, S. J., O'Connor, D. H., Fox, K., & Svoboda, K. (2014). Structural plasticity within the barrel cortex during initial phases of whisker-dependent learning. *J Neurosci, 34*(17), 6078–6083. doi:10.1523/JNEUROSCI.4919-12.2014

Kuiken, T. A., Li, G., Lock, B. A., Lipschutz, R. D., Miller, L., Stubblefield, K., & Englehart, K. (2009). Targeted muscle reinnervation for real-time myoelectric control of multifunction artificial arms. *JAMA 301*(6), 619–628.

Kuiken, T. A., Marasco, P. D., Lock, B. A., Harden, R. N., & Dewald, J. P. (2007). Redirection of cutaneous sensation from the hand to the chest skin of human amputees with targeted reinnervation. *Proc Natl Acad Sci USA, 104*(50), 20061–20066. doi:10.1073/pnas.0706525104

Kukillaya, R. P., & Holmes, P. (2009). A model for insect locomotion in the horizontal plane: Feedforward activation of fast muscles, stability, and robustness. *J Theoret Biol, 261*, 210–226.

Kumar, A. A., Hennek, J. W., Smith, B. S., Kumar, S., Beattie, P., Jain, S., . . . Whitesides, G. M. (2015). From the bench to the field in low-cost diagnostics: Two case studies. *Angew Chem Int Ed Engl, 54*(20), 5836–5853. doi:10.1002/anie.201411741

Kuratani, S. (2013). Evolution. A muscular perspective on vertebrate evolution. *Science, 341*(6142), 139–140. doi:10.1126/science.1241451

Kuratani, S., Nobusada, Y., Horigome, N., & Shigetani, Y. (2001). Embryology of the lamprey and evolution of the vertebrate jaw: Insights from molecular and developmental perspectives. *Philosophical Transactions of the Royal Society B: Biological Sciences,* 356(1414), 1615–1632. doi:10.1098/rstb.2001.0976

Kurlansky, M. (2016). *Paper: Paging through history.* New York: Norton.

Kurth, J. A., & Kier, W. M. (2014). Scaling of the hydrostatic skeleton in the earthworm Lumbricus terrestris. *J Exp Biol, 217*(Pt 11), 1860–1867. doi:10.1242/jeb.098137

Kutch, J. J., & Valero-Cuevas, F. J. (2012). Challenges and new approaches to proving the existence of muscle synergies of neural origin. *PLoS Comput Biol, 8*(5), e1002434. doi:10.1371/journal.pcbi.1002434

Kwakkel, G., & Meskers, C. G. M. (2014). Effects of robotic therapy of the arm after stroke. *Lancet Neurol, 13*(2), 132–133. doi:10.1016/s1474-4422(13)70285-0

Kwan, K. Y., Lam, M. M., Krsnik, Z., Kawasawa, Y. I., Lefebvre, V., & Sestan, N. (2008). SOX5 postmitotically regulates migration, postmigratory differentiation, and projections of subplate and deep-layer neocortical neurons. *Proc Natl Acad Sci USA, 105*(41), 16021–16026. doi:10.1073/pnas.0806791105

Kwok, R. (2013). Once more with feeling. *Nature, 497,* 176–178.

Lahiri, S., Shen, K., Klein, M., Tang, A., Kane, E., Gershow, M., . . . Samuel, A. D. T. (2011). Two alternating motor programs drive navigation in Drosophila larva. *PLoS ONE, 6*(8), 1–12. doi:10.1371/journal.pone.0023180.g001, 10.1371/journal.pone.0023180.g002

Lancaster, M. A., Renner, M., Martin, C. A., Wenzel, D., Bicknell, L. S., Hurles, M. E., . . . Knoblich, J. A. (2013). Cerebral organoids model human brain development and microcephaly. *Nature, 501*(7467), 373–379. doi:10.1038/nature12517

Lander, A. D. (2007). Morpheus unbound: Reimagining the morphogen gradient. *Cell, 128*(2), 245–256. doi:10.1016/j.cell.2007.01.004

Latash, M. L. (2012). The bliss (not the problem) of motor abundance (not redundancy). *Exp Brain Res, 217*(1), 1–5. doi:10.1007/s00221-012-3000-4

Latash, M. L., Scholz, J. P., & Schöner, G. (2007). Toward a new theory of motor synergies. *Motor Control, 11,* 276–308.

Latash, M. L., & Turvey, M. T. (Eds.). (1996). *Dexterity and its development.* Mahwah, NJ: Lawrence Erlbaum Associates.

Lauder, G. V. (2015). Fish locomotion: Recent advances and new directions. *Ann Rev Mar Sci, 7,* 521–545. doi:10.1146/annurev-marine-010814-015614

Laudet, V. (2011). The origins and evolution of vertebrate metamorphosis. *Curr Biol, 21*(18), R726–R737. doi:10.1016/j.cub.2011.07.030

Lawn, J. E., & Kinney, M. (2014). Preterm birth: Now the leading cause of child death worldwide. *Sci Transl Med, 6*(263), 1–3.

Leadbeater, E., & Chittka, L. (2007). Social learning in insects—from miniature brains to consensus building. *Curr Biol, 17*(16), R703–R713. doi:10.1016/j.cub.2007.06.012

Lee, M. S., Cau, A., Naish, D., & Dyke, G. (2014). Sustained miniaturization and anatomical innovation in the dinosaurian ancestors of birds. *Science, 345,* 562–566.

Lee, N. K., Sowa, H., Hinoi, E., Ferron, M., Ahn, J. D., Confavreux, C., . . . Karsenty, G. (2007). Endocrine regulation of energy metabolism by the skeleton. *Cell, 130*(3), 456–469. doi:10.1016/j.cell.2007.05.047

Leeder, A. C., Palma-Guerrero, J., & Glass, N. L. (2011). The social network: Deciphering fungal language. *Nat Rev Microbiol, 9,* 440–451.

Lefort, S., Gray, A. C., & Turrigiano, G. G. (2013). Long-term inhibitory plasticity in visual cortical layer 4 switches sign at the opening of the critical period. *Proc Natl Acad Sci USA, 110*(47), E4540–E4547. doi:10.1073/pnas.1319571110

Lemaire, P. (2011). Evolutionary crossroads in developmental biology: The tunicates. *Development, 138*(11), 2143–2152. doi:10.1242/dev.048975

Lerch, J. P., van der Kouwe, A. J., Raznahan, A., Paus, T., Johansen-Berg, H., Miller, K. L., . . . Sotiropoulos, S. N. (2017). Studying neuroanatomy using MRI. *Nat Neurosci, 20*(3), 314–326. doi:10.1038/nn.4501

Levin, J. E., & Miller, J. P. (1996). Broadband neural encoding in the cricket cereal sensory system enhanced by stochastic resonance. *Nature, 380*(6570), 165. doi:10.1038/380165a0

Levin, M. F., Kleim, J. A., & Wolf, S. L. (2009). What do motor "recovery" and "compensation" mean in patients following stroke? *Neurorehabilitation and Neural Repair 23*, 313–319.

Levin, M. F., Weiss, P. L., & Keshner, E. A. (2015). Emergence of virtual reality as a tool for upper limb rehabilitation: Incorporation of motor control and motor learning principles. *Phys Ther, 95*(3), 415–425. doi:10.2522/ptj.20130579

Levine, A. J., Hinckley, C. A., Hilde, K. L., Driscoll, S. P., Poon, T. H., Montgomery, J. M., & Pfaff, S. L. (2014). Identification of a cellular node for motor control pathways. *Nat Neurosci, 17*(4), 586–593. doi:10.1038/nn.3675

Levine, J. N., Gu, Y., & Cang, J. (2015). Seeing anew through interneuron transplantation. *Neuron, 86*(4), 858–860. doi:10.1016/j.neuron.2015.05.003

Leviton, A., Allred, E. N., Dammann, O., Engelke, S., Fichorova, R. N., Hirtz, D., . . . Investigators, E. S. (2013). Systemic inflammation, intraventricular hemorrhage, and white matter injury. *J Child Neurol, 28*(12), 1637–1645. doi:10.1177/0883073812463068

Leviton, A., Gressens, P., Wolkenhauer, O., & Dammann, O. (2015). Systems approach to the study of brain damage in the very preterm newborn. *Front Syst Neurosci, 9*, 58. doi:10.3389/fnsys.2015.00058

Levy, G., Flash, T., & Hochner, B. (2015). Arm coordination in octopus crawling involves unique motor control strategies. *Curr Biol, 25*(9), 1195–1200. doi:10.1016/j.cub.2015.02.064

Lewis, D., Mirnics, K., Hashimoto, T., & Volk, D. (2004). Gene expression and cortical circuit abnormalities in schizophrenia: Identifying pathophysiological mechanisms. *J Neurochem, 90 Suppl 1*, 131 (Abstract).

Lewis, D. A., Hashimoto, T., & Volk, D. (2005). Cortical inhibitory neurons and schizophrenia. *Nat Rev Neurosci, 6*(4), 312. doi:10.1038/nrn1648

Lewis, P. M., Ackland, H. M., Lowery, A. J., & Rosenfeld, J. V. (2015). Restoration of vision in blind individuals using bionic devices: A review with a focus on cortical visual prostheses. *Brain Res, 1595*, 51–73. doi:10.1016/j.brainres.2014.11.020

Li, R., & Bowerman, B. (2010). Symmetry breaking in biology. *Cold Spring Harbor Perspectives in Biology, 2*(3), a003475. doi:10.1101/cshperspect.a003475

Li, Z., Song, J., Mantini, G., Lu, M., Fang, H., Falconi, C., . . . Wang, Z. (2009). Quantifying the traction force of a single cell by aligned silicon nanowire array. *Nano Lett, 9*(10), 3575–3580. doi:10.1021/nl901774m

Liang, H., & Mahadevan, L. (2009). The shape of a long leaf. *Proc Natl Acad Sci, 106*(52), 22049. doi:10.1073/pnas.0911954106

Liang, H., & Mahadevan, L. (2011). Growth, geometry, and mechanics of a blooming lily. *Proc Natl Acad Sci USA, 108*(14), 5516–5521. doi:10.1073/pnas.1007808108

Liberman, A. M., Cooper, F. S., Shankweiler, D. P., & Studdert-Kennedy, M. (1967). Perception of the speech code. *Psychol Rev, 74*, 431–461.

Liberman, A. M., & Mattingly, I. G. (1985). The motor theory of speech perception revised. *Cognition, 21*(1), 1–36. doi:10.1016/0010-0277(85)90021-6

Lichtman, J., Livet, J., & Sanes, J. R. (2008). A technicolour approach to the connectome. *Nat Rev Neurosci, 9*, 417–422.

Lichtman, J. W., Pfister, H., & Shavit, N. (2014). The big data challenges of connectomics. *Nat Neurosci, 17*(11), 1448–1454. doi:10.1038/nn.3837

Lichtman, J. W., & Smith, S. J. (2008). Seeing circuits assemble. *Neuron, 60*, 441–448.

Liddelow, S. A., & Barres, B. A. (2016). Not everything is scary about a glial scar. *Nature, 532*, 182–183.

Liebovitch, L. S., Jirsa, V. K., & Shehadeh, L. A. (2006). *Structure of genetic regulatory networks: Evidence for scale free networks*. Vienna: World Scientific.

Lienert, F., Lohmueller, J. J., Garg, A., & Silver, P. A. (2014). Synthetic biology in mammalian cells: Next generation research tools and therapeutics. *Nat Rev Mol Cell Biol, 15*(2), 95–107. doi:10.1038/nrm3738

Lillicrap, T. P., & Scott, S. H. (2013). Preference distributions of primary motor cortex neurons reflect control solutions optimized for limb biomechanics. *Neuron, 77*(1), 168–179. doi:10.1016/j.neuron.2012.10.041

Limperopoulos, C., Chilingaryan, G., Sullivan, N., Guizard, N., Robertson, R. L., & du Plessis, A. J. (2014). Injury to the premature cerebellum: Outcome is related to remote cortical development. *Cereb Cortex, 24*(3), 728–736. doi:10.1093/cercor/bhs354

Lin, H., Ouyang, H., Zhu, J., Huang, S., Liu, Z., Chen, S., . . . Liu, Y. (2016). Lens regeneration using endogenous stem cells with gain of visual function. *Nature, 531*(7594), 323–328. doi:10.1038/nature17181

Lin, H. T., Leisk, G. G., & Trimmer, B. (2011). GoQBot: A caterpillar-inspired soft-bodied rolling robot. *Bioinspir Biomim, 6*(2), 026007. doi:10.1088/1748-3182/6/2/026007

Lin, H. T., & Trimmer, B. A. (2010). The substrate as a skeleton: Ground reaction forces from a soft-bodied legged animal. *J Exp Biol, 213*(Pt 7), 1133–1142. doi:10.1242/jeb.037796

Lin, M. Z., & Schnitzer, M. J. (2016). Genetically encoded indicators of neuronal activity. *Nat Neurosci, 19*(9), 114–-1153. doi:10.1038/nn.4359

Lin, P. Y., Hagan, K., Fenoglio, A., Grant, P. E., & Franceschini, M. A. (2016). Reduced cerebral blood flow and oxygen metabolism in extremely preterm neonates with low-grade germinal matrix-intraventricular hemorrhage. *Sci Rep, 6*, 25903. doi:10.1038/srep25903

Lind, J. U., Busbee, T. A., Valentine, A. D., Pasqualini, F. S., Yuan, H., Yadid, M., . . . Parker, K. K. (2016). Instrumented cardiac microphysiological devices via multimaterial three-dimensional printing. *Nat Mater, 16*, 303–308. doi:10.1038/nmat4782

Lipkind, D., Marcus, G. F., Bemis, D. K., Sasahara, K., Jacoby, N., Takahasi, M., . . . Tchernichovski, O. (2013). Stepwise acquisition of vocal combinatorial capacity in songbirds and human infants. *Nature, 498*(7452), 104–108. doi:10.1038/nature12173

Lipton, J. O., & Sahin, M. (2014). The neurology of mTOR. *Neuron, 84*(2), 275–291. doi:10.1016/j.neuron.2014.09.034

Lisman, J. (2015). The challenge of understanding the brain: Where we stand in 2015. *Neuron, 86*(4), 864–882. doi:10.1016/j.neuron.2015.03.032

Lisman, J. E., & Jensen, O. (2013). The θ-γ neural code. *Neuron, 77*(6), 1002. doi:10.1016/j.neuron.2013.03.007

Litcofsky, K. D., Afeyan, R. B., Krom, R. J., Khalil, A. S., & Collins, J. J. (2012). Iterative plug-and-play methodology for constructing and modifying synthetic gene networks. *Nat Methods, 9*(11), 1077–1080. doi:10.1038/nmeth.2205

Liu, C. H., Keshavan, M. S., Tronick, E., & Seidman, L. J. (2015). Perinatal risks and childhood premorbid indicators of later psychosis: Next steps for early psychosocial interventions. *Schizophr Bull, 41*(4), 801–816. doi:10.1093/schbul/sbv047

Liu, J., Fu, T. M., Cheng, Z., Hong, G., Zhou, T., Jin, L., . . . Lieber, C. M. (2015). Syringe-injectable electronics. *Nat Nanotechnol, 10*(7), 629–636. doi:10.1038/nnano.2015.115

Liu, J. C., Xie, C., Dai, X., Jin, L., Zhou, W., & Lieber, C. M. (2013). Multifunctional three-dimensional macroporous nanoelectronic networks for smart materials. *Proc Natl Acad Sci, 110*(17, April 23), 6694–6699. doi:10.1073/pnas.1305209110

Liu, Z., & Keller, P. J. (2016). Emerging imaging and genomic tools for developmental systems biology. *Dev Cell, 36*(6), 597–610. doi:10.1016/j.devcel.2016.02.016

Liubicich, D. M., Serano, J. M., Pavlopoulos, A., Kontarakis, Z., Protas, M. E., Kwan, E., . . . Patel, N. H. (2009). Knockdown of Parhyale Ultrabithorax recapitulates evolutionary changes in crustacean appendage morphology. *Proc Natl Acad Sci USA, 106*(33), 13892–13896. doi:10.1073/pnas.0903105106

Loeb, G. E. (2012). Optimal isn't good enough. *Biol Cybern, 106*(11–12), 757–765. doi:10.1007/s00422-012-0514-6

Long, M. A., Jin, D. Z., & Fee, M. S. (2010). Support for a synaptic chain model of neuronal sequence generation. *Nature, 468*(7322), 394–399. doi:10.1038/nature09514

Lopez-Bendito, G., & Molnar, Z. (2003). Thalamocortical development: How are we going to get there? *Nat Rev Neurosci, 4*(4), 276–289. doi:10.1038/nrn1075

Lopez-Rios, J., Duchesne, A., Speziale, D., Andrey, G., Peterson, K. A., Germann, P., . . . Zeller, R. (2014). Attenuated sensing of SHH by Ptch1 underlies evolution of bovine limbs. *Nature, 511*(7507), 46–51. doi:10.1038/nature13289

LoTurco, J. J., & Booker, A. B. (2013). Neuronal migration disorders. In J. L. Rubenstein & P. Rakic (Eds.), *Comprehensive developmental neuroscience* (pp. 481–494). New York: Elsevier.

Louie, K. (2013). Exploiting exploration: Past outcomes and future actions. *Neuron, 80*(1), 6–9. doi:10.1016/j.neuron.2013.09.016

Loukola, O., Perry, C., Coscos, L., & Chittka, L. (2017). Bumblebees show cognitive flexibility by improving on an observed complex behavior. *Science, 355*(6327), 833–836. doi:10.1126/science.aag2360

Lowe, T., Garwood, R. J., Simonsen, T. J., Bradley, R. S., & Withers, P. J. (2013). Metamorphosis revealed: Time-lapse three-dimensional imaging inside a living chrysalis. *J Roy Soc Int, 10,* 1–6. doi:10.1098/rsif.2013.0304, 10.5061/dryad.b451g

Lozupone, C. A., Stombaugh, J. I., Gordon, J. I., Jansson, J. K., & Knight, R. (2012). Diversity, stability and resilience of the human gut microbiota. *Nature, 489*(7415), 220–230. doi:10.1038/nature11550

Lu, P., Kadoya, K., & Tuszynski, M. H. (2014). Axonal growth and connectivity from neural stem cell grafts in models of spinal cord injury. *Curr Opin Neurobiol, 27,* 103–109. doi:10.1016/j.conb.2014.03.010

Lu, P., Wang, Y., Graham, L., McHale, K., Gao, M., Wu, D., . . . Tuszynski, M. H. (2012). Long-distance growth and connectivity of neural stem cells after severe spinal cord injury. *Cell, 150*(6), 1264–1273. doi:10.1016/j.cell.2012.08.020

Lui, J. H., Hansen, D. V., & Kriegstein, A. R. (2011a). Development and evolution of the human neocortex. *Cell, 146*(1), 18–36. doi:10.1016/j.cell.2011.06.030

Lui, J. H., Hansen, D. V., & Kriegstein, A. R. (2011b). Development and evolution of the human neocortex. *Cell, 146*(2), 332. doi:10.1016/j.cell.2011.07.005

Lumpkin, E. A., & Caterina, M. J. (2007). Mechanisms of sensory transduction in the skin. *Nature, 445*(7130), 858–865. doi:10.1038/nature05662

Lynch, G. F., Okubo, T. S., Hanuschkin, A., Hahnloser, R. H., & Fee, M. S. (2016). Rhythmic continuous-time coding in the songbird analog of vocal motor cortex. *Neuron, 90*(4), 877–892. doi:10.1016/j.neuron.2016.04.021

Lyttle, D., Gill, J., Shaw, K., Thomas, P., & Chiel, H. (2017). Robustness, flexibility, and sensitivity in a multifunctional motor control model. *Biol Cybern, 111*(1), 25–47. doi:10.1007/s00422-016-0704-8

Ma, K. Y., Chirarattananon, P., Fuller, S. B., & Wood, R. J. (2013). Controlled flight of a biologically inspired, insect-scale robot. *Science, 340*(6132), 603–607. doi:10.1126/science.1231806

Ma, L., & Gibson, D. A. (2013). Axon growth and branching. In J. Rubenstein and P. Rakic (Eds.), *Cellular migration and formation of neuronal connections: Comprehensive developmental neuroscience* (Vol. 2, pp. 51–68). Amsterdam: Elsevier. doi:10.1016/B978-0-12-397266-8.00056-9

Maciejasz, P., Eschweiler, J., Gerlach-Hahn, K., Jansen-Troy, A., & Leonhardt, S. (2014). A survey on robotic devices for upper limb rehabilitation. *J Neuroeng Rehabil, 11*(1), 3. doi:10.1186/1743-0003-11-3

MacIver, M. A., Schmitz, L., Mugan, U., Murphey, T. D., & Mobley, C. D. (2017). Massive increase in visual range preceded the origin of terrestrial vertebrates. *Proc Natl Acad Sci USA, 114*(12), E2375–E2384. doi:10.1073/pnas.1615563114

Macosko, E. Z., Pokala, N., Feinberg, E. H., Chalasani, S. H., Butcher, R. A., Clardy, J., & Bargmann, C. I. (2009). A hub-and-spoke circuit drives pheromone attraction and social behaviour in C. elegans. *Nature, 458*(7242), 1171–1175. doi:10.1038/nature07886

Maesani, A., Pradeep, R. F., & Floreano, D. (2014). Artificial evolution by viability rather than competition. *PLoS ONE, 9*(1), 1–12. doi:10.1371/journal.pone0086831
Magdalon, E. C., Michaelsen, S. M., Quevedo, A. A., & Levin, M. F. (2011). Comparison of grasping movements made by healthy subjects in a 3-dimensional immersive virtual versus physical environment. *Acta Psychol (Amst), 138*(1), 126–134. doi:10.1016/j.actpsy.2011.05.015
Majidi, C., Shepherd, R. F., Kramer, R. K., Whitesides, G. M., & Wood, R. J. (2013). Influence of surface traction on soft robot undulation. *Int J Robot Res, 32*(13), 1577–1584. doi:10.1177/0278364913498432
Mallarino, R., & Abzhanov, A. (2012). Paths less traveled: Evo-devo approaches to investigating animal morphological evolution. *Annu Rev Cell Dev Biol, 28,* 743–763. doi:10.1146/annurev-cellbio-101011-155732
Mammoto, T., & Ingber, D. (2010). Mechanical control of tissue and organ development. *Development, 137,* 1407–1420.
Mammoto, T., Mammoto, A., & Ingber, D. E. (2013). Mechanobiology and developmental control. *Annu Rev Cell Dev Biol, 29,* 27–61. doi:10.1146/annurev-cellbio-101512-122340
Mante, V., Sussillo, D., Shenoy, K. V., & Newsome, W. T. (2013). Context-dependent computation by recurrent dynamics in prefrontal cortex. *Nature, 503*(7474), 78–84. doi:10.1038/nature12742
Mao, A. S., & Mooney, D. J. (2015). Regenerative medicine: Current therapies and future directions. *Proc Natl Acad Sci USA, 112*(47), 14452–14459. doi:10.1073/pnas.1508520112
Marasco, P. D., Kim, K., Colgate, J. E., Peshkin, M. A., & Kuiken, T. A. (2011). Robotic touch shifts perception of embodiment to a prosthesis in targeted reinnervation amputees. *Brain, 134*(Pt 3), 747–758. doi:10.1093/brain/awq361
Marder, E. (2011). Variability, compensation, and modulation in neurons and circuits. *Proc Natl Acad Sci USA, 108 Suppl 3,* 15542–15548. doi:10.1073/pnas.1010674108
Marder, E. (2012). Neuromodulation of neuronal circuits: back to the future. *Neuron, 76*(1), 1–11. doi:10.1016/j.neuron.2012.09.010
Marder, E., & Bucher, D. (2007). Understanding circuit dynamics using the stomatogastric nervous system of lobsters and crabs. *Annu Rev Physiol, 69,* 291–316. doi:10.1146/annurev.physiol.69.031905.161516
Marder, E., Goeritz, M. L., & Otopalik, A. G. (2015). Robust circuit rhythms in small circuits arise from variable circuit components and mechanisms. *Curr Opin Neurobiol, 31,* 156–163. doi:10.1016/j.conb.2014.10.012
Marder, E., O'Leary, T., & Shruti, S. (2014). Neuromodulation of circuits with variable parameters: Single neurons and small circuits reveal principles of state-dependent and robust neuromodulation. *Annu Rev Neurosci, 37,* 329–346. doi:10.1146/annurev-neuro-071013-013958
Marder, E., & Taylor, A. L. (2011). Multiple models to capture the variability in biological neurons and networks. *Nat Neurosci, 14,* 133–138.
Marent, T. (2006). *Rainforest* (1st American ed.). New York: Dorling Kindersley.
Margolis, D. J., Lutcke, H., & Helmchen, F. (2014). Microcircuit dynamics of map plasticity in barrel cortex. *Curr Opin Neurobiol, 24*(1), 76–81. doi:10.1016/j.conb.2013.08.019
Margolis, D. J., Lutcke, H., Schulz, K., Haiss, F., Weber, B., Kugler, S., . . . Helmchen, F. (2012). Reorganization of cortical population activity imaged throughout long-term sensory deprivation. *Nat Neurosci, 15*(11), 1539–1546. doi:10.1038/nn.3240
Mariani, J., Coppola, G., Zhang, P., Abyzov, A., Provini, L., Tomasini, L., . . . Vaccarino, F. M. (2015). FOXG1-dependent dysregulation of GABA/glutamate neuron differentiation in autism spectrum disorders. *Cell, 162*(2), 375–390. doi:10.1016/j.cell.2015.06.034
Marin, O., Valiente, M., Ge, X., & Tsai, L. H. (2010). Guiding neuronal cell migrations. *Cold Spring Harb Perspect Biol, 2*(2), a001834. doi:10.1101/cshperspect.a001834
Marín-Padilla, M. (2011). *The human brain: Prenatal development and structure.* Heidelberg: Springer.
Markram, H., Lubke, J., Frotscher, M., & Sakmann, B. (1997). Regulation of synaptic efficacy by coincidence of postsynaptic APs and EPSPs. *Science, 275*(5297), 213–215.

Marta, C.-C., Maria, G., Li, Y., Franziska, B., Suzanne Van Der, V., Christin, K., . . . Tatiana, K. (2017). Gamma oscillations organize top-down signalling to hypothalamus and enable food seeking. *Nature, 542*(7640). doi:10.1038/nature21066

Martinez, R. V., Branch, J. L., Fish, C. R., Jin, L., Shepherd, R. F., Nunes, R. M., . . . Whitesides, G. M. (2013). Robotic tentacles with three-dimensional mobility based on flexible elastomers. *Adv Mater, 25*(2), 205–212. doi:10.1002/adma.201203002

Martinez, R. V., Fish, C. R., Chen, X., & Whitesides, G. M. (2012). Elastomeric origami: Programmable paper-elastomer composites as pneumatic actuators. *Adv Func Mater 36,* 1376–1384. doi:10.1002/adfm.201102978

Masland, R. H. (2012). The neuronal organization of the retina. *Neuron, 76*(2), 266–280. doi:10.1016/j.neuron.2012.10.002

Mayr, E. (1982). *The growth of biological thought diversity, evolution, and inheritance.* Cambridge, MA: Belknap Press of Harvard University Press.

McCall, J. G., Kim, T.-i., Shin, G., Huang, X., Jung, Y. H., Hasani, R. A., . . . Rogers, J. A. (2013). Fabrication and application of flexible, multimodal light-emitting devices for wireless optogenetics. (Protocol) (Report). *Nature Protocols, 8*(12), 2413.

McDonnell, M. D., Boahen, K., Ijspeert, A., & Sejnowski, T. (2014). Engineering intelligent electronic systems based on computational neuroscience. *Proc IEEE, 102,* 646–651.

McDonnell, M. D., & Ward, L. M. (2011). The benefits of noise in neural systems: Bridging theory and experiment. *Nat Rev Neurosci, 12,* 415–425.

McFall-Ngai, M., Hadfield, M. G., Bosch, T. C., Carey, H. V., Domazet-Loso, T., Douglas, A. E., . . . Wernegreen, J. J. (2013). Animals in a bacterial world, a new imperative for the life sciences. *Proc Natl Acad Sci USA, 110*(9), 3229–3236. doi:10.1073/pnas.1218525110

McFall-Ngai, M., Heath-Heckman, E. A., Gillette, A. A., Peyer, S. M., & Harvie, E. A. (2012). The secret languages of coevolved symbioses: Insights from the Euprymna scolopes-Vibrio fischeri symbiosis. *Semin Immunol, 24*(1), 3–8. doi:10.1016/j.smim.2011.11.006

McFall-Ngai, M. J. (2014). The importance of microbes in animal development: Lessons from the squid-vibrio symbiosis. *Annu Rev Microbiol, 68,* 177–194. doi:10.1146/annurev-micro-091313-103654

McGraw, M. B. (1943). *The neuromuscular maturation of the human infant.* New York: Columbia University Press.

Mchedlishvili, L., Mazurov, V., Grassme, K. S., Goehler, K., Robl, B., Tazaki, A., . . . Tanaka, E. M. (2012). Reconstitution of the central and peripheral nervous system during salamander tail regeneration. *Proc Natl Acad Sci USA, 109*(34), E2258–E2266. doi:10.1073/pnas.1116738109

McKenney, P. T., & Pamer, E. G. (2015). From hype to hope: The gut microbiota in enteric infectious disease. *Cell, 163*(6), 1326–1332. doi:10.1016/j.cell.2015.11.032

McLean, W. J., Yin, X., Lu, L., Lenz, D. R., McLean, D., Langer, R., . . . Edge, A. S. (2017). Clonal expansion of Lgr5-positive cells from mammalian cochlea and high-purity generation of sensory hair cells. *Cell Rep, 18*(8), 1917–1929. doi:10.1016/j.celrep.2017.01.066

McMahon, T. A. (1984). *Muscles, reflexes, and locomotion.* Princeton, NJ: Princeton University Press.

McNaughton, B. L., Battaglia, F. P., Jensen, O., Moser, E. I., & Moser, M. B. (2006). Path integration and the neural basis of the "cognitive map." *Nat Rev Neurosci, 7*(8), 663–678. doi:10.1038/nrn1932

Melin, J., & Quake, S. R. (2007). Microfluidic large-scale integration: The evolution of design rules for biological automation. *Annu Rev Biophys Biomol Struct, 36,* 213–231. doi:10.1146/annurev.biophys.36.040306.132646

Menelaou, E., & McLean, D. (2012). A gradient in endogenous rhythmicity and oscillatory drive matches recruitment order in an axial motor pool. *J Neurosci, 32*(32), 10925–10939. doi:10.1523/JNEUROSCI.1809-12.2012

Menon, V. (2013). Developmental pathways to functional brain networks: Emerging principles. *Trends Cogn Sci, 17*(12), 627–640. doi:10.1016/j.tics.2013.09.015

Menzel, R. (2012). The honeybee as a model for understanding the basis of cognition. *Nat Rev Neurosci, 13*(11), 758–768. doi:10.1038/nrn3357

Merabet, L. B. (2011). Building the bionic eye: An emerging reality and opportunity. *Prog Brain Res, 192*, 3–15. doi:10.1016/B978-0-444-53355-5.00001-4

Merians, A. S., & Fluet, G. G. (2014). Rehabilitation applications using virtual reality for persons with residual impairments following stroke. In P. L. Weiss (Ed.), *Virtual reality for physical and motor rehabilitation* (pp. 119–144). New York: Springer Science and Business Media. doi:10.1007/978-1-4939-0968-1_7

Merians, A. S., Fluet, G. G., Qiu, Q., Saleh, S., Lafond, I., Davidow, A., & Adamovich, S. (2011). Robotically facilitated virtual rehabilitation of arm transport integrated with finger movement in persons with hemiparesis. *J Neuroeng Rehabil, 8*, 1–10.

Merlin, C., Heinze, S., & Reppert, S. M. (2012). Unraveling navigational strategies in migratory insects. *Curr Opin Neurobiol, 22*, 353–361.

Metin, C., Vallee, R., Rakic, p., & Bhide, P. (2008). Modes and mishaps of neuronal migration in the mammalian brain. *J Neurosci, 12*, 11746–11752.

Michel, K. B., Heiss, E., Aerts, P., & Van Wassenbergh, S. (2015). A fish that uses its hydrodynamic tongue to feed on land. *Proc Biol Sci, 282*, 1–7. doi:10.1098/rspb.2015.0057

Mijailovich, S. M., Stojanovic, B., Kojic, M., Liang, A., Wedeen, V., & Gilbert, R. (2010). Derivation of a finite-element model of lingual deformation during swallowing from the mechanics of mesoscale myofiber tracts obtained by MRI. *J App Physio, 109*(5), 1500–1514. doi:10.1152/japplphysiol.00493.2010

Miles, G. B., & Sillar, K. T. (2011). Neuromodulation of vertebrate locomotor control networks. *Physiology (Bethesda), 26*(6), 393–411. doi:10.1152/physiol.00013.2011

Milinkovitch, M. C., Manukyan, L., Debry, A., Di-Poi, N., Martin, S., Singh, D., . . . Zwicker, M. (2013). Crocodile head scales are not developmental units but emerge from physical cracking. *Science, 339*(6115), 78–81. doi:10.1126/science.1226265

Miller, A. J. (2002). Oral and pharyngeal reflexes in the mammalian nervous system: Their diverse range in complexity and the pivotal role of the tongue. *Crit Rev Oral Biol Med, 13*(5), 409–425.

Miller, L. A., Goldman, D. I., Hedrick, T. L., Tytell, E. D., Wang, Z. J., Yen, J., & Alben, S. (2012). Using computational and mechanical models to study animal locomotion. *Integr Comp Biol, 52*(5), 553–575. doi:10.1093/icb/ics115

Millet, L. J., & Gillette, M. U. (2012). New perspectives on neuronal development via microfluidic environments. *Trends Neurosci*, 35(12), 752–761. doi:10.1016/j.tins.2012.09.001

Minev, I. R., Musienko, P., Hirsch, A., . . . Lacour, S. P. (2015). Electronic dura mater for long-term multimodal neural interfaces. *Science, 347*, 159–163.

Mirollo, R., & Strogatz, S. H. (1990). Synchronization of pulse-coupled biological oscillators. *SIAM Journal on Applied Mathematics, 50*(6), 1645–1662.

Mischiati, M., Lin, H. T., Herold, P., Imler, E., Olberg, R., & Leonardo, A. (2015). Internal models direct dragonfly interception steering. *Nature, 517*(7534), 333–338. doi:10.1038/nature14045

Mizutari, K., Fujioka, M., Hosoya, M., Bramhall, N., Okano, H. J., Okano, H., & Edge, A. S. (2013). Notch inhibition induces cochlear hair cell regeneration and recovery of hearing after acoustic trauma. *Neuron, 77*(1), 58–69. doi:10.1016/j.neuron.2012.10.032

Mlot, N. J., Tovey, C. A., & Hu, D. L. (2011). Fire ants self-assemble into waterproof rafts to survive floods. *Proc Natl Acad Sci, 108*(19), 7669. doi:10.1073/pnas.1016658108

Mogdans, J., & Bleckmann, H. (2012). Coping with flow: Behavior, neurophysiology and modeling of the fish lateral line system. *Biol Cybern, 106*(11–12), 627–642. doi:10.1007/s00422-012-0525-3

Molnar, Z., Garel, S., Lopez-Bendito, G., Maness, P., & Price, D. J. (2012). Mechanisms controlling the guidance of thalamocortical axons through the embryonic forebrain. *Eur J Neurosci, 35*(10), 1573–1585. doi:10.1111/j.1460-9568.2012.08119.x

Molyneaux, B. J., Arlotta, P., Menezes, J. R., & Macklis, J. D. (2007). Neuronal subtype specification in the cerebral cortex. *Nat Rev Neurosci, 8*(6), 427–437. doi:10.1038/nrn2151

Mongeau, J. M., Demir, A., Lee, J., Cowan, N. J., & Full, R. J. (2013). Locomotion- and mechanics-mediated tactile sensing: Antenna reconfiguration simplifies control during high-speed navigation in cockroaches. *J Exp Biol, 216*(Pt 24), 4530–4541. doi:10.1242/jeb.083477

Mooney, R. (2009). Neurobiology of song learning. *Curr Opin Neurobiol, 19*(6), 654–660. doi:10.1016/j.conb.2009.10.004

Moore, J. D., Deschenes, M., Furuta, T., Huber, D., Smear, M. C., Demers, M., & Kleinfeld, D. (2013). Hierarchy of orofacial rhythms revealed through whisking and breathing. *Nature, 497*(7448), 205–210. doi:10.1038/nature12076

Moore, J. D., Kleinfeld, D., & Wang, F. (2014). How the brainstem controls orofacial behaviors comprised of rhythmic actions. *Trends Neurosci, 37*(7), 370–380. doi:10.1016/j.tins.2014.05.001

Moore, T. Y., Organ, C. L., Edwards, S. V., Biewener, A. A., Tabin, C. J., Jenkins, F. A., Jr., & Cooper, K. L. (2015). Multiple phylogenetically distinct events shaped the evolution of limb skeletal morphologies associated with bipedalism in the jerboas. *Curr Biol, 25*(21), 2785–2794. doi:10.1016/j.cub.2015.09.037

Morgan, C., Darrah, J., Gordon, A. M., Harbourne, R., Spittle, A., Johnson, R., & Fetters, L. (2016). Effectiveness of motor interventions in infants with cerebral palsy: A systematic review. *Dev Med Child Neurol, 58*(9), 900–909. doi:10.1111/dmcn.13105

Morgan, J. L., & Lichtman, J. W. (2013). Why not connectomics? *Nat Methods, 10*(6), 494–500. doi:10.1038/nmeth.2480

Mori, S. (2002). Principles, methods, and applications of diffusion tensor imaging-15. In A. Toga and J. Mazziota (Eds.), *Brain mapping: The methods* (2nd ed.) (pp. 379–397). Boston: Academic Press.

Mori, S. (2013). *Introduction to diffusion tensor imaging and higher order models*. Burlington: Elsevier Science.

Mori, S., & Zhang, J. (2006). Principles of diffusion tensor imaging and its applications to basic neuroscience research. *Neuron, 51*(5), 527–539. doi:10.1016/j.neuron.2006.08.012

Morin, S. A., Shepherd, R. F., Kwok, S. W., Stokes, A. A., Nemiroski, A., & Whitesides, G. M. (2012). Camouflage and display for soft machines. *Science, 337*(6096), 828–832. doi:10.1126/science.1222149

Morishita, H., & Hensch, T. K. (2008). Critical period revisited: impact on vision. *Curr Opin Neurobiol, 18*(1), 101–107. doi:10.1016/j.conb.2008.05.009

Morrison, P., & Morrison, P. (1982). *Powers of ten: A book about the relative size of things in the universe and the effect of adding another zero*. Redding, CT: Scientific American Library.

Morrison, R. J., Hollister, S. J., Niedner, M. F., Mahani, M. G., Park, A. H., Meehta, D. K., . . . Green, G. E. (2015). Mitigation of tracheobronchomalacia with 3D-printed personalized medical devices in pediatric patients. *Sci Transl Med, 7*(285), 1–11.

Mortazavi, F., Oblak, A. L., Morrison, W. Z., Schmahmann, J. D., Stanley, H. E., Wedeen, V. J., & Rosene, D. L. (2017). Geometric navigation of axons in a cerebral pathway: Comparing dMRI with tract tracing and immunohistochemistry. *Cereb Cortex*, February, 1–14. doi:10.1093/cercor/bhx034

Mosadegh, B., Mazzeo, A. D., Shepherd, R. F., Morin, S. A., Gupta, U., Sani, I. Z., . . . Whitesides, G. M. (2014). Control of soft machines using actuators operated by a Braille display. *Lab Chip, 14*(1), 189–199. doi:10.1039/c3lc51083b

Mosadegh, B., Polygerinos, P., Keplinger, C., Wennstedt, S., Shepherd, R. F., Gupta, U., . . . Whitesides, G. M. (2014). Pneumatic networks for soft robotics that actuate rapidly. *Adv Func Mater, 24*(15), 2163–2170. doi:10.1002/adfm.201303288

Moser, T. (2015). Optogenetic stimulation of the auditory pathway for research and future prosthetics. *Curr Opin Neurobiol, 34*, 29–36. doi:10.1016/j.conb.2015.01.004

Moss, F., Ward, L. M., & Sannita, W. G. (2004). Stochastic resonance and sensory information processing: a tutorial and review of application. *Clin Neurophysiol, 115*, 267–281.

Muglia, L. J., & Katz, M. (2010). The enigma of spontaneous preterm birth. *N Engl J Med, 362,* 529–535.

Muller, D., & Nikonenko, I. (2013). Dendritic spines. In J. Rubenstein and P. Rakic (Eds.), *Neural circuit development and function in the brain: Comprehensive developmental neuroscience* (pp. 95–108). Amsterdam: Academic Press. doi:10.1016/b978-0-12-397267-5.00145-x

Murphey, D. K., Herman, A. M., & Arenkiel, B. R. (2014). Dissecting inhibitory brain circuits with genetically-targeted technologies. *Front Neural Circuits, 8,* 124. doi:10.3389/fncir.2014.00124

Murphy, S. V., & Atala, A. (2014). 3D bioprinting of tissues and organs. *Nat Biotechnol, 32*(8), 773–785. doi:10.1038/nbt.2958

Murphy, T. H., & Corbett, D. (2009). Plasticity during stroke recovery: From synapse to behavior. *Nat Rev Neurosci, 10,* 861–872.

Musienko, P., Heutschi, J., Friedli, L., van den Brand, R., & Courtine, G. (2012). Multi-system neurorehabilitative strategies to restore motor functions following severe spinal cord injury. *Exp Neurol, 235*(1), 100_109. doi:10.1016/j.expneurol.2011.08.025

Mussa-Ivaldi, F. A., & Bizzi, E. (2000). Motor learning through the combination of primitives. *Philos Trans R Soc Lond B Biol Sci, 355*(1404), 1755_1769. doi:10.1098/rstb.2000.0733

Muth, J. T., Vogt, D. M., Truby, R. L., Menguc, Y., Kolesky, D. B., Wood, R. J., & Lewis, J. A. (2014). Embedded 3D printing of strain sensors within highly stretchable elastomers. *Adv Mater, 26*(36), 6307_6312. doi:10.1002/adma.201400334

Nacu, E., Gromberg, E., Oliveira, C. R., Drechsel, D., & Tanaka, E. M. (2016). FGF8 and SHH substitute for anterior-posterior tissue interactions to induce limb regeneration. *Nature, 533*(7603), 407–410. doi:10.1038/nature17972

Nacu, E., & Tanaka, E. M. (2011). Limb regeneration: A new development? *Annu Rev Cell Dev Biol, 27,* 409–440. doi:10.1146/annurev-cellbio-092910-154115

Nagpal, R., Patel, A., & Gibson, M. C. (2008). Epithelial topology. *Bioessays, 30*(3), 260–266. doi:10.1002/bies.20722

Natale, L., Paikan, A., Randazzo, M., & Domenichelli, D. E. (2016). The iCub software architecture: Evolution and lessons learned. *Front Robot AI,* 3. doi:10.3389/frobt.2016.00024

Nathan, J. M., Craig, A. T., & David, L. H. (2011). Fire ants self-assemble into waterproof rafts to survive floods. *Proc Natl Acad Sci, 108*(19), 7669. doi:10.1073/pnas.1016658108

Nathanael, J., Vincent, C., & Johanna, R. (2014). Robotic exoskeletons: A perspective for the rehabilitation of arm coordination in stroke patients. *Front Hum Neurosci, 8*(NA), NA-NA. doi:10.3389/fnhum.2014.00947

Nave, K. A. (2010). Myelination and support of axonal integrity by glia. *Nature, 468*(7321), 244–252. doi:10.1038/nature09614

Nawroth, J. C., Lee, H., Feinberg, A. W., Ripplinger, C. M., McCain, M. L., Grosberg, A., . . . Parker, K. K. (2012). A tissue-engineered jellyfish with biomimetic propulsion. *Nat Biotechnol, 30*(8), 792–797. doi:10.1038/nbt.2269

Nef, T., Guidali, M., & Riener, R. (2009). ARMin III—Arm therapy exoskeleton with an ergonomic shoulder actuation. *Applied Bionics and Biomechanics, 6*(2), 127–142. doi:10.1155/2009/962956

Nelson, C. M., & Gleghorn, J. P. (2012). Sculpting organs: Mechanical regulation of tissue development. *Annu Rev Biomed Eng, 14,* 129–154. doi:10.1146/annurev-bioeng-071811-150043

Neutens, C., Adriaens, D., Christiaens, J., De Kegel, B., Dierick, M., Boistel, R., & Van Hoorebeke, L. (2014). Grasping convergent evolution in syngnathids: A unique tale of tails. *J Anat, 224*(6), 710–723. doi:10.1111/joa.12181

Newman, S. A. (2012). Physico-genetic determinants in the evolution of development. *Science, 338*(6104), 217–219. doi:10.1126/science.1222003

Newman, S. A., & Bhat, R. (2008). Dynamical patterning modules: Physico-genetic determinants of morphological development and evolution. *Phys Biol,* 5(1), 015008. doi:10.1088/1478-3975/5/1/015008

Newman, S. A., Forgacs, G., & Muller, G. B. (2006). Before programs: The physical origination of multicellular forms. *Int J Dev Biol, 50*(2-3), 289-299. doi:052049sn [pii], 10.1387/ijdb.052049sn

Nguyen, Q. T., & Kleinfeld, D. (2005). Positive feedback in a brainstem tactile sensorimotor loop. *Neuron, 45*(3), 447–457. doi:10.1016/j.neuron.2004.12.042

Nicosia, V., Vertes, P. E., Schafer, W. R., Latora, V., & Bullmore, E. (2013). Phase transition in the economically modeled growth of a cellular nervous system. *Proc Natl Acad Sci, 110*, 7880–7885.

Nielson, J. L., Haefeli, J., Salegio, E. A., Liu, A. W., Guandique, C. F., Stuck, E. D., . . . Ferguson, A. R. (2015). Leveraging biomedical informatics for assessing plasticity and repair in primate spinal cord injury. *Brain Res, 1619*, 124–138. doi:10.1016/j.brainres.2014.10.048

Nikolic, D., Fries, P., & Singer, W. (2013). Gamma oscillations: Precise temporal coordination without a metronome. *Trends Cogn Sci, 17*(2), 54–55. doi:10.1016/j.tics.2012.12.003

Nilsson, D. E., Warrant, E. J., Johnsen, S., Hanlon, R., & Shashar, N. (2012). A unique advantage for giant eyes in giant squid. *Curr Biol, 22*(8), 683–688. doi:10.1016/j.cub.2012.02.031

Nishikawa, K., Biewener, A. A., Aerts, P., Ahn, A. N., Chiel, H. J., Daley, M. A., . . . Szymik, B. (2007). Neuromechanics: An integrative approach for understanding motor control. *Integr Comp Biol, 47*(1), 16–54. doi:10.1093/icb/icm024

Nishiyama, J., & Yasuda, R. (2015). Biochemical computation for spine structural plasticity. *Neuron, 87*(1), 63–75. doi:10.1016/j.neuron.2015.05.043

Noctor, S. C., Cunningham, C. L., & Kriegstein, A. R. (2013). Radial migration in the developing cerebral cortex. In J. Rubenstein and P. Rakic (Eds.), *Comprehensive developmental neuroscience: Cellular migration and formation of neural connections* (Chapter 16, pp. 299–316). New York: Academic Press.

Nonaka, T. (2013). Motor variability but functional specificity: The case of a C4 tetraplegic mouth calligrapher. *Ecol Psychol, 25*(2), 131–154. doi:10.1080/10407413.2013.780492

Nonaka, T., Bril, B., & Rein, R. (2010). How do stone knappers predict and control the outcome of flaking? Implications for understanding early stone tool technology. *Journal of Human Evolution, 59*(2), 155–167. doi:10.1016/j.jhevol.2010.04.006

Nonaka, T., & Goldfield, E. C. (2017; under review). Mother-infant interaction in the emergence of a tool-using skill at mealtime: A process of affordance selection. *Ecological Psychology*.

Nonaka, T., & Sasaki, M. (2009). When a toddler starts handling multiple detached objects: Descriptions of a toddler's niche through everyday actions. *Ecological Psychology, 21*(2), 155–183. doi:10.1080/10407410902877207

Noorduin, W. L., Grinthal, A., Mahadevan, L., & Aizenberg, J. (2013). Rationally designed complex, hierarchical microarchitectures. *Science, 340*(6134), 832–837. doi:10.1126/science.1234621

Norell, M. A., & Xu, X. (2005). Feathered dinosaurs. *Ann Rev Earth Planetary Sci, 33*(1), 277–299. doi:10.1146/annurev.earth.33.092203.122511

Novak, B., & Tyson, J. (2008). Design principles of biochemical oscillators. *Nat Rev Mol Cell Biol, 9*, 981–991.

Novak, D., & Riener, R. (2015). Control strategies and artificial intelligence in rehabilitation robotics. *AI Magazine, Winter*, 23–33.

Nyholm, S. V., & Graf, J. (2012). Knowing your friends: Invertebrate innate immunity fosters beneficial bacterial symbioses. *Nat Rev Microbiol, 10*(12), 815–827. doi:10.1038/nrmicro2894

Nyholm, S. V., & McFall-Ngai, M. (2004). The winnowing: Establishing the squid-vibrio symbiosis. *Nat Rev Microbiol, 2*(8), 632–642. doi:10.1038/nrmicro957

Nyholm, S. V., & McFall-Ngai, M. (2014). Animal development in a microbial world. In A. Minelli & T. Pradeu (Eds.), *Towards a theory of development*. New York: Oxford University Press.

Ober, T. J., Foresti, D., & Lewis, J. A. (2015). Active mixing of complex fluids at the microscale. *Proc Natl Acad Sci USA, 112*(40), 12293–12298. doi:10.1073/pnas.1509224112

Oh, S. W., Harris, J. A., Ng, L., Winslow, B., Cain, N., Mihalas, S., . . . Zeng, H. (2014). A mesoscale connectome of the mouse brain. *Nature, 508*(7495), 207–214. doi:10.1038/nature13186

Okada, Y., Hamalainen, M., Pratt, K., Mascarenas, A., Miller, P., Han, M., . . . Paulson, D. (2016). BabyMEG: A whole-head pediatric magnetoencephalography system for human brain development research. *Rev Sci Instrum, 87*(9), 094301. doi:10.1063/1.4962020

Okubo, T. S., Mackevicius, E. L., Payne, H. L., Lynch, G. F., & Fee, M. S. (2015). Growth and splitting of neural sequences in songbird vocal development. *Nature, 528*, 352–357. doi:10.1038/nature15741

Olkowicz, S., Kocourek, M., Lucan, R. K., Portes, M., Fitch, W. T., Herculano-Houzel, S., & Nemec, P. (2016). Birds have primate-like numbers of neurons in the forebrain. *Proc Natl Acad Sci USA, 113*(26), 7255–7260. doi:10.1073/pnas.1517131113

Olusanya, B. O., Neumann, K. J., & Saunders, J. E. (2014). The global burden of disabling hearing impairment: A call to action. *Bulletin of the World Health Organization, 92*(5), 367. doi:10.2471/BLT.13.128728

Olveczky, B. P., & Gardner, T. J. (2011). A bird's eye view of neural circuit formation. *Curr Opin Neurobiol, 21*(1), 124–131. doi:10.1016/j.conb.2010.08.001

Omura, T., Omura, K., Tedeschi, A., Riva, P., Painter, M. W., Rojas, L., . . . Woolf, C. J. (2015). Robust axonal regeneration occurs in the injured CAST / Ei mouse CNS. *Neuron, 86*(5), 1215–1227. doi:10.1016/j.neuron.2015.05.005

Onal, C., Wood, R. J., & Rus, D. (2013). An origami-inspired approach to worm robots. *IEEE / ASME Trans Mechatronics, 18*, 430–438.

Ong, J. M., & da Cruz, L. (2012). The bionic eye: A review. *Clin Exper Ophthalmol, 40*(1), 6–17. doi:10.1111/j.1442-9071.2011.02590.x

Orefice, L. L., Zimmerman, A. L., Chirila, A. M., Sleboda, S. J., Head, J. P., & Ginty, D. D. (2016). Peripheral mechanosensory neuron dysfunction underlies tactile and behavioral deficits in mouse models of ASDs. *Cell, 166*(2), 299–313. doi:10.1016/j.cell.2016.05.033

O'Shea, T. M., Allred, E. N., Dammann, O., Hirtz, D., Kuban, K. C., Paneth, N., . . . Leviton, A. (2009). The ELGAN study of the brain and related disorders in extremely low gestational age newborns. *Early Hum Dev, 85*(11), 719–725. doi:10.1016/j.earlhumdev.2009.08.060

Overduin, S. A., d'Avella, A., Carmena, J. M., & Bizzi, E. (2012). Microstimulation activates a handful of muscle synergies. *Neuron, 76*(6), 1071–1077. doi:10.1016/j.neuron.2012.10.018

Overduin, S. A., d'Avella, A., Carmena, J. M., & Bizzi, E. (2014). Muscle synergies evoked by microstimulation are preferentially encoded during behavior. *Front Comput Neurosci, 8*, 20. doi:10.3389/fncom.2014.00020

Overvelde, J. T., Kloek, T., D'Haen J, J., & Bertoldi, K. (2015). Amplifying the response of soft actuators by harnessing snap-through instabilities. *Proc Natl Acad Sci USA, 112*(35), 10863–10868. doi:10.1073/pnas.1504947112

Oxman, N. (2012). Programming matter. *Architectural Design, 82*(2), 88–95. doi:10.1002/ad.1384

Ozbolat, I. T. (2015). Bioprinting scale-up tissue and organ constructs for transplantation. *Trends Biotechnol, 33*(7), 395–400. doi:10.1016/j.tibtech.2015.04.005

Packer, A. M., Russell, L. E., Dalgleish, H., & Hausser, M. (2015). Simultaneous all-optical manipulation and recording of neural circuit activity with cellular resolution *in vivo*. *Nat Methods, 12*(2), 140–146. doi:10.1038/nmeth.3217

Palatinus, Z., Dixon, J. A., & Kelty-Stephen, D. G. (2013). Fractal fluctuations in quiet standing predict the use of mechanical information for haptic perception. *Ann Biomed Eng, 41*(8), 1625–1634. doi:10.1007/s10439-012-0706-1

Palmer, C., Bik, E. M., DiGiulio, D. B., Relman, D. A., Brown, P. O., & Ruan, Y. (2007). Development of the human infant intestinal microbiota. *PLoS Biology, 5*(7), 1556–1573. doi:10.1371/journal.pbio.0050177

Paolicelli, R. C., Bolasco, G., Pagani, F., Maggi, L., Scianni, M., Panzanelli, P., . . . Gross, C. T. (2011). Synaptic pruning by microglia is necessary for normal brain development. *Science, 333*(6048), 1456–1458. doi:10.1126/science.1202529

Pardee, K., Green, A. A., Takahashi, M. K., Braff, D., Lambert, G., Lee, J. W., . . . Collins, J. J. (2016). Rapid, low-cost detection of Zika virus using programmable biomolecular components. *Cell, 165*(5), 1255–1266. doi:10.1016/j.cell.2016.04.059

Park, S. I., Brenner, D. S., Shin, G., Morgan, C. D., Copits, B. A., Chung, H. U., . . . Rogers, J. A. (2015). Soft, stretchable, fully implantable miniaturized optoelectronic systems for wireless optogenetics. *Nat Biotechnol, 33*(12), 1280–1286. doi:10.1038/nbt.3415

Park, S.-J., Gazzola, M., Park, K. S., Park, S., DiSanto, V., Blevins, E. L., . . . Parker, K. K. (2016). Phototactic guidance of a tissue-engineered soft-robotic ray. *Science, 353*(6295), 158–162.

Park, Y.-L., Chen, B. R., Perez-Arancibia, N. O., Young, D., Stirling, L., Wood, R. J., . . . Nagpal, R. (2014). Design and control of a bio-inspired soft wearable robotic device for ankle-foot rehabilitation. *Bioinspir Biomim, 9*(1), 016007. doi:10.1088/1748-3182/9/1/016007

Park, Y.-L., Majidi, C., Kramer, R., Bérard, P., & Wood, R. J. (2010). Hyperelastic pressure sensing with a liquid-embedded elastomer. *J Micromechan Microeng, 20*(12), 125029. doi:10.1088/0960-1317/20/12/125029

Park, Y.-L., Santos, J., Galloway, K. G., Goldfield, E., & Wood, R. J. (2014). *A soft wearable robotic device for active knee motions using flat pneumatic artificial muscles.* Paper presented at the IEEE International Conference on Robotics and Automation (ICRA), Hong Kong, China.

Parkhurst, C. N., Yang, G., Ninan, I., Savas, J. N., Yates, J. R., 3rd, Lafaille, J. J., . . . Gan, W. B. (2013). Microglia promote learning-dependent synapse formation through brain-derived neurotrophic factor. *Cell, 155*(7), 1596–1609. doi:10.1016/j.cell.2013.11.030

Parpura, V. (2012). Bionanoelectronics: Getting close to the action. *Nat Nanotechnol, 7*(3), 143–145. doi:10.1038/nnano.2012.22

Patrick, J. D., Jeff, H. D., Eugene, G., & David, K. (2008). Finding coherence in spontaneous oscillations. *Nat Neurosci, 11*(9), 991. doi:10.1038/nn0908-991

Paydarfar, D., Gilbert, R. J., Poppel, C. S., & Nassab, P. F. (1995). Respiratory phase resetting and airflow changes induced by swallowing in humans. *J Physiol, 483*(Pt 1), 273–288.

Peitgen, H.-O., Jurgens, H., & Saupe, D. (1992). *Chaos and fractals: New frontiers of science.* New York: Springer-Verlag.

Peña Ramirez, J., Aihara, K., Fey, R. H. B., & Nijmeijer, H. (2014). Further understanding of Huygens' coupled clocks: The effect of stiffness. *Physica D: Nonlinear Phenomena, 270,* 11–19. doi:10.1016/j.physd.2013.12.005

Pennycott, A., Wyss, D., Vallery, H., Klamroth-Marganska, V., & Riener, R. (2012). Towards more effective robotic gait training for stroke rehabilitation: A review. *J Neuroengin Rehabil, 9,* 1–13.

Perdikis, D., Huys, R., & Jirsa, V. (2011). Complex processes from dynamical architectures with time-scale hierarchy. *PLoS ONE, 6,* 1–12.

Perez-Arancibia, N. O., Duhamel, P.-E. J., Ma, K. Y., & Wood, R. J., (2015). Model-free control of a hovering flapping-wing microrobot. *J of Intelligent & Robotic Sys, 77,* 95–111. doi:10.1007/s10846-014-0096-8)

Pérez-Arancibia, N. O., Ma, K. Y., Galloway, K. C., Greenberg, J. D., & Wood, R. J. (2011). First controlled vertical flight of a biologically inspired microrobot. *Bioinspiration and Biomemetics 6*(3), 036009. doi:10.1088/1748-3182/6/3/036009

Perko, L. (2001). *Differential equations and dynamical systems* (3rd ed.). New York: Springer.

Petersen, K. H., Nagpal, R., & Werfel, J. K. (2011). TERMES: An autonomous robotic system for three-dimensional collective construction. In H. Durrant-Whyte, N. Roy, & P. Abbeel (Eds.), *Robotics: Science and systems,* vol. 7. Cambridge, MA: MIT Press.

Petkova, V., & Ehrsson, H. H. (2008). If I were you: Perceptual illusion of body swapping. *PLoS ONE, 3*(12), 1–9. doi:10.1371/journal.pone.0003832.g001, 10.1371/journal.pone.0003832.g002

Pettinger, P. (1998). *Bill Evans: How my heart sings.* New Haven, CT: Yale University Press.

Peyrache, A., Lacroix, M. M., Petersen, P. C., & Buzsaki, G. (2015). Internally organized mechanisms of the head direction sense. *Nat Neurosci, 18*(4), 569–575. doi:10.1038/nn.3968

Pezzulo, G., & Cisek, P. (2016). Navigating the affordance landscape: Feedback control as a process model of behavior and cognition. *Trends Cogn Sci, 20*(6), 414–424. doi:10.1016/j.tics.2016.03.013

Pfeifer, R., Lungarella, M., & Iida, F. (2007). Self-organization, embodiment, and biologically-inspired robotics. *Science, 318,* 1088–1093.

Pfeiffer, B. E., & Foster, D. J. (2015). Autoassociative dynamics in the generation of sequences of hippocampal place cells. *Science, 349*(6244), 180. doi:10.1126/science.aaa9633

Piao, X., Hill, R., Bodell, A., & Chang, B. (2004). G protein-coupled receptor-dependent development of human frontal cortex. *Science, 303*(5666), 2033–2036. doi:10.1126/science.1092780

Pinto, C. M., & Golubitsky, M. (2006). Central pattern generators for bipedal locomotion. *J Math Biol, 53*(3), 474–489. doi:10.1007/s00285-006-0021-2

Pokroy, B., Epstein, A. K., Persson-Gulda, M. C. M., & Aizenberg, J. (2009). Fabrication of bioinspired actuated nanostructures with arbitrary geometry and stiffness. *Adv Mater, 21*(4), 463–469. doi:10.1002/adma.200801432

Pokroy, B., Kang, S. H., Mahadevan, L., & Aizenberg, J. (2009). Self-organization of a mesoscale bristle into ordered, hierarchical helical assemblies. *Science, 323*(5911).

Poldrack, R. A., & Farah, M. J. (2015). Progress and challenges in probing the human brain. *Nature, 526*(7573), 371–379. doi:10.1038/nature15692

Polygerinos, P., Wang, Z., Overvelde, J. T., Galloway, K. C., Wood, R. J., Bertoldi, K., & Walsh, C. J. (2015). Modeling of soft fiber-reinforced bending actuators. *IEEE Trans Robotics, 31*, 778–789.

Poo, C., & Isaacson, J. S. (2009). Odor representations in olfactory cortex: "Sparse" coding, global inhibition, and oscillations. *Neuron, 62*(6), 850–-861. doi:10.1016/j.neuron.2009.05.022

Porter, M. M., Adriaens, D., Hatton, R. L., Meyers, M. A., & McKittrick, J. (2015). Biomechanics: Why the seahorse tail is square. *Science, 349*(6243), 48–53, aaa6683. doi:10.1126/science.aaa6683

Portugues, R., Feierstein, C. E., Engert, F., & Orger, M. B. (2014). Whole-brain activity maps reveal stereotyped, distributed networks for visuomotor behavior. *Neuron, 81*(6), 1328–1343. doi:10.1016/j.neuron.2014.01.019

Portugues, R., Haesemeyer, M., Blum, M. L., & Engert, F. (2015). Whole-field visual motion drives swimming in larval zebrafish via a stochastic process. *J Exp Biol, 218*(Pt 9), 1433–1443. doi:10.1242/jeb.118299

Portugues, R., Severi, K. E., Wyart, C., & Ahrens, M. B. (2013). Optogenetics in a transparent animal: Circuit function in the larval zebrafish. *Curr Opin Neurobiol, 23*(1), 119–126. doi:10.1016/j.conb.2012.11.001

Potts, R. (2012). Evolution and environmental change in early human prehistory. *Ann Rev Anthropol, 41*(1), 151–167. doi:10.1146/annurev-anthro-092611-145754

Power, J., Fair, D., Schlaggar, B., & Pertersen, S. E. (2010). The development of human functional brain networks. *Neuron, 67*, 735–748.

Powers of ten: A film dealing with the relative size of things in the universe and the effect of adding another zero. (2000). Directed by Eames Demetrios and Shelley Mills. Santa Monica, CA.

Preyer, W. T. (1885). *Specielle physiologie des embryo: Untersuchungen über die lebenserscheinungen vor der geburt.* Leipzig, n.p.

Prinz, A. A., Bucher, D., & Marder, E. (2004). Similar network activity from disparate circuit parameters. *Nat Neurosci, 7*(12), 1345–1352. doi:10.1038/nn1352

Prinz, M., Erny, D., & Hagemeyer, N. (2017). Ontogeny and homeostasis of CNS myeloid cells. *Nat Immunol, 18*(4), 385–392. doi:10.1038/ni.3703

Priplata, A., Niemi, J., Salen, M., Harry, J., Lipsitz, L. A., & Collins, J. J. (2002). Noise-enhanced human balance control. *Phys Rev Lett, 89*(23), 238101. doi:10.1103/PhysRevLett.89.238101

Proctor, J., & Holmes, P. (2010). Reflexes and preflexes: On the role of sensory feedback on rhythmic patterns in insect locomotion. *BiolCybernet, 102*, 513–531.

Proctor, J., Kukillaya, R. P., & Holmes, P. (2010). A phase-reduced neuro-mechanical model for insect locomotion: Feedforward stability and proprioceptive feedback. *Philos Trans R Soc A, 368*, 5087–5104.

Prud'homme, B., Minervino, C., Hocine, M., Cande, J. D., Aouane, A., Dufour, H. D., . . . Gompel, N. (2011). Body plan innovation in treehoppers through the evolution of an extra wing-like appendage. *Nature, 473*(7345), 83–86. doi:10.1038/nature09977

Purnell, B. (2013). Introduction: Getting into shape. *Science, 340,* 1183.
Qin, D., Xia, Y., & Whitesides, G. M. (2010). Soft lithography for micro- and nanoscale patterning. *Nature Protocols, 5*(3), 491. doi:10.1038/nprot.2009.234
Qiu, A., Mori, S., & Miller, M. I. (2015). Diffusion tensor imaging for understanding brain development in early life. *Annu Rev Psychol, 66,* 853–876. doi:10.1146/annurev-psych-010814-015340
Quesada, R., Triana, E., Vargas, G., Douglass, J. K., Seid, M. A., Niven, J. E., . . . Wcislo, W. T. (2011). The allometry of CNS size and consequences of miniaturization in orb-weaving and cleptoparasitic spiders. *Arthropod Structure and Development, 40,* 521–529.
Rabbitt, S. M., Kazdin, A. E., & Scassellati, B. (2015). Integrating socially assistive robotics into mental healthcare interventions: Applications and recommendations for expanded use. *Clin Psychol Rev, 35,* 35–46. doi:10.1016/j.cpr.2014.07.001
Raff, R. A. (1996). *The shape of life: Genes, development and the evolution of animal form.* Chicago: University of Chicago Press.
Raichle, M. E. (2010). Two views of brain function. *Trends Cogn Sci, 14*(4), 180–190. doi:10.1016/j.tics.2010.01.008
Raineteau, O., & Schwab, M. E. (2001). Plasticity of motor systems after incomplete spinal cord injury. *Nat Rev Neurosci, 2,* 263–273.
Rakic, P. (1988). Specification of cerebral cortical areas. *Science, 241,* 170–176.
Rakic, P. (2007). The radial edifice of cortical architecture: From neuronal silhouettes to genetic engineering. *Brain Res Rev, 55,* 204–219.
Rakic, P. (2009). Evolution of the neocortex: a perspective from developmental biology. *Nat Rev Neurosci, 10*(10), 724–735. doi:10.1038/nrn2719
Rakic, P., Ayoub, A. E., Breunig, J. J., & Dominguez, M. H. (2009). Decision by division: Making cortical maps. *Trends Neurosci, 32*(5), 291–301. doi:10.1016/j.tins.2009.01.007
Rakoff-Nahoum, S., Kong, Y., Kleinstein, S. H., Subramanian, S., Ahern, P. P., Gordon, J., & Medzhitov, R. (2015). Analysis of gene–environment interactions in postnatal development of the mammalian intestine. *Proc Natl Acad Sci, 112*(7), 1929. doi:10.1073/pnas.1424886112
Raman, R., Cvetkovic, C., & Bashir, R. (2017). A modular approach to the design, fabrication, and characterization of muscle-powered biological machines. *Nat Protoc, 12*(3), 519–533. doi:10.1038/nprot.2016.185
Raman, R., Cvetkovic, C., Uzel, S. G., Platt, R. J., Sengupta, P., Kamm, R. D., & Bashir, R. (2016). Optogenetic skeletal muscle-powered adaptive biological machines. *Proc Natl Acad Sci USA, 113*(13), 3497–3502. doi:10.1073/pnas.1516139113
Ramdya, P., Lichocki, P., Cruchet, S., Frisch, L., Tse, W., Floreano, D., & Benton, R. (2015). Mechanosensory interactions drive collective behaviour in Drosophila. *Nature, 519*(7542), 233–236. doi:10.1038/nature14024
Randel, D. M. (Ed.) (2003). *The Harvard dictionary of music* (4th ed.). Cambridge, MA: Belknap Press of Harvard University Press.
Ransohoff, R. M., & Stevens, B. (2011). Neuroscience. How many cell types does it take to wire a brain? *Science, 333*(6048), 1391–1392. doi:10.1126/science.1212112
Raphael, G., Tsianos, G. A., & Loeb, G. E. (2010). Spinal-like regulator facilitates control of a two-degree-of-freedom wrist. *J Neurosci, 30*(28), 9431–9444. doi:10.1523/JNEUROSCI.5537-09.2010
Rash, B. G., & Rakic, P. (2014). Neuroscience: Genetic resolutions of brain convolutions. *Science, 343*(6172), 744–745. doi:10.1126/science.1250246
Raspopovic, S., Capogrosso, M., Petrini, F. M., Bonizzato, M., Rigosa, J., Di Pino, G., . . . Micera, S. (2014). Restoring natural sensory feedback in real-time bidirectional hand prostheses. *Sci Transl Med, 6*(222), 1–10. doi:10.1126/scitranslmed.3006820
Rauscent, A., Einum, J., Le Ray, D., Simmers, J., & Combes, D. (2009). Opposing aminergic modulation of distinct spinal locomotor circuits and their functional coupling during amphibian metamorphosis. *J Neurosci, 29*(4), 1163–1174. doi:10.1523/JNEUROSCI.5255-08.2009

Rauscent, A., Le Ray, D., Cabirol-Pol, M. J., Sillar, K. T., Simmers, J., & Combes, D. (2006). Development and neuromodulation of spinal locomotor networks in the metamorphosing frog. *J Physiol Paris, 100*(5-6), 317–327. doi:10.1016/j.jphysparis.2007.05.009
Reed, E. (1988). *James J. Gibson and the psychology of perception.* New Haven, CT: Yale University Press.
Reid, C. R., Latty, T., Dussutour, A., & Beekman, M. (2012). Slime mold uses an externalized spatial "memory" to navigate in complex environments. *Proc Natl Acad Sci, 109*(43), 17490–17494.
Reid, L. B., Rose, S. E., & Boyd, R. N. (2015). Rehabilitation and neuroplasticity in children with unilateral cerebral palsy. *Nat Rev Neurol, 11*(7), 390–400. doi:10.1038/nrneurol.2015.97
Rein, R., Nonaka, T., & Bril, B. (2014). Movement pattern variability in stone knapping: Implications for the development of percussive traditions. *PLoS ONE, 9*(11), e113567. doi:10.1371/journal.pone.0113567
Reinitz, J. (2012). Pattern formation. *Nature, 482*, 464.
Reinkensmeyer, D. J., Bonato, P., Boninger, M. L., Chan, L., Cowan, R. E., Fregly, B. J., & Rodgers, M. M. (2012). Major trends in mobility technology research and development: Overview of the results of the NSF-WTEC European study. *J Neuroeng Rehabil, 9*, 22. doi:10.1186/1743-0003-9-22
Reinkensmeyer, D. J., Burdet, E., Casadio, M., Krakauer, J. W., Kwakkel, G., Lang, C. E., . . . Schweighofer, N. (2016). Computational neurorehabilitation: Modeling plasticity and learning to predict recovery. *J Neuroeng Rehabil, 13*(1), 42. doi:10.1186/s12984-016-0148-3
Reiser, M. B., & Dickinson, M. H. (2013). Visual motion speed determines a behavioral switch from forward flight to expansion avoidance in Drosophila. *J Exp Biol, 216*(Pt 4), 719–732. doi:10.1242/jeb.074732
Ren, K., Chen, Y., & Wu, H. (2014). New materials for microfluidics in biology. *Curr Opin Biotechnol, 25*, 78–85. doi:10.1016/j.copbio.2013.09.004
Renier, N., Adams, Eliza l., Kirst, C., Wu, Z., Azevedo, R., Kohl, J., . . . Tessier-Lavigne, M. (2016). Mapping of brain activity by automated volume analysis of immediate early genes. *Cell, 165*(7), 1789–1802. doi:10.1016/j.cell.2016.05.007
Resnik, L., Klinger, S. L., & Etter, K. (2014). The DEKA Arm: Its features, functionality, and evolution during the Veterans Affairs Study to optimize the DEKA Arm. *Prosthetics and Orthotics International, 38*(6), 492–504. doi:10.1177/0309364613506913
Revzen, S., Burden, S. A., Moore, T. Y., Mongeau, J. M., & Full, R. J. (2013). Instantaneous kinematic phase reflects neuromechanical response to lateral perturbations of running cockroaches. *Biol Cybern, 107*(2), 179–200. doi:10.1007/s00422-012-0545-z
Revzen, S., Koditschek, D., & Full, R. J. (2009). Towards testable neuromechnical control architectures for running. In D. Sternad (Ed.), *Progress in motor control: A multidisciplinary perspective.* New York: Springer.
Rey, H. A. (1952). *The stars: A new way to see them.* New York: Houghton Mifflin.
Rey, M., & Rey, H. A. (1966). *Curious George goes to the hospital.* Boston: Houghton Mifflin Harcourt.
Richards, T., & Gomes, S. (2015). How to build a microbial eye. *Nature, 523*, 166–167.
Richardson, M. J., Harrison, S. J., Kallen, R. W., Walton, A., Eiler, B. A., Saltzman, E., & Schmidt, R. C. (2015). Self-organized complementary joint action: Behavioral dynamics of an interpersonal collision-avoidance task. *J Exp Psychol Hum Percept Perform, 41*(3), 665–679. doi:10.1037/xhp0000041
Rico-Guevara, A., Fan, T. H., & Rubega, M. A. (2015). Hummingbird tongues are elastic micropumps. *Proc Biol Sci, 282*(1813), 20151014. doi:10.1098/rspb.2015.1014
Rico-Guevara, A., & Rubega, M. A. (2011). The hummingbird tongue is a fluid trap, not a capillary tube. *Proc Natl Acad Sci USA, 108*(23), 9356–9360. doi:10.1073/pnas.1016944108
Ridaura, V. K., Faith, J. J., Rey, F. E., Cheng, J., Duncan, A. E., Kau, A. L., . . . Gordon, J. I. (2013). Gut microbiota from twins discordant for obesity modulate metabolism in mice. *Science, 341*(6150), 1241214. doi:10.1126/science.1241214

Riedl, J., & Louis, M. (2012). Behavioral neuroscience: Crawling is a no-brainer for fruit fly larvae. *Curr Biol, 22*(20), R867–R869. doi:10.1016/j.cub.2012.08.018

Rieffel, J., Valero-Cuevas, F., & Lipson, H. (2010). Morphological communication: Exploiting coupled dynamics in a complex mechanical structure to achieve locomotion. *J Royal Society Interface, 7*, 613–621.

Righetti, L., Buchli, J., & Ijspeert, A. (2006). Dynamic Hebbian learning in adaptive frequency oscillators. *Physica D, 216*, 269–281.

Riley, M. A., Shockley, K., & Van Orden, G. (2012). Learning from the body about the mind. *Top Cogn Sci, 4*(1), 21–34. doi:10.1111/j.1756-8765.2011.01163.x

Rio, K., Bonneaud, S., & Warren, W. H. (2012). Speed coordination in pedestrian groups: Linking individual locomotion with crowd behavior. *J Vis, 12*(9), 190. doi:10.1167/12.9.190

Rio, K., & Warren, W. H. (2014). The visual coupling between neighbors in real and virtual crowds. *Transportation Research Procedia, 2*, 132–140. doi:10.1016/j.trpro.2014.09.017

Rio, K. W., Rhea, C. K., & Warren, W. H. (2014). Follow the leader: Visual control of speed in pedestrian following. *J Vis, 14*(2). doi:10.1167/14.2.4

Robert, M. T., Guberek, R., Sveistrup, H., & Levin, M. F. (2013). Motor learning in children with hemiplegic cerebral palsy and the role of sensation in short-term motor training of goal-directed reaching. *Dev Med Child Neurol, 55*(12), 1121–1128. doi:10.1111/dmcn.12219

Roberts, T. F., Tschida, K. A., Klein, M. E., & Mooney, R. (2010). Rapid spine stabilization and synaptic enhancement at the onset of behavioural learning. *Nature, 463*(7283), 948–952. doi:10.1038/nature08759

Rogers, J., Someya, T., & Huang, Y. (2010). Materials and mechanics for stretchable electronics. *Science, 327*, 1603–1607.

Rogier, E. W., Frantz, A. L., Bruno, M. E., Wedlund, L., Cohen, D. A., Stromberg, A. J., & Kaetzel, C. S. (2014). Secretory antibodies in breast milk promote long-term intestinal homeostasis by regulating the gut microbiota and host gene expression. *Proc Natl Acad Sci, 111*(8), 3074. doi:10.1073/pnas.1315792111

Ronsse, R., Lenzi, T., Vitiello, N., Koopman, B., van Asseldonk, E., De Rossi, S. M., . . . Ijspeert, A. J. (2011). Oscillator-based assistance of cyclical movements: Model-based and model-free approaches. *Med Biol Eng Comput, 49*(10), 1173–1185. doi:10.1007/s11517-011-0816-1

Roos, G., Van Wassenbergh, S., Leysen, H., Herrel, A., Adriaens, D., & Aerts, P. (2009). Ontogeny of feeding kinematics in the seahorse *Hippocampus reidi* from newly born to adult. *Integrative and Comparative Biology, 49*, E146–E146.

Rosenbaum, P. (2007). A report: The definition and classification of cerebral palsy, April 2006. *Dev Med Child Neurol, 49 Suppl 109*, 8–14.

Rosenthal, S., Biswas, J., & Veloso, M. (2010). *An effective personal mobile robot agent through symbiotic human-robot interaction*. Paper presented at the Proceedings of the 9th International Conference on Autonomous Agents and Multiagent Systems, May 2010, Toronto, Ontario, Canada.

Rosenthal, S., Veloso, M., & Dey, A. (2012a). Acquiring accurate human responses to robots' questions. *Int J of Soc Robotics, 4*(2), 117–129. doi:10.1007/s12369-012-0138-y

Rosenthal, S., Veloso, M., & Dey, A. (2012b). Is someone in this office available to help me? *J Intell Robot Syst, 66*(1), 205–221. doi:10.1007/s10846-011-9610-4

Rosenzweig, E. S., Brock, J. H., Culbertson, M. D., Lu, P., Moseanko, R., Edgerton, V. R., . . . Tuszynski, M. H. (2009). Extensive spinal decussation and bilateral termination of cervical corticospinal projections in rhesus monkeys. *J Comp Neurol, 513*(2), 151–163. doi:10.1002/cne.21940

Roth, E., Sponberg, S., & Cowan, N. J. (2014). A comparative approach to closed-loop computation. *Curr Opin Neurobiol, 25*, 54–62. doi:10.1016/j.conb.2013.11.005

Rouse, E. J., Mooney, L. M., & Herr, H. M. (2014). Clutchable series-elastic actuator: Implications for prosthetic knee design. *Int J Robot Res, 33*(13), 1611–1625. doi:10.1177/0278364914545673

Roux, L., Stark, E., Sjulson, L., & Buzsáki, G. (2014). In vivo optogenetic identification and manipulation of GABAergic interneuron subtypes. *Curr Opin Neurobiol, 26,* 88–95. doi:10.1016/j.conb.2013.12.013
Rubene, D., Hastad, O., Tauson, R., Wall, H., & Odeen, A. (2010). The presence of UV wavelengths improves the temporal resolution of the avian visual system. *J Exp Biol, 213*(Pt 19), 3357–3363. doi:10.1242/jeb.042424
Rubenstein, M., Ahler, C., Hoff, N., Cabrera, A., & Nagpal, R. (2014). Kilobot: A low cost robot with scalable operations designed for collective behaviors. *Robotics and Autonomous Systems, 62*(7), 966–975. doi:10.1016/j.robot.2013.08.006
Rubenstein, M., Cornejo, A., & Nagpal, R. (2014). Robotics: Programmable self-assembly in a thousand-robot swarm. *Science, 345*(6198), 795–799. doi:10.1126/science.1254295
Runeson, S. (1977). On the possibility of "smart" perceptual mechanisms. *Scandinavian Journal of Psychology, 18*(1), 172–179. doi:10.1111/j.1467-9450.1977.tb00274.x
Salmaso, N., Jablonska, B., Scafidi, J., Vaccarino, F. M., & Gallo, V. (2014). Neurobiology of premature brain injury. *Nat Neurosci, 17*(3), 341–346. doi:10.1038/nn.3604
Salter, M. W., & Beggs, S. (2014). Sublime microglia: Expanding roles for the guardians of the CNS. *Cell, 158*(1), 15–24. doi:10.1016/j.cell.2014.06.008
Saltzman, E., & Byrd, D. (2000). Demonstrating effects of parameter dynamics on gestural timing. *J Acoust Soc Amer, 107*(5), 2904 (Abstract). doi:10.1121/1.428807
Saltzman, E., & Holt, K. (2014). Movement forms: A graph-dynamic perspective. *Ecol Psychol, 26*(1–2), 60–68. doi:10.1080/10407413.2014.874891
Saltzman, E., & Kelso, J. A. S. (1987). Skilled actions: A task-dynamic approach. *Psychol Rev, 94*(1), 84–106.
Saltzman, E., Nam, H., Goldstein, L., & Byrd, D. (2006). The distinctions between state, parameter and graph dynamics in sensorimotor control and coordination. In M. L. Latash & F. Lestienne (Eds.), *Motor Control and Learning* (pp. 63–87). New York: Springer.
Saltzman, E., Nam, H., Krivokapic, J., & Goldstein, L. (2008). *A task-dynamic toolkit for modeling the effects of prosodic structure on articulation.* Paper presented at the Speech Prosody 2008, Campinas, Brazil.
Saltzman, E. L., & Munhall, K. G. (1992). Skill acquisition and development: The roles of state-, parameter, and graph dynamics. *J Mot Behav, 24*(1), 49–57.
Sampson, T. R., & Mazmanian, S. K. (2015). Control of brain development, function, and behavior by the microbiome. *Cell Host Microbe, 17*(5), 565–576. doi:10.1016/j.chom.2015.04.011
Sanchez, T., Welch, D., Nicastro, D., & Dogic, Z. (2011). Cilia-like beating of active microtubule bundles. *Science, 333,* 456–459.
Sane, S. P., & Dickinson, M. H. (2002). The aerodynamic effects of wing rotation and a revised quasi-steady model of flapping flight. *J Exper Biol, 205*(Pt 8), 1087.
Sane, S. P., & McHenry, M. J. (2009). The biomechanics of sensory organs. *Integrative and Comparative Biology,* i8–i23.
Sanes, J. R., & Zipursky, S. L. (2010). Design principles of insect and vertebrate visual systems. *Neuron, 66*(1), 15–36. doi:10.1016/j.neuron.2010.01.018
Sanger, T. D., & Kalaska, J. F. (2014). Crouching tiger, hidden dimensions. *Nat Neurosci, 17*(3), 338–340. doi:10.1038/nn.3663
Santello, M., Baud-Bovy, G., & Jorntell, H. (2013). Neural bases of hand synergies. *Front Comput Neurosci, 7,* 23. doi:10.3389/fncom.2013.00023
Sargent, B., Scholz, J., Reimann, H., Kubo, M., & Fetters, L. (2015). Development of infant leg coordination: Exploiting passive torques. *Infant Behavior and Development, 40,* 108–121. doi:10.1016/j.infbeh.2015.03.002
Sarkar, A., Lehto, S. M., Harty, S., Dinan, T. G., Cryan, J. F., & Burnet, P. W. (2016). Psychobiotics and the manipulation of bacteria-gut-brain signals. *Trends Neurosci, 39*(11), 763–781. doi:10.1016/j.tins.2016.09.002

Sarvestani, K., Kozlov, A., Harischandra, N., Grillner, S., & Ekeberg, O. (2013). A computational model of visually guided locomotion in lamprey. *Biol Cybern, 107*(5), 497–512. doi:10.1007/s00422-012-0524-4

Sathish, S., Sayeeda, H., Tanya, Y., Rashidul, H., Mustafa, M., Mohammed, A. A., . . . Jeffrey, I. G. (2014). Persistent gut microbiota immaturity in malnourished Bangladeshi children. *Nature, 510*(7505), 417. doi:10.1038/nature13421

Sato, Y., Hiratsuka, Y., Kawamata, I., Murata, S., & Nomura, S.-I. M. (2017). Micrometer-sized molecular robot changes its shape in response to signal molecules. *Sci Robot, 2*(4), eaal3735. doi:10.1126/scirobotics.aal3735

Satterlie, R. A. (2011). Do jellyfish have central nervous systems? *J Exp Biol, 214*(Pt 8), 1215–1223. doi:10.1242/jeb.043687

Satterlie, R. A. (2015). The search for ancestral nervous systems: An integrative and comparative approach. *J Exp Biol, 218*(Pt 4), 612–617. doi:10.1242/jeb.110387

Saunders, A., Oldenburg, I. A., Berezovskii, V. K., Johnson, C. A., Kingery, N. D., Elliott, H. L., . . . Sabatini, B. L. (2015). A direct GABAergic output from the basal ganglia to frontal cortex. *Nature, 521*(7550), 85–89. doi:10.1038/nature14179

Savin, T., Kurpios, N. A., Shyer, A. E., Florescu, P., Liang, H., Mahadevan, L., & Tabin, C. J. (2011). On the growth and form of the gut. *Nature, 476*(7358), 57–62. doi:10.1038/nature10277

Sawczuk, A., & Mosier, K. M. (2001). Neural control of tongue movement with respect to respiration and swallowing. *Crit Rev Oral Biol Med, 12*(1), 18–37.

Saxena, T., & Bellamkonda, R. V. (2015). Implantable electronics: A sensor web for neurons. *Nat Mater, 14*(12), 1190–1191. doi:10.1038/nmat4454

Scarry, R. (1979). *What do people do all day?* Abridged ed. New York: Random House.

Scassellati, B., Admoni, H., & Mataric, M. (2012). Robots for use in autism research. *Annu Rev Biomed Eng, 14*, 275–294. doi:10.1146/annurev-bioeng-071811-150036

Schafer, D. P., Lehrman, E. K., Kautzman, A. G., Koyama, R., Mardinly, A. R., Yamasaki, R., . . . Stevens, B. (2012). Microglia sculpt postnatal neural circuits in an activity and complement-dependent manner. *Neuron, 74*(4), 691–705. doi:10.1016/j.neuron.2012.03.026

Schafer, D. P., Lehrman, E. K., & Stevens, B. (2013). The "quad-partite" synapse: Microglia-synapse interactions in the developing and mature CNS. *Glia, 61*(1), 24–36. doi:10.1002/glia.22389

Schafer, D. P., & Stevens, B. (2013). Phagocytic glial cells: Sculpting synaptic circuits in the developing nervous system. *Curr Opin Neurobiol, 23*(6), 1034–1040. doi:10.1016/j.conb.2013.09.012

Schafer, W. (2016). Nematode nervous systems. *Curr Biol, 26*(20), R955–R959. doi:10.1016/j.cub.2016.07.044

Schaller, V., Weber, C., Semmrich, C., Frey, E., & Bausch, A. R. (2010). Polar patterns of driven filaments. *Nature, 467*(7311), 73–77. doi:10.1038/nature09312

Scheck, S. M., Boyd, R. N., & Rose, S. E. (2012). New insights into the pathology of white matter tracts in cerebral palsy from diffusion magnetic resonance imaging: A systematic review. *Dev Med Child Neurol, 54*(8), 684–696. doi:10.1111/j.1469-8749.2012.04332.x

Schmidt, R. C., Shaw, B. K., & Turvey, M. T. (1993). Coupling dynamics in interlimb coordination. *J Exp Psychol: Hum Percep Perform, 19*, 397–415.

Schneider, D. M., Nelson, A., & Mooney, R. (2014). A synaptic and circuit basis for corollary discharge in the auditory cortex. *Nature, 513*(7517), 189–194. doi:10.1038/nature13724

Schöll, E. (2010). Neural control: Chaos control sets the pace. *Nat Phys, 6*(3), 161–162. doi:10.1038/nphys1611

Scholpp, S., & Lumsden, A. (2010). Building a bridal chamber: Development of the thalamus. *Trends Neurosci, 33*(8), 373–380. doi:10.1016/j.tins.2010.05.003

Scholz, J. P., Danion, F., Latash, M. L., & Schöner, G. (2002). Understanding finger coordination through analysis of the structure of force variability. *BiolCybern, 86*(1), 29–39.

Scholz, J. P., & Schöner, G. (1999). The uncontrolled manifold concept: Identifying control variables for a functional task. *Exper Brain Res, 126*(3), 289–306.

Schöner, G., & Scholz, J. P. (2007). Analyzing variance in multi-degree-of-freedom movements: Uncovering structure versus extracting correlations. *Motor Control, 11*(3), 259–275. doi:10.1123/mcj.11.3.259

Schöner, M. G., Schöner, C. R., Simon, R., Grafe, T. U., Puechmaille, S. J., Ji, L. L., & Kerth, G. (2015). Bats are acoustically attracted to mutualistic carnivorous plants. *Curr Biol, 25*(14), 1911–1916. doi:10.1016/j.cub.2015.05.054

Schrodel, T., Prevedel, R., Aumayr, K., Zimmer, M., & Vaziri, A. (2013). Brain-wide 3D imaging of neuronal activity in Caenorhabditis elegans with sculpted light. *Nat Methods, 10*(10), 1013–1020. doi:10.1038/nmeth.2637

Schroter, M., Paulsen, O., & Bullmore, E. T. (2017). Micro-connectomics: Probing the organization of neuronal networks at the cellular scale. *Nat Rev Neurosci, 18*(3), 131–146. doi:10.1038/nrn.2016.182

Schulze, A., Gomez-Marin, A., Rajendran, V., Ahammad, P., Jayaraman, V., & Louis, M. (2012). Using optogenetics to explore the sensory representation of dynamic odor stimuli in Drosophila larvae. *Journal of Neurogenetics, 26*, 9–10.

Schwab, M. E. (2010). Functions of Nogo proteins and their receptors in the nervous system. *Nat Rev Neurosci, 11*(12), 799–811. doi:10.1038/nrn2936

Schwab, M. E., & Strittmatter, S. M. (2014). Nogo limits neural plasticity and recovery from injury. *Curr Opin Neurobiol, 27*, 53–60. doi:10.1016/j.conb.2014.02.011

Schwarz, D. A., Lebedev, M. A., Hanson, T. L., Dimitrov, D. F., Lehew, G., Meloy, J., . . . Nicolelis, M. A. (2014). Chronic, wireless recordings of large-scale brain activity in freely moving rhesus monkeys. *Nat Methods, 11*(6), 670–676. doi:10.1038/nmeth.2936

Scott, S. H., & Kalaska, J. F. (1997). Reaching movements with similar hand paths but different arm orientations. I. Activiity of individual cells in motor cortex. *J Neurophysiol, 77*, 826–852.

Seelig, J. D., & Jayaraman, V. (2015). Neural dynamics for landmark orientation and angular path integration. *Nature, 521*(7551), 186–191. doi:10.1038/nature14446

Seeman, N. C. (1999). DNA engineering and its application to nanotechnology. *Trends Biotechnol, 17*(11), 437–443. doi:10.1016/S0167-7799(99)01360-8

Seidman, L. J., & Nordentoft, M. (2015). New targets for prevention of schizophrenia: Is it time for interventions in the premorbid phase? *Schizophr Bull, 41*(4), 795–800. doi:10.1093/schbul/sbv050

Sejnowski, T. J., Churchland, P. S., & Movshon, J. A. (2014). Putting big data to good use in neuroscience. *Nat Neurosci, 17*(11), 1440–1441. doi:10.1038/nn.3839

Sekar, A., Bialas, A. R., de Rivera, H., Davis, A., Hammond, T. R., Kamitaki, N., . . . McCarroll, S. A. (2016). Schizophrenia risk from complex variation of complement component 4. *Nature, 530*(7589), 177–183. doi:10.1038/nature16549

Selen, L., Beek, P., & Dieën, J. (2005). Can co-activation reduce kinematic variability? A simulation study. *Biol Cybern, 93*(5), 373–381. doi:10.1007/s00422-005-0015-y

Selen, L., Beek, P., & van Dieën, J. (2007). Fatigue-induced changes of impedance and performance in target tracking. *Experimental Brain Research, 181*(1), 99–108. doi:10.1007/s00221-007-0909-0

Sergio, L. E., & Kalaska, J. F. (2003). Systematic changes in motor cortex cell activity with arm posture during directional isometric force generation. *J Neurophysiol, 89*(1), 212.

Serwane, F., Mongera, A., Rowghanian, P., Kealhofer, D. A., Lucio, A. A., Hockenbery, Z. M., & Campas, O. (2017). In vivo quantification of spatially varying mechanical properties in developing tissues. *Nat Methods, 14*(2), 181–186. doi:10.1038/nmeth.4101

Seth, R.-N., Yong, K., Steven, H. K., Sathish, S., Philip, P. A., Jeffrey, I. G., & Ruslan, M. (2015). Analysis of gene–environment interactions in postnatal development of the mammalian intestine. *Proc Natl Acad Sci, 112*(7), 1929. doi:10.1073/pnas.1424886112

Shadmehr, R., & Krakauer, J. W. (2008). A computational neuroanatomy for motor control. *Exp Brain Res, 185*(3), 359–381. doi:10.1007/s00221-008-1280-5

Shadmehr, R., & Mussa-Ivaldi, A. (1994). Adaptive representation of dynamics during learning of a motor task. *J Neurosci, 14*, 3208–3224.

Shah, P. K., Gerasimenko, Y., Shyu, A., Lavrov, I., Zhong, H., Roy, R. R., & Edgerton, V. R. (2012). Variability in step training enhances locomotor recovery after a spinal cord injury. *Eur J Neurosci, 36*(1), 2054–2062. doi:10.1111/j.1460-9568.2012.08106.x

Shapiro, M. D., Marks, M. E., Peichel, C. L., Blackman, B. K., Nereng, K. S., Jonsson, B., . . . Kingsley, D. M. (2004). Genetic and developmental basis of evolutionary pelvic reduction in threespine sticklebacks. *Nature, 428*, 717–723.

Sharon, G., Sampson, T. R., Geschwind, D. H., & Mazmanian, S. K. (2016). The central nervous system and the gut microbiome. *Cell, 167*(4), 915–932. doi:10.1016/j.cell.2016.10.027

Shaw, P., Law, J., & Lee, M. (2013). A comparison of learning strategies for biologically constrained development of gaze control on an iCub robot. *Autonomous Robots, 37*(1), 97–110. doi:10.1007/s10514-013-9378-4

Shenoy, K. V., & Carmena, J. M. (2014). Combining decoder design and neural adaptation in brain-machine interfaces. *Neuron, 84*(4), 665–680. doi:10.1016/j.neuron.2014.08.038

Shenoy, K. V., & Nurmikko, A. V. (2012). Brain models enabled by next-generation neurotechnology. *IEEE Pulse, March / April*, 31–36.

Shenoy, K. V., Sahani, M., & Churchland, M. M. (2013). Cortical control of arm movements: A dynamical systems perspective. *Annu Rev Neurosci, 36*, 337–359. doi:10.1146/annurev-neuro-062111-150509

Shepherd, R., Wise, A., & Fallon, J. (2013). Cochlear implants. *10*, 315–331. doi:10.1016/b978-0-7020-5310-8.00016-8

Shepherd, R. F., Ilievski, F., Choi, W., Morin, S. A., Stokes, A. A., Mazzeo, A. D., . . . Whitesides, G. M. (2011). Multigait soft robot. *Proc Natl Acad Sci USA, 108*(51), 20400–20403. doi:10.1073/pnas.1116564108

Shepherd, R. K., Shivdasani, M. N., Nayagam, D. A., Williams, C. E., & Blamey, P. J. (2013). Visual prostheses for the blind. *Trends Biotechnol, 31*(10), 562–571. doi:10.1016/j.tibtech.2013.07.001

Sherman, D. L., & Brophy, P. J. (2005). Mechanisms of axon ensheathment and myelin growth. *Nat Rev Neurosci, 6*(9), 683–690. doi:10.1038/nrn1743

Shigetani, Y., Sugahara, F., & Kuratani, S. (2005). A new evolutionary scenario for the vertebrate jaw. *BioEssays: News and Reviews in Molecular, Cellular and Developmental Biology 27*(3), 331–338.

Shih, C. T., Sporns, O., Yuan, S. L., Su, T. S., Lin, Y. J., Chuang, C. C., . . . Chiang, A. S. (2015). Connectomics-based analysis of information flow in the Drosophila brain. *Curr Biol, 25*(10), 1249–1258. doi:10.1016/j.cub.2015.03.021

Shim, J., Grosberg, A., Nawroth, J. C., Parker, K. K., & Bertoldi, K. (2012). Modeling of cardiac muscle thin films: Pre-stretch, passive and active behavior. *J Biomechan, 45*, 832–841.

Shin, J. W., & Mooney, D. J. (2016). Improving stem cell therapeutics with mechanobiology. *Cell Stem Cell, 18*(1), 16–19. doi:10.1016/j.stem.2015.12.007

Shmuelof, L., & Krakauer, J. W. (2011). Are we ready for a natural history of motor learning? *Neuron, 72*(3), 469–476. doi:10.1016/j.neuron.2011.10.017

Shubin, N. H., Daeschler, E. B., & Jenkins, F. A., Jr. (2014). Pelvic girdle and fin of Tiktaalik roseae. *Proc Natl Acad Sci USA, 111*(3), 893–899. doi:10.1073/pnas.1322559111

Shulz, D. E., & Feldman, D. E. (2013). Spike timing-dependent plasticity. In J. Rubenstein and P. Rakic (Eds.), *Neural circuit development and function in the brain: Comprehensive developmental neuroscience* (155–181). Amsterdam: Academic Press. doi:10.1016/b978-0-12-397267-5.00029-7

Siegel, M., Buschman, T. J., & Miller, E. K. (2015). Cortical information flow during flexible sensorimotor decisions. *Science, 348*(6241), 1352–1355.

Siegle, J. H., Pritchett, D. L., & Moore, C. I. (2014). Gamma-range synchronization of fast-spiking interneurons can enhance detection of tactile stimuli. *Nat Neurosci, 17*(10), 1371–1379. doi:10.1038/nn.3797

Silasi, G., & Murphy, T. H. (2014). Stroke and the connectome: How connectivity guides therapeutic intervention. *Neuron, 83*(6), 1354–1368. doi:10.1016/j.neuron.2014.08.052

Sillar, K. T., Combes, D., Ramanathan, S., Molinari, M., & Simmers, J. (2008). Neuromodulation and developmental plasticity in the locomotor system of anuran amphibians during metamorphosis. *Brain Res Rev, 57*(1), 94–102. doi:10.1016/j.brainresrev.2007.07.018

Sillar, K. T., Combes, D., & Simmers, J. (2014). Neuromodulation in developing motor microcircuits. *Curr Opin Neurobiol, 29*, 73–81. doi:10.1016/j.conb.2014.05.009

Simon, M., Woods, W., Serebrenik, Y., et al. (2010). Visceral-locomotory pistoning in crawling caterpillars. *Curr Biol, 20*, 1458–1463.

Simone-Finstrom, M., & Spivak, M. (2010). Propolis and bee health: the natural history and significance of resin use by honey bees. *Apidologie, 41*(3), 295–311. doi:10.1051/apido/2010016

Skotheim, J. M., & Mahadevan, L. (2005). Physical limits and design principles for plant and fungal movements. *Science, 308*(5726), 1308–1310. doi:10.1126/science.1107976

Slack, J. M. W. (2002). Conrad Hal Waddington: The last Renaissance biologist? *Nature Reviews Genetics, 3*(11), 889.

Sloan, S. A., & Barres, B. A. (2014). Mechanisms of astrocyte development and their contributions to neurodevelopmental disorders. *Curr Opin Neurobiol, 27*, 75–81. doi:10.1016/j.conb.2014.03.005

Slotkin, J. R., Pritchard, C. D., Luque, B., Ye, J., Layer, R. T., Lawrence, M. S., . . . Langer, R. (2017). Biodegradable scaffolds promote tissue remodeling and functional improvement in non-human primates with acute spinal cord injury. *Biomaterials, 123*, 63–76. doi:10.1016/j.biomaterials.2017.01.024

Smith, J. C., Abdala, A. P., Borgmann, A., Rybak, I., & Paton, J. F. (2013). Brainstem respiratory networks: Building blocks and microcircuits. *Trends Neurosci, 36*, 152–162.

Smith, K. K., & Kier, W. M. (1989). Trunks, tongues, and tentacles: Moving with skeletons of muscle. *Am Scientist, 77*, 28–35.

Smith, V. C., Kelty-Stephen, D., Qureshi Amad, M., Mao, W., Cakert, K., Osborne, J., & Paydarfar, D. (2015). Stochastic resonance effects on apnea, bradycardia, and oxygenation: A randomized controlled trial. *Pediatrics, 136*, 1561–1568.

Smyser, C. D., & Neil, J. J. (2015). Use of resting-state functional MRI to study brain development and injury in neonates. *Semin Perinatol, 39*(2), 130–140. doi:10.1053/j.semperi.2015.01.006

Soekadar, S. R., Witkowski, M., Gómez, C., Opisso, E., Medina, J., Cortese, M., . . . Vitiello, N. (2016). Hybrid EEG / EOG-based brain / neural hand exoskeleton restores fully independent daily living activities after quadriplegia. *Sci Robot, 1*(1), eaag3296. doi:10.1126/scirobotics.aag3296

Sommer, F., & Backhed, F. (2013). The gut microbiota—masters of host development and physiology. *Nat Rev Microbiol, 11*(4), 227–238. doi:10.1038/nrmicro2974

Song, F., Xiao, K. W., Bai, K., & Bai, Y. L. (2007). Microstructure and nanomechanical properties of the wing membrane of dragonfly. *Mater Sci Engin: A, 457*(1), 254–260. doi:10.1016/j.msea.2007.01.136

Song, J. W., Mitchell, P. D., Kolasinski, J., Ellen Grant, P., Galaburda, A. M., & Takahashi, E. (2014). Asymmetry of white matter pathways in developing human brains. *Cereb Cortex 25*(9), 2883–2893, doi:10.1093/cercor/bhu084

Sonnenburg, J. L., & Backhed, F. (2016). Diet-microbiota interactions as moderators of human metabolism. *Nature, 535*(7610), 56–64. doi:10.1038/nature18846

Sonoda, K., Asakura, A., Minoura, M., Elwood, R. W., & Gunji, Y. P. (2012). Hermit crabs perceive the extent of their virtual bodies. *Biol Lett, 8*(4), 495–497. doi:10.1098/rsbl.2012.0085

Soska, K. C., & Adolph, K. E. (2014). Postural position constrains multimodal object exploration in infants. *Infancy, 19*(2), 138–161. doi:10.1111/infa.12039

Sosnik, R., Hauptmann, B., Karni, A., & Flash, T. (2004). When practice leads to co-articulation: The evolution of geometrically defined movement primitives. *Exp Brain Res, 156*(4), 422–438. doi:10.1007/s00221-003-1799-4

Southwell, D. G., Froemke, R. C., Alvarez-Buylla, A., Stryker, M., & Gandhi, S. P. (2010). Cortical plasticity induced by inhibitory neuron transplantation. *Science, 327*, 1145–1148.

Southwell, D. G., Paredes, M. F., Galvao, R. P., Jones, D. L., Froemke, R. C., Sebe, J. Y., . . . Alvarez-Buylla, A. (2012). Intrinsically determined cell death of developing cortical interneurons. *Nature, 491*(7422), 109–113. doi:10.1038/nature11523

Spaulding, S., & Breazeal, C. (2015). Affect and inference in Bayesian knowledge tracing with a robot tutor (pp. 219–220). *Proceedings of the Tenth Annual ACM / IEEE International Conference on Human-Robot Interaction.*

Spergel, D. N. (2015). The dark side of cosmology: Dark matter and dark energy. *Science, 347,* 1100–1102.

Spira, M. E., & Hai, A. (2013). Multi-electrode array technologies for neuroscience and cardiology. *Nat Nanotechnol, 8*(2), 83–94. doi:10.1038/nnano.2012.265

Sponberg, S., Libby, T., Mullens, C. H., & Full, R. J. (2011). Shifts in a single muscle's control potential of body dynamics are determined by mechanical feedback. *Philos Trans R Soc B, 366,* 1606–1620.

Sporns, O. (2011). *Networks of the brain.* Cambridge, MA: MIT Press.

Sporns, O. (2012). *Discovering the human connectome.* Cambridge, MA: MIT Press.

Sporns, O. (2013a). Making sense of brain network data. *Nat Methods, 10*(6), 491–493. doi:10.1038/nmeth.2485

Sporns, O. (2013b). Network attributes for segregation and integration in the human brain. *Curr Opin Neurobiol, 23*(2), 162–171. doi:10.1016/j.conb.2012.11.015

Sporns, O. (2014). Contributions and challenges for network models in cognitive neuroscience. *Nat Neurosci, 17*(5), 652–660. doi:10.1038/nn.3690

Sporns, O., & Betzel, R. F. (2016). Modular brain networks. *Annu Rev Psychol, 67,* 613–640. doi:10.1146/annurev-psych-122414-033634

Sporns, O., & Honey, C. J. (2013). Topographic dynamics in the resting brain. *Neuron, 78*(6), 955–956. doi:10.1016/j.neuron.2013.05.037

Sporns, O., Tononi, G., & Kötter, R. (2005). The human connectome: A structural description of the human brain (Review). *PLOS Comp Biol, 1*(4), e42. doi:10.1371/journal.pcbi.0010042

Squires, T. M., & Quake, S. R. (2005). Microfluidics: Fluid physics at the nanoliter scale. *Rev Mod Phys, 77,* 977–1026.

Sreetharan, P. S., Whitney, J. P., Strauss, M. D., & Wood, R. J. (2012). Monolithic fabrication of millimeter-scale machines. *J Micromech and Microeng, 22*(5), 055027. doi:10.1088/0960-1317/22/5/055027

Sreetharan, P. S., & Wood, R. J. (2010). Passive aerodynamic drag balancing in a flapping-wing robotic insect. *J Mech Des, 132*(5), 051006. doi:10.1115/1.4001379

Sreetharan, P. S., & Wood, R. J. (2011). Passive torque regulation in an underactuated flapping wing robotic insect. *Auton Robots, 31*(2–3), 225–234. doi:10.1007/s10514-011-9242-3

Srinivasan, M. V. (2010). Honey bees as a model for vision, perception, and cognition. *Annu Rev Entomol, 55,* 267–284. doi:10.1146/annurev.ento.010908.164537

Srinivasan, R., Li, Q., Zhou, X., Lu, J., Lichtman, J., & Wong, S. T. (2010). Reconstruction of the neuromuscular junction connectome. *Bioinformatics, 26*(12), i64–i70. doi:10.1093/bioinformatics/btq179

Srivastava, M., Simakov, O., Chapman, J., Fahey, B., Gauthier, M. E., Mitros, T., . . . Rokhsar, D. S. (2010). The Amphimedon queenslandica genome and the evolution of animal complexity. *Nature, 466*(7307), 720–726. doi:10.1038/nature09201

Standen, E. M., Du, T. Y., & Larsson, H. C. (2014). Developmental plasticity and the origin of tetrapods. *Nature, 513*(7516), 54–58. doi:10.1038/nature13708

Starkey, M. L., & Schwab, M. E. (2014). How plastic is the brain after a stroke? *Neuroscientist, 20*(4), 359–371. doi:10.1177/1073858413514636

Steck, K., Wittlinger, M., & Wolf, H. (2009). Estimation of homing distance in desert ants, Cataglyphis fortis, remains unaffected by disturbance of walking behaviour. *J Exp Biol, 212*(18), 2893–2901. doi:10.1242/jeb.030403

Stegmaier, J., Amat, F., Lemon, W. C., McDole, K., Wan, Y., Teodoro, G., . . . Keller, P. J. (2016). Real-time three-dimensional cell segmentation in large-scale microscopy data of developing embryos. *Dev Cell, 36*(2), 225–240. doi:10.1016/j.devcel.2015.12.028

Steinberg, E. E., Christoffel, D. J., Deisseroth, K., & Malenka, R. C. (2015). Illuminating circuitry relevant to psychiatric disorders with optogenetics. *Curr Opin Neurobiol, 30*, 9–16. doi:10.1016/j.conb.2014.08.004

Stephen, D., Hsu, W., Young, D., Saltzman, E., Holt, K., Newman, D., . . . Goldfield, E. (2012). Multifractal fluctuations in joint angles during spontaneous kicking reveal multiplicativity-driven coordination. *Chaos, Solitons, and Fractals, 45*, 1201–1219.

Stephenson-Jones, M., Samuelsson, E., Ericsson, J., Robertson, B., & Grillner, S. (2011). Evolutionary conservation of the basal ganglia as a common vertebrate mechanism for action selection. *Curr Biol, 21*(13), 1081–1091. doi:10.1016/j.cub.2011.05.001

Stocker, R. (2012). Marine microbes see a sea of gradients. *Science, 338*(6107), 628–633. doi:10.1126/science.1208929

Stoll, B. J., Hansen, N. I., Bell, E. F., Shankaran, S., Laptook, A. R., Walsh, M. C., . . . Human Development Neonatal Research Network. (2010). Neonatal outcomes of extremely preterm infants from the NICHD Neonatal Research Network. *Pediatrics, 126*(3), 443–456. doi:10.1542/peds.2009-2959

Stoner, R., Chow, M. L., Boyle, M. P., Sunkin, S. M., Mouton, P. R., Roy, S., . . . Courchesne, E. (2014). Patches of disorganization in the neocortex of children with autism. *N Engl J Med, 370*(13), 1209–1219. doi:10.1056/NEJMoa1307491

Straw, A. D., Lee, S., & Dickinson, M. H. (2010). Visual control of altitude in flying Drosophila. *Curr Biol, 20*(17), 1550–1556. doi:10.1016/j.cub.2010.07.025

Studholme, C. (2015). Mapping the developing human brain in utero using quantitative MR imaging techniques. *Semin Perinatol, 39*(2), 105–112. doi:10.1053/j.semperi.2015.01.003

Subramanian, S., Blanton, L. V., Frese, Steven A., Charbonneau, M., Mills, David A., & Gordon, Jeffrey I. (2015). Cultivating healthy growth and nutrition through the gut microbiota. *Cell, 161*(1), 36–48. doi:10.1016/j.cell.2015.03.013

Sumbre, G., Fiorito, G., Flash, T., & Hochner, B. (2005). Neurobiology: Motor control of flexible octopus arms. *Nature, 433*(7026), 595.

Sumbre, G., Fiorito, G., Flash, T., & Hochner, B. (2006). Octopuses use a human-like strategy to control precise point-to-point arm movements. *Curr Biol, 16*(8), 767–772. doi:10.1016/j.cub.2006.02.069

Sun, T., & Hevner, R. F. (2014). Growth and folding of the mammalian cerebral cortex: From molecules to malformations. *Nat Rev Neurosci, 15*(4), 217–232. doi:10.1038/nrn3707

Sun, Y., Jallerat, Q., Szymanski, J. M., & Feinberg, A. W. (2015). Conformal nanopatterning of extracellular matrix proteins onto topographically complex surfaces. *Nat Methods, 12*(2), 134–136. doi:10.1038/nmeth.3210

Sung, C., Demaine, E., Demaine, M., & Rus, D. (2013, August 4–7). *Joining unfoldings of 3-D surfaces.* Paper presented at the IDETC / CIE, Portland, OR.

Supekar, K. S., Musen, M. A., & Menon, V. (2009). Development of large-scale functional Brain networks in children. *Neuroimage, Suppl 1, 47*, S109.

Sussillo, D. (2014). Neural circuits as computational dynamical systems. *Curr Opin Neurobiol, 25*, 156–163. doi:10.1016/j.conb.2014.01.008

Sussillo, D., & Barak, O. (2013). Opening the black box: Low-dimensional dynamics in high-dimensional recurrent neural networks. *Neural Computation, 25*, 626–649.

Sussillo, D., Churchland, M. M., Kaufman, M. T., & Shenoy, K. V. (2015). A neural network that finds a naturalistic solution for the production of muscle activity. *Nat Neurosci, 18*(7), 1025–1033. doi:10.1038/nn.4042

Sussillo, D., Stavisky, S. D., Kao, J. C., Ryu, S. I., & Shenoy, K. V. (2016). Making brain-machine interfaces robust to future neural variability. *Nat Commun, 7*, 13749. doi:10.1038/ncomms13749

Sutton, R. S., & Barto, A. (1998). *Reinforcement learning: An introduction*. Cambridge, MA: MIT Press.

Suver, M. P., Mamiya, A., & Dickinson, M. H. (2012). Octopamine neurons mediate flight-induced modulation of visual processing in Drosophila. *Curr Biol, 22*(24), 2294–2302. doi:10.1016/j.cub.2012.10.034

Svoboda, K., & Yasuda, R. (2006). Principles of two-photon excitation microscopy and its applications to neuroscience. *Neuron, 50*(6), 823–839. doi:10.1016/j.neuron.2006.05.019

Swanson, L. W., & Lichtman, J. W. (2016). From cajal to connectome and beyond. *Annu Rev Neurosci, 39*, 197–216. doi:10.1146/annurev-neuro-071714-033954

Taber, L. A. (2014). Morphomechanics: Transforming tubes into organs. *Curr Opin Genet Dev, 27*, 7–13. doi:10.1016/j.gde.2014.03.004

Tabot, G. A., Dammann, J. F., Berg, J. A., Tenore, F. V., Boback, J. L., Vogelstein, R. J., & Bensmaia, S. J. (2013). Restoring the sense of touch with a prosthetic hand through a brain interface. *Proc Natl Acad Sci U S A, 110*(45), 18279–18284. doi:10.1073/pnas.1221113110

Takahashi, E., Dai, G., Rosen, G. D., Wang, R., Ohki, K., Folkerth, R. D., . . . Ellen Grant, P. (2011). Developing neocortex organization and connectivity in cats revealed by direct correlation of diffusion tractography and histology. *Cereb Cortex, 21*(1), 200–211. doi:10.1093/cercor/bhq084

Takahashi, E., Dai, G., Wang, R., Ohki, K., Rosen, G. D., Galaburda, A. M., . . . Wedeen, V. J. (2010). Development of cerebral fiber pathways in cats revealed by diffusion spectrum imaging. *Neuroimage, 49*, 1231–1240.

Takahashi, E., Folkerth, R. D., Galaburda, A. M., & Grant, P. E. (2012). Emerging cerebral connectivity in the human fetal brain: An MR tractography study. *Cereb Cortex, 22*(2), 455–464. doi:10.1093/cercor/bhr126

Takahashi, E., Hayashi, E., Schmahmann, J. D., & Ellen Grant, P. (2014). Development of cerebellar connectivity in human fetal brains revealed by high angular resolution diffusion tractography. *Neuroimage, 96*, 326–333. doi:10.1016/j.neuroimage.2014.03.022

Takahashi, E., Song, J. W., Folkerth, R. D., Grant, P. E., & Schmahmann, J. D. (2013). Detection of postmortem human cerebellar cortex and white matter pathways using high angular resolution diffusion tractography: A feasibility study. *Neuroimage, 68*, 105–111. doi:10.1016/j.neuroimage.2012.11.042

Takatoh, J., Nelson, A., Zhou, X., Bolton, M. M., Ehlers, M. D., Arenkiel, B. R., . . . Wang, F. (2013). New modules are added to vibrissal premotor circuitry with the emergence of exploratory whisking. *Neuron, 77*(2), 346–360. doi:10.1016/j.neuron.2012.11.010

Takeoka, A., Vollenweider, I., Courtine, G., & Arber, S. (2014). Muscle spindle feedback directs locomotor recovery and circuit reorganization after spinal cord injury. *Cell, 159*(7), 1626–1639. doi:10.1016/j.cell.2014.11.019

Takesian, A. E., & Hensch, T. K. (2013). Balancing plasticity / stability across brain development. *Prog Brain Res, 207*, 3–34. doi:10.1016/B978-0-444-63327-9.00001-1

Tallinen, T., Chung, J. Y., Biggins, J. S., & Mahadevan, L. (2014). Gyrification from constrained cortical expansion. *Proc Natl Acad Sci USA, 111*(35), 12667–12672. doi:10.1073/pnas.1406015111

Tallinen, T., Chung, J. Y., Rousseau, F., Girard, N., Lefèvre, J., & Mahadevan, L. (2016). On the growth and form of cortical convolutions. *Nat Phys 12*, 588–593. doi:10.1038/nphys3632

Talpalar, A. E., Bouvier, J., Borgius, L., Fortin, G., Pierani, A., & Kiehn, O. (2013). Dual-mode operation of neuronal networks involved in left-right alternation. *Nature, 500*(7460), 85–88. doi:10.1038/nature12286

Tan, F., Walshe, P., Viani, L., & Al-Rubeai, M. (2013). Surface biotechnology for refining cochlear implants. *Trends Biotechnol, 31*(12), 678–687. doi:10.1016/j.tibtech.2013.09.001

Tanaka, E. M. (2016). The molecular and cellular choreography of appendage regeneration. *Cell, 165*(7), 1598–1608. doi:10.1016/j.cell.2016.05.038

Tanaka, E. M., & Ferretti, P. (2009). Considering the evolution of regeneration in the central nervous system. *Nat Rev Neurosci, 10*(10), 713–723. doi:10.1038/nrn2707

Tanaka, H., Whitney, J. P., & Wood, R. J. (2011). Effect of flexural and torsional wing flexibility on lift generation in hoverfly flight. *Integr Comp Biol, 51*(1), 142–150. doi:10.1093/icb/icr051

Tang-Schomer, M. D., White, J. D., Tien, L. W., Schmitt, L. I., Valentin, T. M., Graziano, D. J., . . . Kaplan, D. L. (2014). Bioengineered functional brain-like cortical tissue. *Proc Natl Acad Sci USA, 111*(38), 13811–13816. doi:10.1073/pnas.1324214111

Tapia, J. C., Wylie, J. D., Kasthuri, N., Hayworth, K. J., Schalek, R., Berger, D. R., . . . Lichtman, J. W. (2012). Pervasive synaptic branch removal in the mammalian neuromuscular system at birth. *Neuron, 74*(5), 816–829. doi:10.1016/j.neuron.2012.04.017

Tapus, A., Maja, M., & Scassellatti, B. (2007). The grand challenges in socially assistive robotics. *IEEE Robotics and Automation Magazine 14*, 35–42.

Tau, G. Z., & Peterson, B. S. (2010). Normal development of brain circuits. *Neuropsychopharmacology, 35*, 147–168.

Taube, J. S. (2007). The head direction signal: Origins and sensory-motor integration. *Annu Rev Neurosci, 30*, 181–207. doi:10.1146/annurev.neuro.29.051605.112854

Taylor, A. H., Hunt, G. R., Holzhaider, J. C., & Gray, R. D. (2007). Spontaneous metatool use by New Caledonian crows. *Curr Biol, 17*(17), 1504–1507. doi:10.1016/j.cub.2007.07.057

Taylor, G. K., & Krapp, H. G. (2007). Sensory systems and flight stability: What do insects measure and why? *Advances in Insect Physiology, 34*, 231–316. doi:10.1016/S0065-2806(07)34005-8

Tedeschi, A. (2011). Tuning the orchestra: Transcriptional pathways controlling axon regeneration. *Front Mol Neurosci, 4*, 60. doi:10.3389/fnmol.2011.00060

Tennenbaum, M., Liu, Z., Hu, D., & Fernandez-Nieves, A. (2016). Mechanics of fire ant aggregations. *Nat Mater, 15*(1), 54–59. doi:10.1038/nmat4450

Teoh, Z. E., Fuller, S. B., Chirarattananon, P., Perez-Arancibia, N. O., Greenberg, J. D., & Wood, R. J. (2012). *A hovering flapping-wing microrobot with altitude control and passive upright stability.* Paper presented at the IEEE / RSJ International Conference on Intelligent Robots and Systems, Vilamoura, Algarve, Portugal.

Tetsuya, I., Jun, C., & Friesen, W. O. (2014). Biological clockwork underlying adaptive rhythmic movements. *Proc Natl Acad Sci, 111*(3), 978. doi:10.1073/pnas.1313933111

Teulier, C., Lee, D. K., & Ulrich, B. D. (2015). Early gait development in human infants: Plasticity and clinical applications. *Dev Psychobiol, 57*(4), 447–458. doi:10.1002/dev.21291

Teulier, C., Sansom, J. K., Muraszko, K., & Ulrich, B. D. (2012). Longitudinal changes in muscle activity during infants' treadmill stepping. *J Neurophysiol, 108*(3), 853–862. doi:10.1152/jn.01037.2011

Thakor, N. (2013). Translating the brain-machine interface. *Sci Transl Med*, 5(210), 1–7.

Thelen, E. (1979). Rhythmical stereotypies in normal human infants. *Animal Behaviour, 27*, 699–715. doi:10.1016/0003-3472(79)90006-X

Thelen, E. (1981). Rhythmical behavior in infancy: An ethological perspective. *Dev Psychol, 17*(3), 237–257.

Thelen, E. (1989). The (re)discovery of motor development: Learning new things from an old field. *Dev Psychol, 25*(6), 946–949. doi:10.1037/0012-1649.25.6.946

Thelen, E. (1995). Motor development: A new synthesis. *American Psychologist, 50*(2), 79–95.

Thelen, E. (1996). Motor development: A new synthesis. *Annual Progress in Child Psychiatry & Child Development*, 32–66.

Thelen, E. (2000). Grounded in the world: Developmental origins of the embodied mind. *Infancy, 1*(1), 3–28.

Thelen, E., & Adolph, K. E. (1992). Arnold L. Gessell: the paradox of nature and nurture. (APA Centennial Feature). *Dev Psychol, 28*(3), 368.

Thelen, E., Corbetta, D., Kamm, K., Spencer, J. P., Schneider, K., & Zernicke, R. F. (1993). The transition to reaching: Mapping intention and intrinsic dynamics. *Child Dev, 64*(4), 1058–1098.

Thelen, E., & Fisher, D. M. (1983). From spontaneous to instrumental behavior: Kinematic analysis of movement changes during very early learning. *Child Dev, 54*, 129–140.

Thelen, E., Fisher, D. M., & Ridley-Johnson, R. (2002). The relationship between physical growth and a newborn reflex. *Infant Behav Dev, 25*(1), 72–85.

Thelen, E., Schöner, G., Scheier, C., & Smith, L. B. (2001). The dynamics of embodiment: A field theory of infant perseverative reaching. *Behav Brain Sci, 24*(1), 1–34; discussion 34–86.

Thelen, E., & Smith, L. (1994). *A dynamic systems approach to the development of cognition and action*. Cambridge, MA: MIT Press.

Thelen, E., & Ulrich, B. D. (1991). Hidden skills: A dynamic systems analysis of treadmill stepping during the first year. *Monogr Soc Res Child Dev, 56*(1), 1–98; discussion 99–104.

Therrien, A. S., & Bastian, A. J. (2015). Cerebellar damage impairs internal predictions for sensory and motor function. *Curr Opin Neurobiol, 33*, 127–133. doi:10.1016/j.conb.2015.03.013

Thompson, D. A. W. (1942). *On growth and form* (2nd ed.). Cambridge, UK: Cambridge University Press.

Thompson, W. R., Rubin, C. T., & Rubin, J. (2012). Mechanical regulation of signaling pathways in bone. *Gene, 503*, 179–193.

Thuret, S., Moon, L. D., & Gage, F. H. (2006). Therapeutic interventions after spinal cord injury. *Nat Rev Neurosci, 7*(8), 628–643. doi:10.1038/nrn1955

Tibbits, S. (2012). Design to self-assembly. *Architectural Des, 82*, 68–73.

Tim, D. W., Gen, S., & Berhane, A. (1994). Australopithecus ramidus, a new species of early hominid from Aramis, Ethiopia. *Nature, 371*(6495), 306. doi:10.1038/371306a0

Tolley, M. T., Shepherd, R. F., Mosadegh, B., Galloway, K. C., Wehner, M., Karpelson, M., . . . Whitesides, G. M. (2014). A resilient, untethered soft robot. *Soft Robotics, 1*(3), 213–223. doi:10.1089/soro.2014.0008

Tomchek, S., & Dunn, W. (2007). Sensory processing in children with and without autism: A comparative study using the short sensory profile. *American Journal of Occupational Therapy, 61*(2), 190–200.

Toyoizumi, T., Miyamoto, H., Yazaki-Sugiyama, Y., Atapour, N., Hensch, T. K., & Miller, K. D. (2013). A theory of the transition to critical period plasticity: Inhibition selectively suppresses spontaneous activity. *Neuron, 80*(1), 51–63. doi:10.1016/j.neuron.2013.07.022

Tresch, M. C., Saltiel, P., & Bizzi, E. (1999). The construction of movement by the spinal cord. *Nat Neurosci, 2*(2), 162–167. doi:10.1038/5721

Trevisan, M., Mindlin, G., & Goller, F. (2006). Nonlinear model predicts diverse respiratory patterns of birdsong. *Phys Rev Lett, 96*(5), 1–4. doi:10.1103/PhysRevLett.96.058103

Trimmer, B. (2013). A journal of soft robotics: Why now? *Soft Robotics, 1*, 1–4.

Trimmer, B., & Issberner, J. (2007). Kinematics of soft-bodied, legged locomotion in Manduca sexta larvae. *Biol Bull, 212*, 130–142.

Truby, R. L., & Lewis, J. A. (2016). Printing soft matter in three dimensions. *Nature, 540*(7633), 371–378. doi:10.1038/nature21003

Tsai, H.-H., Li, H., Fuentealba, L. C., Molofsky, A. V., Taveira-Marques, R., Zhuang, H., . . . Rowitch, D. H. (2012). Regional astrocyte allocation regulates CNS synaptogenesis and repair. *Science, 337*(6092), 358–362. doi:10.1126/science.1222381

Tsai, H.-H., Niu, J., Munji, R., Davalos, D., Chang, J., Zhang, H., . . . Fancy, S. P. (2016). Oligodendrocyte precursors migrate along vasculature in the developing nervous system. *Science, 351*, 379–384.

Tschida, K., & Mooney, R. (2012). The role of auditory feedback in vocal learning and maintenance. *Curr Opin Neurobiol, 22*(2), 320–327. doi:10.1016/j.conb.2011.11.006

Tschida, K. A., & Mooney, R. (2012). Deafening drives cell-type-specific changes to dendritic spines in a sensorimotor nucleus important to learned vocalizations. *Neuron, 73*(5), 1028–1039. doi:10.1016/j.neuron.2011.12.038

Turing, A. M. (2004). *The essential Turing: Seminal writings in computing, logic, philosophy, artificial intelligence, and artificial life, plus the secrets of Enigma*. Oxford: Oxford University Press.

Turner, J. S. (2000). *The extended organism: The physiology of animal-built structures*. Cambridge, MA: Harvard University Press.

Turner, J. S. (2010). Termites as models of swarm cognition. *Swarm Intell, 5*(1), 19–43. doi:10.1007/s11721-010-0049-1

Turney, S. G., & Lichtman, J. W. (2012). Reversing the outcome of synapse elimination at developing neuromuscular junctions in vivo: Evidence for synaptic competition and its mechanism. *PLoS Biol, 10*(6), e1001352. doi:10.1371/journal.pbio.1001352

Turrigiano, G. (2011). Too many cooks? Intrinsic and synaptic homeostatic mechanisms in cortical circuit refinement. *Annu Rev Neurosci, 34,* 89–103. doi:10.1146/annurev-neuro-060909-153238

Turrigiano, G. (2012). Homeostatic synaptic plasticity: Local and global mechanisms for stabilizing neuronal function. *Cold Spring Harb Perspect Biol, 4*(1), a005736. doi:10.1101/cshperspect.a005736

Turvey, M. T. (1990). Coordination. *Am Psychol, 45,* 938–953.

Turvey, M. T. (2007). Action and perception at the level of synergies. *Hum Movement Sci, 26*(4), 657–697. doi:10.1016/j.humov.2007.04.002

Turvey, M. T., & Carello, C. (2011). Obtaining information by dynamic (effortful) touching. *Philos Trans R Soc Lond B Biol Sci, 366*(1581), 3123–3132. doi:10.1098/rstb.2011.0159

Turvey, M. T., Carello, C., Fitzpatrick, P., Pagano, C., & Kadar, E. (1996). Spinors and selective dynamic touch. *J Exp Psychol: Hum Percep Perform, 22*(5), 1113–1126.

Turvey, M. T., & Fonseca, S. T. (2014). The medium of haptic perception: A tensegrity hypothesis. *J Mot Behav, 46*(3), 143–187. doi:10.1080/00222895.2013.798252

Turvey, M. T., Harrison, S. J., Frank, T. D., & Carello, C. (2012). Human odometry verifies the symmetry perspective on bipedal gaits. *J Exp Psychol: HumPercep Perform, 38*(4), 1014–1025.

Tuszynski, M. H., & Steward, O. (2012). Concepts and methods for the study of axonal regeneration in the CNS. *Neuron, 74*(5), 777–791. doi:10.1016/j.neuron.2012.05.006

Tye, K. M., & Deisseroth, K. (2012). Optogenetic investigation of neural circuits underlying brain disease in animal models. *Nat Rev Neurosci, 13*(4), 251–266. doi:10.1038/nrn3171

Tymofiyeva, O., Hess, C. P., Ziv, E., Tian, N., Bonifacio, S. L., McQuillen, P. S., . . . Xu, D. (2012). Towards the "baby connectome": Mapping the structural connectivity of the newborn brain. *PLoS ONE, 7*(2), e31029. doi:10.1371/journal.pone.0031029

Tytell, E. D., Holmes, P., & Cohen, A. H. (2011). Spikes alone do not behavior make: Why neuroscience needs biomechanics. *Curr Opin Neurobiol, 21*(5), 816–822. doi:10.1016/j.conb.2011.05.017

Tytell, E. D., Hsu, C. Y., & Fauci, L. J. (2014). The role of mechanical resonance in the neural control of swimming in fishes. *Zoology (Jena), 117*(1), 48–56. doi:10.1016/j.zool.2013.10.011

Tytell, E. D., Hsu, C. Y., Williams, T. L., Cohen, A. H., & Fauci, L. J. (2010). Interactions between internal forces, body stiffness, and fluid environment in a neuromechanical model of lamprey swimming. *Proc Natl Acad Sci USA, 107*(46), 19832–19837. doi:10.1073/pnas.1011564107

Uddin, L. Q., & Menon, V. (2009). The anterior insula in autism: Under-connected and under-examined. *Neuroscience and Biobehavioral Reviews,* 33(8), 1198–1203. doi:10.1016/j.neubiorev.2009.06.002

Uddin, L. Q., Supekar, K. S., Ryali, S., & Menon, V. (2011). Dynamic reconfiguration of structural and functional connectivity across core neurocognitive brain networks with development. *J Neurosci, 31*(50), 18578–18589. doi:10.1523/JNEUROSCI.4465-11.2011

Uhlhaas, P. J., & Singer, W. (2010). Abnormal neural oscillations and synchrony in schizophrenia. *Nat Rev Neurosci, 11*(2), 100–113. doi:10.1038/nrn2774

Uhlhaas, P. J., & Singer, W. (2011). The development of neural synchrony and large-scale cortical networks during adolescence: Relevance for the pathophysiology of schizophrenia and neurodevelopmental hypothesis. *Schizophr Bull,* 37(3), 514–523. doi:10.1093/schbul/sbr034

Uhlhaas, P. J., & Singer, W. (2015). Oscillations and neuronal dynamics in schizophrenia: The search for basic symptoms and translational opportunities. *Biol Psychiatry, 77*(12), 1001–1009. doi:10.1016/j.biopsych.2014.11.019

Underwood, E. (2016). Barcoding the brain. *Science, 351,* 799–800.

Valero-Cuevas, F. J., Anand, V. V., Saxena, A., & Lipson, H. (2007). Beyond parameter estimation: Extending biomechanical modeling by the explicit exploration of model topology. *IEEE Trans Bio-Med Eng, 54*(11), 1951–1964.

Van den Brand, R., Heutschi, J., Barraud, Q., DiGiovanna, J., Bartholdi, K., Huerlimann, M., . . . Courtine, G. (2012). Restoring voluntary control of locomotion after paralyzing spinal cord injury. *Science, 336*(6085), 1182–1185. doi:10.1126/science.1217416

Van den Heuvel, M., & Fornito, A. (2014). Brain networks in schizophrenia. *Neuropsychol Rev, 24*(1), 32–48. doi:10.1007/s11065-014-9248-7

Van den Heuvel, M., Kahn, R. S., Goni, J., & Sporns, O. (2012). High-cost, high-capacity backbone for global brain communication. *Proc Natl Acad Sci, 109*, 11372–11377.

Van den Heuvel, M. P., & Sporns, O. (2013). Network hubs in the human brain. *Trends Cogn Sci, 17*(12), 683–696. doi:10.1016/j.tics.2013.09.012

Van der Steen, M. M., & Bongers, R. M. (2011). Joint angle variability and co-variation in a reaching with a rod task. *Exp Brain Res, 208*(3), 411–422. doi:10.1007/s00221-010-2493-y

Van Dijk, K. R., Hedden, T., Venkataraman, A., Evans, K. C., Lazar, S. W., & Buckner, R. L. (2010). Intrinsic functional connectivity as a tool for human connectomics: Theory, properties, and optimization. *J Neurophysiol, 103*(1), 297–321. doi:10.1152/jn.00783.2009

Van Kordelaar, J., van Wegen, E. E., Nijland, R. H., Daffertshofer, A., & Kwakkel, G. (2013). Understanding adaptive motor control of the paretic upper limb early poststroke: The EXPLICIT-stroke program. *Neurorehabil Neural Repair, 27*(9), 854–863. doi:10.1177/1545968313496327

Van Ooyen, A. (2011). Using theoretical models to analyse neural development. *Nat Rev Neurosci, 12*(6), 311–326. doi:10.1038/nrn3031

Van Wassenbergh, S., Leysen, H., Adriaens, D., & Aerts, P. (2013). Mechanics of snout expansion in suction-feeding seahorses: Musculoskeletal force transmission. *J Exp Biol, 216*(Pt 3), 407–417. doi:10.1242/jeb.074658

Varlet, M., Marin, L., Raffard, S., Schmidt, R. C., Capdevielle, D., Boulenger, J. P., . . . Bardy, B. G. (2012). Impairments of social motor coordination in schizophrenia. *PLoS ONE, 7*(1), e29772. doi:10.1371/journal.pone.0029772

Vasung, L., Fischi-Gomez, E., & Huppi, P. S. (2013). Multimodality evaluation of the pediatric brain: DTI and its competitors. *Pediatr Radiol, 43*(1), 60–68. doi:10.1007/s00247-012-2515-y

Vaziri, A., & Mahadevan, L. (2008). Localized and extended deformations of elastic shells. *Proc Natl Acad Sci, 105*, 7913–7918.

Vertes, P. E., Alexander-Bloch, A., Gogtay, N., Giedd, J. N., Rapoport, J. L., & Bullmore, E. T. (2012). Simple models of human brain functional networks. *Proc Natl Acad Sci, 109*, 5868–5873.

Vinther, J., Stein, M., Longrich, N. R., & Harper, D. A. (2014). A suspension-feeding anomalocarid from the Early Cambrian. *Nature, 507*(7493), 496–499. doi:10.1038/nature13010

Vogel, G. (2013). How do organs know they have reached the right size? *Science, 340*, 1156–1157.

Vogel, S. (2003). *Comparative biomechanics*. Princeton, NJ: Princeton Univerity Press.

Vollrath, M. A., Kwan, K. Y., & Corey, D. P. (2007). The micromachinery of mechanotransduction in hair cells. *Annu Rev Neurosci, 30*, 339–365. doi:10.1146/annurev.neuro.29.051605.112917

Volpe, J. J. (2008). *Neurology of the newborn* (5th ed.). Philadelphia: Saunders.

Volpe, J. J. (2009). The encephalopathy of prematurity—brain injury and impaired brain development inextricably intertwined. *Semin Pediatr Neurol, 16*(4), 167–178. doi:10.1016/j.spen.2009.09.005

Volpe, J. J. (2012). Neonatal encephalopathy: An inadequate term for hypoxic-ischemic encephalopathy. *Ann Neurol, 72*(2), 156–166. doi:10.1002/ana.23647

Volpe, J. J., Kinney, H. C., Jensen, F. E., & Rosenberg, P. A. (2011). The developing oligodendrocyte: Key cellular target in brain injury in the premature infant. *Int J Dev Neurosci, 29*(4), 423–440. doi:10.1016/j.ijdevneu.2011.02.012

Vuong, H. E., Yano, J. M., Fung, T. C., & Hsiao, E. Y. (2017). The microbiome and host behavior. *Annu Rev Neurosci*. doi:10.1146/annurev-neuro-072116-031347

Waddington, C. H. (1957). *The strategy of the genes: A discussion of some aspects of theoretical biology.* New York: Macmillan.

Wahl, A. S., Omlor, W., Rubio, J. C., Chen, J. L., Zheng, H., Schroter, A., . . . Schwab, M. E. (2014). Asynchronous therapy restores motor control by rewiring of the rat corticospinal tract after stroke. *Science, 344*(6189), 1250–1255. doi:10.1594/

Wake, H., Moorhouse, A. J., Miyamoto, A., & Nabekura, J. (2013). Microglia: Actively surveying and shaping neuronal circuit structure and function. *Trends Neurosci, 36*(4), 209–217. doi:10.1016/j.tins.2012.11.007

Wallace, A. (2006). D'Arcy Thompson and the theory of transformations. *Nature Reviews Genetics, 7*(5), 401. doi:10.1038/nrg1835

Walsh, M. K., & Lichtman, J. (2003). In vivo time-lapse imaging of synaptic takeover associated with naturally occurring synapse elimination. *Neuron, 37,* 67–73.

Wang, L., Conner, J. M., Nagahara, A. H., & Tuszynski, M. H. (2016). Rehabilitation drives enhancement of neuronal structure in functionally relevant neuronal subsets. *Proc Natl Acad Sci USA, 113*(10), 2750–2755.

Wang, N., Tytell, J., & Ingber, D. E. (2009). Mechanotransduction at a distance: Mechanically coupling the extracellular matrix with the nucleus. *Nat Rev Mol Cell Biol, 10,* 75–82.

Wang, W. C., & McLean, D. L. (2014). Selective responses to tonic descending commands by temporal summation in a spinal motor pool. *Neuron, 83*(3), 708–721. doi:10.1016/j.neuron.2014.06.021

Wang, X. (2016). The ying and yang of auditory nerve damage. *Neuron, 89*(4), 680–682. doi:10.1016/j.neuron.2016.02.007

Wang, Z., Chen, L. M., Negyessy, L., Friedman, R. M., Mishra, A., Gore, J. C., & Roe, A. W. (2013). The relationship of anatomical and functional connectivity to resting-state connectivity in primate somatosensory cortex. *Neuron, 78*(6), 1116–1126. doi:10.1016/j.neuron.2013.04.023

Warden, M. R., Selimbeyoglu, A., Mirzabekov, J. J., Lo, M., Thompson, K. R., Kim, S. Y., . . . Deisseroth, K. (2012). A prefrontal cortex-brainstem neuronal projection that controls response to behavioural challenge. *Nature, 492*(7429), 428–432. doi:10.1038/nature11617

Warp, E., Agarwal, G., Wyart, C., et al. (2012). Emergence of patterned activity in the developing zebrafish spinal cord. *Curr Biol, 22,* 93–102.

Warren, W., & Rio, K. (2015). The visual coupling between neighbors in a virtual crowd. *J Vis, 15*(12), 747. doi:10.1167/15.12.747

Warren, W. H. (1984). Perceiving affordances: Visual guidance of stair climbing. *J Exp Psychol: Hum Percep Perform, 10,* 683–703.

Warren, W. H. (1988). Action modes and laws of control for the visual guidance of action. In O. G. Meijer & K. Roth (Eds.), *Complex movement behaviour: "The" motor-action controversy* (pp. 339–380). Amsterdam: North-Holland; Elsevier Science.

Warren, W. H. (2006). The dynamics of perception and action. *Psychol Rev, 113*(2), 358–389. doi:10.1037/0033-295X.113.2.358

Warren, W. H., & Fajen, B. R. (2008). Behavioral dynamics of visually-guided locomotion. In A. Fuchs & V. Jirsa (Eds.), *Coordination: Neural, behavioral and social dynamics* (pp. 45–76). Berlin: Springer.

Weaver, J. A., Melin, J., Stark, D., Quake, S., R., & Horowitz, M. A. (2010). Static control logic for microfluidic devices using pressure-gain valves. *Nature Physics, 6*(3), 218. doi:10.1038/nphys1513

Webb, B. (2004). Neural mechanisms for prediction: Do insects have forward models? *Trends in Neurosciences, 27*(5), 278–282. doi:10.1016/j.tins.2004.03.004

Wedeen, V. J., Rosene, D. L., Wang, R., Dai, G., Mortazavi, F., Hagmann, P., . . . Tseng, W. Y. (2012). The geometric structure of the brain fiber pathways. *Science, 335*(6076), 1628–1634. doi:10.1126/science.1215280

Wehner, M., Park, Y.-L., Walsh, C., Nagpal, R., Wood, R. J., & Goldfield, E. (2012). *Experimental characterization of components for active soft orthotics.* Paper presented at the IEEE RAS / EMBS International Conference on Biomedical Robotics and Biomechatronics, Rome, Italy.

Wehner, M., Truby, R. L., Fitzgerald, D. J., Mosadegh, B., Whitesides, G. M., Lewis, J. A., & Wood, R. J. (2016). An integrated design and fabrication strategy for entirely soft, autonomous robots. *Nature, 536*(7617), 451–455. doi:10.1038/nature19100

Wehner, R. (1997). The ant's celestial compass system: Spectral and polarization channels. In M. Lehrer (Ed.), *Orientation and Communication in Arthropods* (pp. 145–185). Birkhäuser Verlag.

Weinberg, D. H. (2005). Mapping the large-scale structure of the universe. (ASTRONOMY). *Science, 309*(5734), 564.

Weinkamer, R., & Fratzl, P. (2011). Mechanical adaptation of biological materials—the examples of bone and wood. *Mater Sci Eng: C, 31*(6), 1164–1173. doi:10.1016/j.msec.2010.12.002

Weir, P. T., & Dickinson, M. H. (2012). Flying Drosophila orient to sky polarization. *Curr Biol, 22*(1), 21–27. doi:10.1016/j.cub.2011.11.026

Weiss, P., & Garber, B. (1952). Shape and movement of mesenchymal cells as functions of the physical structure of the medium. Contributions to a quantitative morphology. *Proc Natl Acad Sci, 38*, 264–280.

Weiss, P., Keshner, E., & Levin, M. (2014). *Virtual reality for physical and motor rehabilitation.* New York: Springer.

Wen, Q., Po, M. D., Hulme, E., Chen, S., Liu, X., Kwok, S. W., . . . Samuel, A. D. (2012). Proprioceptive coupling within motor neurons drives C. elegans forward locomotion. *Neuron, 76*(4), 750–761. doi:10.1016/j.neuron.2012.08.039

Weng, S.-J., Wiggins, J. L., Peltier, S. J., Carrasco, M., Risi, S., Lord, C., & Monk, C. S. (2010). Alterations of resting state functional connectivity in the default network in adolescents with autism spectrum disorders. *Brain Res, 1313*, 202–214. doi:10.1016/j.brainres.2009.11.057

Wenger, N., Moraud, E. M., Gandar, J., Musienko, P., Capogrosso, M., Baud, L., . . . Courtine, G. (2016). Spatiotemporal neuromodulation therapies engaging muscle synergies improve motor control after spinal cord injury. *Nat Med, 22*(2), 138–145. doi:10.1038/nm.4025

Wenger, N., Moraud, E. M., Raspopovic, S., Bonizzato, M., DiGiovanna, J., Musienko, P., . . . Courtine, G. (2014). Closed-loop neuromodulation of spinal sensorimotor circuits controls refined locomotion after complete spinal cord injury. *Sci Transl Med, 6*(255), 255ra133. doi:10.1126/scitranslmed.3008325

Wennekamp, S., Mesecke, S., Nedelec, F., & Hiiragi, T. (2013). A self-organization framework for symmetry breaking in the mammalian embryo. *Nat Rev Mol Cell Biol, 14*(7), 452–459. doi:10.1038/nrm3602

Wenner, P. (2012). Motor development: Activity matters after all. *Curr Biol, 22*(2), R47–R48. doi:10.1016/j.cub.2011.12.008

Werfel, J., & Nagpal, R. (2008). Three-dimensional construction with mobile robots and modular blocks. *Int J Robot Res, 27*(3-4), 463–479. doi:10.1177/0278364907084984

Werfel, J., Petersen, K., & Nagpal, R. (2014). Designing collective behavior in a termite-inspired robot construction team. *Science, 343*(6172), 754–758. doi:10.1126/science.1245842

Werker, J. F., & Hensch, T. K. (2015). Critical periods in speech perception: New directions. *Annu Rev Psychol, 66*, 173–196. doi:10.1146/annurev-psych-010814-015104

Wernegreen, J. J. (2012). Endosymbiosis. *Curr Biol, 22*(14), R555–R561. doi:10.1016/j.cub.2012.06.010

West, G. B. (2012). The importance of quantitative systemic thinking in medicine. *The Lancet, 379*, 1551–1559.

West, G. B., & Brown, J. H. (2005). The origin of allometric scaling laws in biology from genomes to ecosystems: Towards a quantitative unifying theory of biological structure and organization. *J Exp Biol, 208*(Pt 9), 1575–1592. doi:10.1242/jeb.01589

White, T., Asfaw, B., Beyene, Y., Haile-Selassie, Y., Lovejoy, C., Suwa, G., & Woldegabriel, G. (2009). *Ardipithecus ramidus* and the paleobiology of early hominids. *Science* (Washington), *326*(5949), 75–86.

Whitesides, G. M. (2006). The origins and future of microfluidics. *Nature, 442*, 368–373.

Whitesides, G. M., & Grzybowski, B. (2002). Self-assembly at all scales. *Science, 295,* 2418–2421.

Whiting, H. T. A., & Whiting, H. T. A. (1983). *Human motor actions: Bernstein reassessed.* Amsterdam: North-Holland.

Whitney, J. P., Sreetharan, P. S., Ma, K. Y., & Wood, R. J. (2011). Pop-up book MEMS. *J Micromech and Microeng, 21*(11), 115021. doi:10.1088/0960-1317/21/11/115021

Whitney, J. P., & Wood, R. J. (2010). Aeromechanics of passive rotation in flapping flight. *J Fluid Mech, 660,* 197–220. doi:10.1017/s002211201000265x

Wilson, A., & Lichtwark, G. (2011). The anatomical arrangement of muscle and tendon enhances limb versatility and locomotor performance. *Philos Trans R Soc Lond B Biol Sci, 366*(1570), 1540–1553. doi:10.1098/rstb.2010.0361

Winold, H., Thelen, E., & Ulrich, B. D. (1994). Coordination and control in the bow arm movements of highly skilled cellists. *Ecol Psychol, 6,* 1–31.

Witter, L., & De Zeeuw, C. I. (2015). Regional functionality of the cerebellum. *Curr Opin Neurobiol, 33,* 150–155. doi:10.1016/j.conb.2015.03.017

Wittlinger, M., Wehner, R., & Wolf, H. (2006). The ant odometer: Stepping on stilts and stumps. *Science, 312,* 1965–1967.

Wodlinger, B., Downey, J. E., Tyler-Kabara, E. C., Schwartz, A. B., Boninger, M. L., & Collinger, J. L. (2015). Ten-dimensional anthropomorphic arm control in a human brain¯machine interface: Difficulties, solutions, and limitations. *J Neural Eng 12*(1), 016011. doi:10.1088/1741-2560/12/1/016011

Wolf, D. H., Satterthwaite, T. D., Calkins, M. E., Ruparel, K., Elliott, M. A., Hopson, R. D., . . . Gur, R. E. (2015). Functional neuroimaging abnormalities in youth with psychosis spectrum symptoms. *JAMA Psychiatry, 72*(5), 456–465. doi:10.1001/jamapsychiatry.2014.3169

Wolf, H. (2011). Odometry and insect navigation. *J Exp Biol, 214*(Pt 10), 1629–1641. doi:10.1242/jeb.038570

Wolf, S. L., Winstein, C. J., Miller, J. P., Taub, E., Uswatte, G., Morris, D., . . . Excite Investigators. (2006). Effect of constraint-induced movement therapy on upper extremity function 3 to 9 months after stroke: The EXCITE Randomized Clinical Trial. *JAMA, 296*(17), 2095–2104. doi:10.1001/jama.296.17.2095

Wolff, P. H. (1960). *The developmental psychologies of Jean Piaget and psychoanalysis.* New York.

Wolff, P. H. (1973). Natural history of sucking patterns in infant goats. *J Compar and Physiol Psychol, 84,* 252–257.

Wolff, P. H. (1987). *The development of behavioral states and the expression of emotions in early infancy:New proposals for investigation.* Chicago: University of Chicago Press.

Wolpert, L., Jessell, T., Lawrence, P., Meyerowitz, E., Robertson, E., & Smith, J. (2007). *Principles of development* (3rd ed.). New York: Oxford University Press.

Womelsdorf, T., Valiante, T. A., Sahin, N. T., Miller, K. J., & Tiesinga, P. (2014). Dynamic circuit motifs underlying rhythmic gain control, gating and integration. *Nat Neurosci, 17*(8), 1031–1039. doi:10.1038/nn.3764

Wong, T. S., Kang, S. H., Tang, S. K., Smythe, E. J., Hatton, B. D., Grinthal, A., & Aizenberg, J. (2011). Bioinspired self-repairing slippery surfaces with pressure-stable omniphobicity. *Nature, 477*(7365), 443–447. doi:10.1038/nature10447

Wood, R., Nagpal, R., & Wei, G.-Y. (2013). Flight of the robobees. *Scient Amer, 308*(3), 60–65. doi:10.1038/scientificamerican0313-60

Wood, R. J. (2008). The first takeoff of a biologically inspired at-scale robotic insect. *IEEE Trans Robotics, 24,* 1–7.

Wood, R. J., Avadhanula, S., Sahai, R., Steltz, E., & Fearing, R. (2008). Microrobot design using fiber reinforced composites. *J Mech Des, 130*(5), 1–10. doi:10.1115/1.2885509

Woolley, S. C., Rajan, R., Joshua, M., & Doupe, A. J. (2014). Emergence of context-dependent variability across a basal ganglia network. *Neuron, 82*(1), 208–223. doi:10.1016/j.neuron.2014.01.039

Wozniak, M. A., & Chen, C. S. (2009). Mechanotransduction in development: A growing role for contractility. *Nat Rev Mol Cell Biol, 10*(1), 34–43. doi:10.1038/nrm2592

Wu, F., Stark, E., Ku, P. C., Wise, K. D., Buzsaki, G., & Yoon, E. (2015). Monolithically integrated muleds on silicon neural probes for high-resolution optogenetic studies in behaving animals. *Neuron, 88*(6), 1136–1148. doi:10.1016/j.neuron.2015.10.032

Wu, H. G., Miyamoto, Y. R., Castro, L. N., Olveczky, B. P., & Smith, M. A. (2014). Temporal structure of motor variability is dynamically regulated and predicts motor learning ability. *Nat Neurosci, 17*(2), 312–321. doi:10.1038/nn.3616

Wu, M. C., Chu, L. A., Hsiao, P. Y., Lin, Y. Y., Chi, C. C., Liu, T. H., . . . Chiang, A. S. (2014). Optogenetic control of selective neural activity in multiple freely moving Drosophila adults. *Proc Natl Acad Sci U S A, 111*(14), 5367-5372. doi:10.1073/pnas.1400997111

Wu, W., Moreno, A. M., Tangen, J. M., & Reinhard, J. (2013). Honeybees can discriminate between Monet and Picasso paintings. *J Comp Physiol A Neuroethol Sens Neural Behav Physiol, 199*(1), 45–55. doi:10.1007/s00359-012-0767-5

Wyatt, L. A., & Keirstead, H. S. (2012). Stem cell-based treatments for spinal cord injury–Chapter 13. *Progress in Brain Research, 201*, 233–252. doi:10.1016/B978-0-444-59544-7.00012-3

Wyss, A. F., Hamadjida, A., Savidan, J., Liu, Y., Bashir, S., Mir, A., . . . Belhaj-Saif, A. (2013). Long-term motor cortical map changes following unilateral lesion of the hand representation in the motor cortex in macaque monkeys showing functional recovery of hand functions. *Restor Neurol Neurosci, 31*(6), 733–760. doi:10.3233/RNN-130344

Xia, Y., & Whitesides, G. (1998). Soft lithography. *Annual Review of Materials Science, 28*, 153.

Xie, C., Liu, J., Fu, T. M., Dai, X., Zhou, W., & Lieber, C. M. (2015). Three-dimensional macroporous nanoelectronic networks as minimally invasive brain probes. *Nat Mater, 14*(12), 1286–1292. doi:10.1038/nmat4427

Xu, G., Takahashi, E., Folkerth, R. D., Haynes, R. L., Volpe, J. J., Grant, P. E., & Kinney, H. C. (2014). Radial coherence of diffusion tractography in the cerebral white matter of the human fetus: Neuroanatomic insights. *Cereb Cortex, 24*(3), 579–592. doi:10.1093/cercor/bhs330

Xu, X., Zhou, Z., Dudley, R., Mackem, S., Chuong, C. M., Erickson, G. M., & Varricchio, D. J. (2014). An integrative approach to understanding bird origins. *Science, 346*(6215), 1253293. doi:10.1126/science.1253293

Yang, W., & Yuste, R. (2017). In vivo imaging of neural activity. *Nat Methods, 14*(4), 349–359. doi:10.1038/nmeth.4230

Yilmaz, M., & Meister, M. (2013). Rapid innate defensive responses of mice to looming visual stimuli. *Curr Biol, 23*(20), 2011–2015. doi:10.1016/j.cub.2013.08.015

Yiu, G., & He, Z. (2006). Glial inhibition of CNS axon regeneration. *Nat Rev Neurosci, 7*(8), 617–627. doi:10.1038/nrn1956

Yoshizawa, M., Goricki, S., Soares, D., & Jeffery, W. R. (2010). Evolution of a behavioral shift mediated by superficial neuromasts helps cavefish find food in darkness. *Curr Biol, 20*(18), 1631–1636. doi:10.1016/j.cub.2010.07.017

Yu, C.-H., & Nagpal, R. (2009). Self-adapting modular robotics: A generalized distributed consensus framework. *2009 IEEE International Conference on Robotics and Automation* (pp. 1881–1888).

Yu, C.-H., & Nagpal, R. (2011). A self-adaptive framework for modular robots in a dynamic environment: Theory and applications. *Int J Robot Res, 30*(8), 1015–1036. doi:10.1177/0278364910384753

Yuste, R. (2011). Dendritic spines and distributed circuits. *Neuron, 71*(5), 772–781. doi:10.1016/j.neuron.2011.07.024

Yuste, R. (2013). Electrical compartmentalization in dendritic spines. *Annu Rev Neurosci, 36*, 429–449. doi:10.1146/annurev-neuro-062111-150455

Yuste, R. (2015). From the neuron doctrine to neural networks. *Nat Rev Neurosci, 16*(8), 487–497. doi:10.1038/nrn3962

Zagorovsky, K., & Chan, W. C. (2013). Bioimaging: illuminating the deep. *Nat Mater, 12*(4), 285–287. doi:10.1038/nmat3608

Zaidi, M. (2007). Skeletal remodeling in health and disease. *Nature Medicine, 13,* 791–801.

Zeil, J., & Hemmi, J. M. (2006). The visual ecology of fiddler crabs. *J Comp Physiol A Neuroethol Sens Neural Behav Physiol, 192*(1), 1–25. doi:10.1007/s00359-005-0048-7

Zeiler, S. R., & Krakauer, J. W. (2013). The interaction between training and plasticity in the post-stroke brain. *Curr Opin Neurol, 26*(6), 609–616. doi:10.1097/WCO.0000000000000025

Zelazo, P. D., Adolph, K. E., & Robinson, S. R. (2013). The road to walking: What learning to walk tells us about development. In P. D. Zelazo (Ed.), *The Oxford handbook of developmental psychology* (Vol. 1: *Body and Mind,* pp. 1–73). New York: Oxford University Press.

Zhang, A., & Lieber, C. M. (2016). Nano-bioelectronics. *Chem Rev, 116*(1), 215–257. doi:10.1021/acs.chemrev.5b00608

Zhang, F., Gradinaru, V., Adamantidis, A. R., Durand, R., Airan, R. D., de Lecea, L., & Deisseroth, K. (2010). Optogenetic interrogation of neural circuits: Technology for probing mammalian brain structures. *Nat Protoc, 5*(3), 439–456. doi:10.1038/nprot.2009.226

Zhang, H. Y., Issberner, J., & Sillar, K. T. (2011). Development of a spinal locomotor rheostat. *Proc Natl Acad Sci USA, 108*(28), 11674–11679. doi:10.1073/pnas.1018512108

Zhang, J., Aggarwal, M., & Mori, S. (2012). Structural insights into the rodent CNS via diffusion tensor imaging. *Trends Neurosci, 35*(7), 412–421. doi:10.1016/j.tins.2012.04.010

Zhang, J., Lanuza, G. M., Britz, O., Wang, Z., Siembab, V. C., Zhang, Y., . . . Goulding, M. (2014). V1 and V2b interneurons secure the alternating flexor-extensor motor activity mice require for limbed locomotion. *Neuron, 82*(1), 138–150. doi:10.1016/j.neuron.2014.02.013

Zhao, B., & Muller, U. (2015). The elusive mechanotransduction machinery of hair cells. *Curr Opin Neurobiol, 34,* 172–179. doi:10.1016/j.conb.2015.08.006

Zhao, H.-P., Feng, X.-Q., Yu, S.-W., Cui, W.-Z., & Zou, F.-Z. (2005). Mechanical properties of silkworm cocoons. *Polymer, 46*(21), 9192–9201. doi:10.1016/j.polymer.2005.07.004

Zhou, K., Wolpert, D. M., & De Zeeuw, C. I. (2014). Motor systems: Reaching out and grasping the molecular tools. *Curr Biol, 24*(7), R269–R271. doi:10.1016/j.cub.2014.02.048

Zhou, Z. (2014). Dinosaur evolution: Feathers up for selection. *Curr Biol, 24*(16), R751–R753. doi:10.1016/j.cub.2014.07.017

Ziegler, M. D., Zhong, H., Roy, R. R., & Edgerton, V. R. (2010). Why variability facilitates spinal learning. *J Neurosci, 30*(32), 10720–10726. doi:10.1523/JNEUROSCI.1938-10.2010

Zilles, K., Palomero-Gallagher, N., & Amunts, K. (2013). Development of cortical folding during evolution and ontogeny. *Trends Neurosci, 36*(5), 275–284. doi:10.1016/j.tins.2013.01.006

Zimmerman, A., Bai, L., & Ginty, D. D. (2014). The gentle touch receptors of mammalian skin. *Science, 346*(6212), 950–954. doi:10.1126/science.1254229

Zoghbi, H. Y., & Bear, M. F. (2012). Synaptic dysfunction in neurodevelopmental disorders associated with autism and intellectual disabilities. *Cold Spring Harb Perspect Biol, 4,*1–22. doi:10.1101/cshperspect.a009886

Zrenner, E. (2013). Fighting blindness with microelectronics. *Sci Transl Med, 5,* 1–7.

Zullo, L., Sumbre, G., Agnisola, C., Flash, T., & Hochner, B. (2009). Nonsomatotopic organization of the higher motor centers in octopus. *Curr Biol, 19*(19), 1632–1636. doi:10.1016/j.cub.2009.07.067

Index